Studies in

THE SCOTTISH LATEGLACIAL ENVIRONMENT

Studies in

THE SCOTTISH LATEGLACIAL ENVIRONMENT

Edited by

J. M. GRAY and J. J. LOWE

Pergamon Press

OXFORD · NEW YORK · TORONTO

SYDNEY · PARIS . FRANKFURT

U.K. Pergamon Press Ltd., Headington Hill Hall, Oxford OX3 0BW, England
U.S.A. Pergamon Press Inc., Maxwell House, Fairview Park, Elmsford, New York 10523, U.S.A.
CANADA Pergamon of Canada Ltd., 75 The East Mall, Toronto, Ontario, Canada
AUSTRALIA Pergamon Press (Aust.) Pty. Ltd., 19a Boundary Street, Rushcutters Bay, N.S.W. 2011, Australia
FRANCE Pergamon Press SARL, 24 rue des Ecoles, 75240 Paris, Cedex 05, France
WEST GERMANY Pergamon Press GmbH, 6242 Kronberg-Taunus, Pferdstrasse 1, Frankfurt-am-Main, West Germany

---

First edition 1977

Library of Congress Cataloging in Publication Data

Main entry under title:

Studies in the Scottish Lateglacial Environment.

Includes bibliographical references and index.
1. Geology, Stratigraphic—Pleistocene. 2. Glacial epoch—Scotland.
I. Gray, J M II. Lowe, J J
QE697.S88 1977 551.7'92'09411 76–41218

ISBN 0–08–020498–8 (Hardcover)

In order to make this volume available as economically and rapidly as possible the author's typescript has been reproduced in its original form. This method unfortunately has its typographical limitations but it is hoped that they in no way distract the reader.

Printed in Great Britain by A. Wheaton & Co., Exeter

PROFESSOR W. W. BISHOP

Late in the production of this volume one of the contributors, Bill Bishop, Head of the Geology Department at Queen Mary College, London, died suddenly and prematurely at the age of 45. As well as his Scottish work, for over a decade he made important contributions to the geological and geomorphological background to fossil man in East Africa. He had recently begun a project in Pakistan and was about to take up his new appointment as Director of the Peabody Museum of Natural History at Yale. His loss will be felt not only in his research fields but also in the numerous committees and organisations in which he so eagerly involved himself. His enthusiasm, liveliness, humour and song will be sadly missed by his many friends and colleagues all over the world.

Contents

Contributors

PROFESSOR W. W. BISHOP
Department of Geology, Queen Mary College, University of London, Mile End Road, London E1 4NS

DR. C. M. CLAPPERTON
Department of Geography, University of Aberdeen, High Street, Old Aberdeen, AB9 2UF

DR. G. R. COOPE
Department of Geological Sciences, University of Birmingham, P.O. Box 363, Birmingham, B15 2TT

DR. R. A. CULLINGFORD
Department of Geography, University of Exeter, Rennes Drive, Exeter, EX4 4RJ

MR. D. K. GRAHAM
Institute of Geological Sciences, Murchison House, West Mains Road, Edinburgh, EH9 3LA

DR. J. M. GRAY
Department of Geography, Queen Mary College, University of London, Mile End Road, London E1 4NS

MR. J. J. LOWE
Geography Section, City of London Polytechnic, Calcutta House, Old Castle Street, London E1 7NT

DR. J. D. PEACOCK
Institute of Geological Sciences, Murchison House, West Mains Road, Edinburgh, EH9 3LA

DR. W. PENNINGTON
School of Biological Sciences, University of Leicester, Adrian Building, University Road, Leicester, LE1 7RH

DR. J. E. ROBINSON
Department of Geology, University College, University of London, Gower Street, London WC1E 6BT

DR. J. B. SISSONS
Department of Geography, University of Edinburgh, High School Yards, Edinburgh, EH1 1NR

DR. D. E. SUGDEN
Department of Geography, University of Aberdeen, High Street, Old Aberdeen, AB9 2UF

MR. D. G. SUTHERLAND
Department of Geography, University of Edinburgh, High School Yards, Edinburgh, EH1 1NR

PROFESSOR Y. VASARI
Oulun Yliopiston, Kasvitieteen Laitos, Department of Botany, University of Oulu, Oulu, Finland

DR. M. J. C. WALKER
Department of Geography, St. David's University College, Lampeter, Dyfed, SA48 7ED

MR. I. WILKINSON
Institute of Geological Sciences, Ring Road, Halton, Leeds, LS15 8TQ

Acknowledgements

We must express our gratitude to a number of individuals and organizations who assisted in various ways in the production of this volume. First and foremost our thanks must go to our supervisors Dr. Brian Sissons and Dr. Walter Newey for instilling in us the excitement of recreating the Scottish Lateglacial landscape. Without their enthusiasm and encouragement this book would never have been produced. Dr. H. J. B. Birks and Dr. J. T. Greensmith kindly provided advice on various aspects of the volume. We are most grateful to the contributors for being so patient with our incessant enquiries and with the unavoidable delays in the publication of their papers, and for the advice they gave so freely. Thanks are also due to those workers who allowed us or the contributors to quote their unpublished results or ideas.

We gratefully acknowledge the following for permission to reproduce illustrations:- the Royal Scottish Geographical Society (Fig. 1, p.2); Dr. J. B. Sissons (Fig. 2, p.2); Centre National de la Recherche Scientifique, Montpellier (publishers of *Pollen et Spores*) (Fig. 4, p.106 and Fig. 5, p.107); the Royal Society (Fig. 4a, p.124, Fig. 4d, p.127 and Fig. 6, p.131); Universitetsforlaget, Oslo (publishers of *Boreas*) (Fig. 4b, p.125 and Fig. 5a, p.128); Societas pro Fauna et Flora Fennica (publishers of *Acta Botanica Fennica*) (Fig. 1, p.144, Fig. 3, p.148, Fig. 7, p.154 and Fig. 9, p.158); Blackwell Scientific Publications Ltd. (publishers of *Journal of Ecology*) and Dr. H. H. Birks (Fig. 5, p.152); the Scottish Academic Press (publishers of *Scottish Journal of Geology*) and Dr. J. B. Sissons (Fig. 2, p.166).

The mammoth task of typing the volume was undertaken by Mrs. Linda Agombar. Without her efficiency, care and patience the task of producing this book would have been a much more harrowing experience. Mr. Alan Gillard, Mrs. Valerie Harwood, Mr. Peter Newman, Mrs. Maureen Newton, Mr. Stephen Pratt, Mrs. Mary Putney and Mr. Don Shewan, all of Queen Mary College Geography Department, also gave invaluable secretarial, cartographic or photographic assistance. Finally we owe much to Mr. H. Holt and the staff at Pergamon Press.

Murray Gray
John Lowe

London, July, 1976

Introduction

In recent years the occurrence in various parts of the world of floods, droughts and record temperatures has reinforced for many people the ever-changing nature of our climate. These recent variations, however, are only ripples in comparison to the oscillations that characterised the Lateglacial period. In general terms 'Lateglacial' usually refers to a very limited period of geological time covering the transition from the cold climate of the Last Glaciation to the temperate conditions that have characterised the last 10,000 years. The importance of this period has been widely recognised, particularly in Europe, where there is much biological and geological evidence for climatic oscillations that resulted in major environmental changes at this time. It was a period of very great transition in Scotland; transition from a cold, polar climate which had resulted in an ice sheet covering virtually the whole country during the preceding Devensian (Last) glacial stage, to the temperate conditions that led to final disappearance of glacier ice from the Scottish uplands around 10,000 years ago. As this book illustrates, however, this transition was not straightforward.

The Scottish Lateglacial environment may appear to be a very restricted subject for a book, but the contents of this volume attest the wealth of information now available for environmental reconstructions, the dramatic nature of environmental changes, and the important implications of studies of the Lateglacial environment for Quaternary studies in general as well as for modern climatic variations. The suggestion that mean July temperature changed by 7°C in as little as 70 years at the beginning of the Lateglacial (Coope and Brophy, 1972) is quite astounding. As far as Scotland is concerned, the evidence for the 3,000 - 4,000 years of Lateglacial time far exceeds that available for the remainder of the Pleistocene (at least 1 million years). Further, unlike most of the Pleistocene, the Lateglacial lies within the range of radiocarbon dating, and thus it is possible to build up a detailed absolute chronology as an independent scale for the measurement of times and rates of biological and geological changes.

The reconstruction of environmental conditions during the Lateglacial is, however, a demanding task, for, as the large number of recent studies has shown, climatic changes during the Lateglacial, and resultant modifications of the interdependent biological, lithological, glaciological and morphological processes, were sudden and, in terms of the overall duration of the Pleistocene, comparatively short-lived. This has resulted in misinterpretations of palaeoenvironmental records, for an unstable equilibrium appears to have existed during much of the Lateglacial between climate on the one hand and physical and biological systems on the other. A multi-disciplinary approach is thus essential to correctly interpret Lateglacial stratigraphic records.

This volume grew from our awareness of the large amount of research in progress on the Scottish Lateglacial environment and from our belief that it would be far more beneficial to the subject if the results of some of this research could be grouped together in one volume rather than published separately in the scientific journals of several disciplines. By bringing together the work of researchers in geography, geology and botany departments and institutions we hope the volume will be of value to students of various disciplines and to Quaternary researchers in other countries.

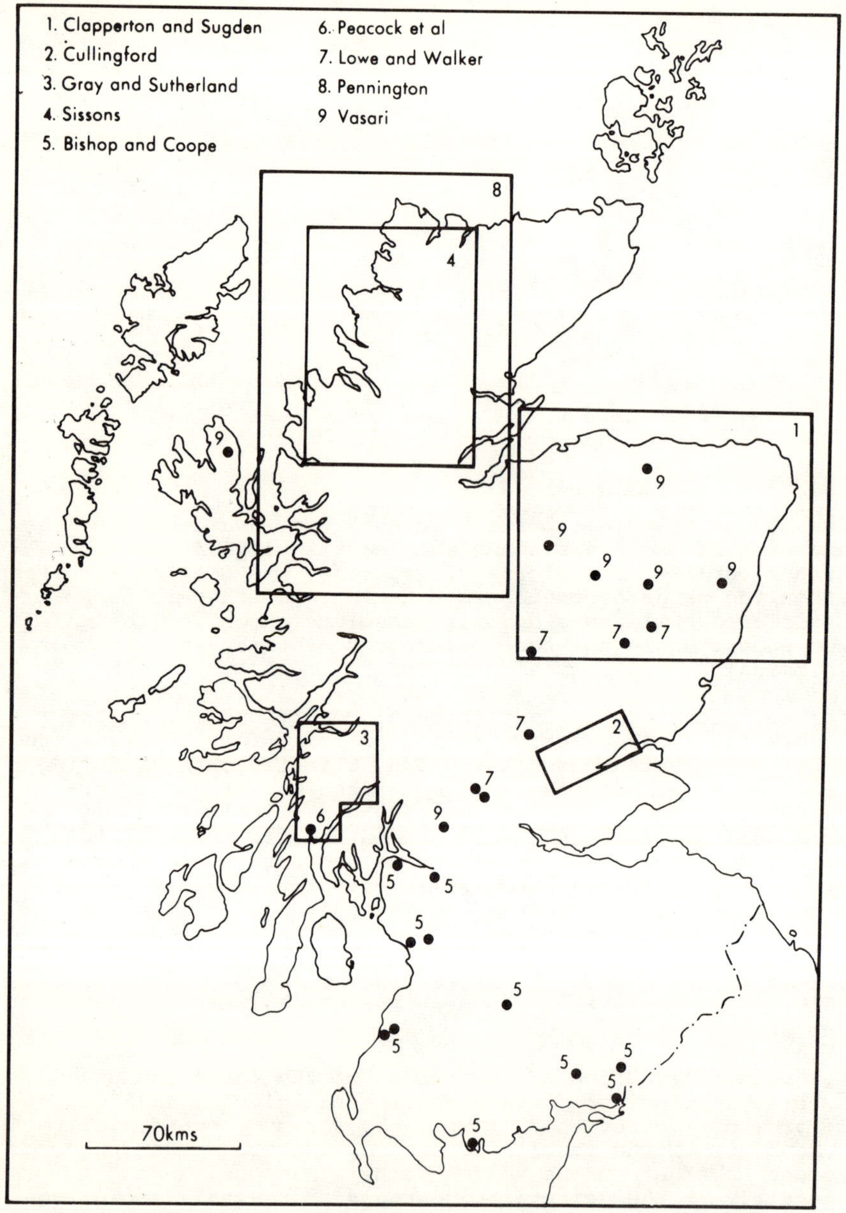

Fig. 1. Study areas and main localities of papers in this volume.

Figure 1 shows the areas and localities considered in the papers in this volume. In the opening chapter CLAPPERTON and SUGDEN review the Late Devensian glaciation in north-east Scotland, thus setting the subsequent papers in a broader context. They discuss the evidence for an unglaciated area in Buchan during the Last Glaciation and for readvances during decay of the Late Devensian ice sheet. CULLINGFORD presents very detailed evidence concerning Lateglacial sea-level changes in the Earn-Tay area and considers their relationship to deglaciation. This area has become a controversial one due to uncertainty regarding the 'Perth Readvance' and Cullingford's work adds significantly to our understanding of this problem. GRAY and SUTHERLAND question the validity of the Oban-Ford Moraine, the only pre-Loch Lomond Readvance moraine remaining virtually unchallenged in the literature. Much of the impetus for the study of the Scottish Lateglacial environment has been due to the large amount of research carried out by SISSONS. In his paper he presents the first results of his recent research on the Loch Lomond Readvance in the little-known north-west Scottish Highlands.

In a double paper BISHOP and COOPE attempt a reconstruction of Lateglacial and early Flandrian environments in south-west Scotland based on their research on the stratigraphy and associated coleopteran assemblages of numerous sites. The beetle evidence has allowed construction of a mean July temperature curve for the Lateglacial and early Flandrian, the first detailed temperature curve published for the Lateglacial of Scotland. PEACOCK, GRAHAM, ROBINSON and WILKINSON have carried out detailed research on the fauna in marine sediments at Lochgilphead. This, together with radiocarbon dates on marine shells, has allowed construction of a relative sea-level curve and conclusions regarding the contemporaneous climate.

LOWE and WALKER describe results of pollen-stratigraphic investigations at six Lateglacial sites in the southern and eastern Grampian Highlands, and discuss the palaeoenvironmental implications of inferred vegetation and soil developments. In recent years PENNINGTON has contributed much to our understanding of the Scottish Lateglacial environment, and in her paper she further discusses analyses of biological and chemical variables from sediment profiles from lochs in northern Scotland. On the basis of absolute pollen counts and radiocarbon dates chronozones are defined for the Lateglacial. VASARI presents a number of radiocarbon dates that are related to pollen profiles from the Scottish Highlands and the Isle of Skye. The subdivision of the Lateglacial and early Flandrian vegetation succession is discussed, and regional contrasts are suggested for Scotland for the pattern of environmental change during the Lateglacial.

Finally we attempt to integrate the new results contained in this volume into a synthesis of the Scottish Lateglacial environment, and we discuss the chronology of the environmental changes. In the past there has been some inconsistency and disagreement as regards the terminology and chronology of the Lateglacial. In the literature this term has been given different meanings and has been subdivided in various ways. 'Lateglacial'* has for example been used as a general term for the Jessen-Godwin pollen zones I, II and III, as a chronostratigraphic unit, as a climatostratigraphic unit, and as a general descriptive term applying to local events or episodes, such as glacial retreat stages, shorelines and sea-level changes, vegetational developments, and inferred 'environments'. It has been applied in a precise way based on chronostratigraphical definitions, and in an imprecise, informal or descriptive way, and this, together with the fact that a number of different forms of the word are used in the literature (sometimes within the same publication) has resulted, if not in confusion, in a lack of clarity. However, it is such a widely used term, applying to a reasonably specific and important period of time, that we feel that it should be retained but defined as strictly as available data will permit for any particular region. The use of the term 'Late Devensian' in place of Lateglacial can only add to any existing confusion, for the Late Devensian covers a much wider span of time (26,000 - 10,000 B.P.; Mitchell *et al.*, 1973) and is not equivalent to the Late Weichselian of continental north-west Europe (13,000 - 10,000 B.P.; Mangerud *et al.*, 1974).

The experience of editing this volume has vividly underlined for us the difficulty that exists in the use of terms and the need for a more rigorous approach to terminology and chronology than in the past. We are of the opinion that, although it is important to erect international chronozones, it is also important to define and subdivide the Lateglacial on the basis of climatostratigraphy for any region. Current evidence suggests only two climatostratigraphic subdivisions of the Scottish Lateglacial, *viz.* the Lateglacial Interstadial and the Loch Lomond Stadial. An attempt is made to use these terms in this volume with the following definitions:-

Lateglacial Interstadial: *The period between the start of the apparently rapid thermal improvement that occurred between about 14,000 and 13,000 B.P. and the beginning of the marked thermal decline that took place around 11,000 B.P.*

Loch Lomond Stadial: *The period between the start of the marked thermal decline that occurred around 11,000 B.P. and the international chronostratigraphic boundary for the beginning of the Flandrian, viz. 10,000 B.P.*

As discussed in our final paper, it is difficult at present to date the climatostratigraphic boundaries of these units more precisely. Nevertheless, we propose these formal climatostratigraphic terms in the conviction that future research will result in precise boundaries that can be accepted with confidence and in a greater understanding of the climatic parameters of the Lateglacial climatic episodes.

In arriving at our conclusions we have questioned the relevance for Scotland of the Jessen-Godwin Lateglacial zonation and recently-proposed chronostratigraphic boundaries. It should therefore be pointed out that our definitions may not be totally accepted by all the authors contributing to the volume, nor strictly adhered to in the text. We have, where possible, however, attempted to use the terms 'Lateglacial', 'Lateglacial Interstadial' and 'Loch Lomond Stadial' as outlined above. In view of some differences in interpretation, however, and in order to place the Lateglacial environment in a wider context, it has been found desirable to discuss events both prior to and immediately succeeding the Lateglacial.

There are of course many gaps in our knowledge of the Scottish Lateglacial environment and we are only too well aware of the deficiencies of this book. At the end of our synthesis we try to outline where the major gaps in our knowledge lie or where further information is required or may soon be forthcoming. The scope for research is manifold and we look forward to the time when the environment is known in much more detail and with much more certainty than presented here. Our book is simply a statement of current knowledge of the Scottish Lateglacial environment providing an integrated survey of available information for students of the Quaternary at both home and abroad. We hope that similar statements will soon be published on the Lateglacial environments of other countries or regions of the world.

---

*Includes alternative forms such as 'Late Glacial', 'Late-Glacial', 'late-Glacial', 'lateglacial', 'Late-glacial' and 'late-glacial'.

The Late Devensian Glaciation of North-East Scotland

CHALMERS M. CLAPPERTON (University of Aberdeen)
and
DAVID E. SUGDEN (University of Aberdeen)

ABSTRACT

This chapter is a review of current interpretations of Late Devensian glacial phenomena in north-east Scotland. Features of glacial erosion and deposition, of meltwater erosion and deposition, and of periglacial activity are briefly described before an assessment is made of the probable chronology. It is concluded that many erosional landforms were at least partly created by the Late Devensian ice sheet which appears also to have deposited all of the exposed tills and to have been temperate (wet)-based in places. The distribution and pattern of glacial landforms and deposits, particularly those of meltwater origin, suggest that the entire area of north-east Scotland was covered by the Late Devensian ice sheet. The latter was composed of three distinct ice streams with a triple junction zone located over central Buchan. Deglaciation led to the formation of glacial lakes and associated sediments in parts of Buchan and it is suggested that as the margins of the three ice masses withdrew the area was exposed to severe periglacial conditions for some time before the Lateglacial Interstadial. A return to periglacial conditions over north-east Scotland during the Loch Lomond Readvance was accompanied by the formation of corrie glaciers and possibly small plateau ice caps in the Cairngorm and Grampian mountains.

INTRODUCTION

Perhaps the best known and most comprehensive account of the glaciation of north-east Scotland is that by Synge (1956). On the basis of his own observations and a synthesis of previous work by T. F. Jamieson, A. Bremner and S. Simpson he suggested a pattern of glacier movement and a chronology for glacial events over the region. Many of the conclusions of this early work were based on such evidence as drift stratigraphy, stratigraphical interpolations and the distribution of supposed indicator erratics; ice marginal positions were postulated from supposed marginal and glacial lake overflow channels and ice terminal positions were derived from concentrations of hummocky accumulations of fluvioglacial sands and gravels. Figure 1 summarises the glacial sequence envisaged by Synge, to which have been added the Alpine chronological correlative stages suggested in his text. Accepting Synge's proposed moraine limits Sissons (1967a) subsequently rearranged the chronology to fit with his assessment of glacial readvances in Scotland as a whole (Fig. 2) but he has recently modified this view (Sissons, 1974c).

In recent years the present writers have mapped in detail the glacial landforms of some key localities in north-east Scotland and have obtained a great deal of significant information from continuous drift exposures in pipeline trenches temporarily opened through the area from Stonehaven to Cruden Bay. This work, and that in a small part of the area by Sissons (1972a) and Sissons and Grant (1972), have encouraged a reassessment of the hitherto well-established views of Synge, particularly in the light of recent developments in glaciological theory (*e.g.* Paterson, 1969) and the application of such theory to glacial geomorphology (*e.g.* Sugden and John, 1976).

This chapter is a review of current opinion on the interpretation of Late Devensian glacial phenomena in north-east Scotland, providing a generalised description and interpretation of the main landform and drift associations and a summary of the most recent chronological evidence.

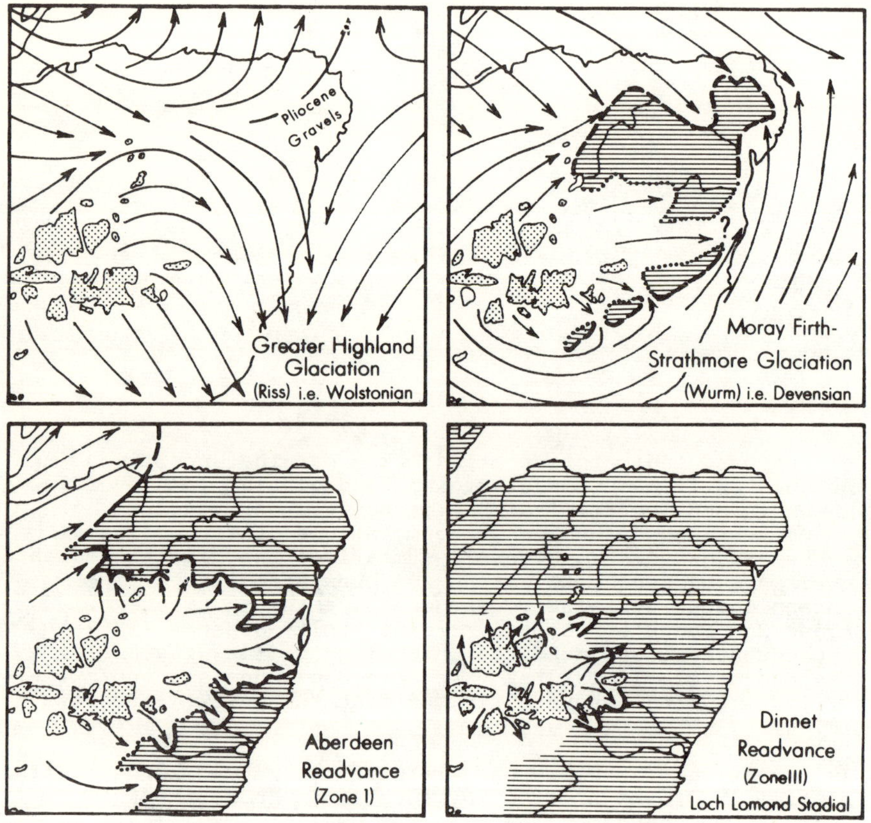

Fig. 1. The glacial sequence in north-east Scotland according to Synge (1956).

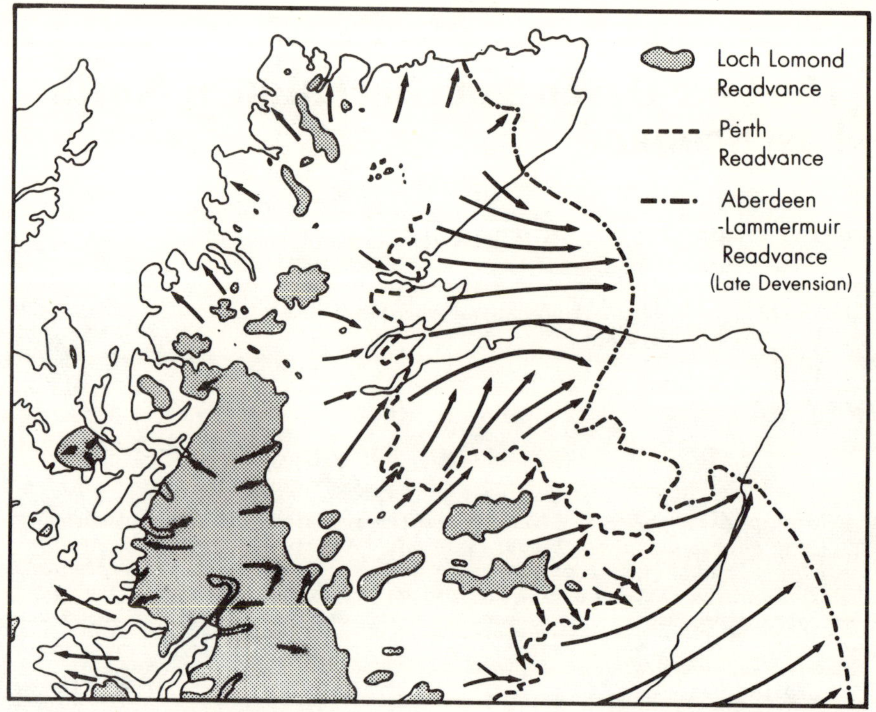

Fig. 2. The glacial sequence in northern Scotland according to Sissons (1967a).

PHYSICAL SETTING

There is not scope in this chapter to write a detailed account of the broad 'preglacial' elements of the landscape but since they are of some significance in understanding the glacial events a brief summary may be helpful.*

The topography of north-east Scotland is broadly composed of dissected plateau-like surfaces that steadily slope upwards in a series of large steps from the coastline towards the Cairngorm Mountains where the highest summit, Ben Macdhui, lies at an altitude of 1311 m. Although a number of separate topographical surfaces of different extent and in various degrees of dissection make up the stepped landscape, three major plateaux have been identified (Fleet, 1938). The highest is formed by ground above 915 m and lies in the Cairngorm and Lochnagar massifs (Fig. 3); it is separated by a distinct break of slope from the main Grampian plateau lying roughly between 600 and 800 m. Fragmented outliers of this surface and possible remnants of other lower ones are sharply bounded by steep slopes from the third main surface, that of the Buchan plateau lying between 100 and 150 m. Differential weathering, slope retreat and the development of a fluvial network in response to intermittent uplift in preglacial times led to the dissection of this landscape into narrow, steep-sided valleys that in places alternate with

*Fuller descriptions have been published by, for example, Odell and Walton (1962) and Clapperton (1969).

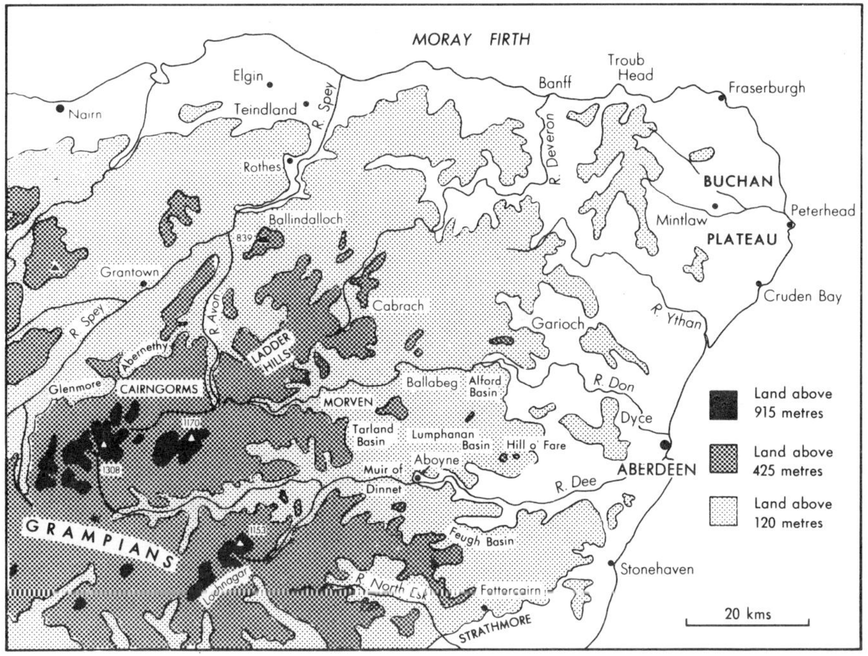

Fig. 3. General relief features of north-east Scotland and localities mentioned in the text.

wide basins surrounded by equally steep slopes. The basins of Tarland, Lumphanan and Feugh in the Dee area, Alford, Garioch and Dyce in the Don area and Glenmore, Abernethy and the Cabrach in the Spey area are particularly fine examples (Fig. 3). It is known that weathering processes decomposed underlying bedrock to depths in excess of 10 m in places, particularly in granite and schist, for many pockets of such decomposed bedrock presently remain beneath a cover of till, for example at Bandley (NJ/605165) in the Alford Basin. Such were the broad elements of the landscape geometry inherited and superficially modified by Pleistocene ice sheets which expanded across the area towards the North Sea basin from major dispersal centres farther west and from local centres such as the Grampian and Cairngorm mountains (Sissons, 1976b).

FEATURES OF GLACIAL EROSION

The landforms of glacial erosion in north-east Scotland are almost certainly the product of more than one ice sheet and cannot be wholly attributed to that of the Late Devensian. However, it is clear from the abundance of fresh bedrock surfaces and from the thick drift sequences deposited by the last ice sheet that a considerable amount of erosion was achieved, particularly where the ice sheet may have been wet-based. Landforms of glacial erosion in the region must therefore be considered in any assessment of the possible dynamic characteristics of the Late Devensian ice sheet.

They can be grouped into three main categories: (a) those of areal scouring which were formed by mechanisms operating at the glacier bed over wide areas of varied underlying topography; (b) those of linear erosion formed by similar mechanisms but at the base of powerful streams of ice discharging through and from the higher ground; (c) those of more localised erosion on the higher ground where small glaciers enlarged hollows in a combination of subglacial and subaerial mechanisms. A brief consideration of these landform types may help establish patterns of ice movement and suggest aspects of the dynamic characteristics of the Late Devensian ice sheet over the region.

Landscapes made up of features of areal scouring are typically composed of bare mamillated surfaces of fresh bedrock with a micro relief that shows a combination of smoothly abraded surfaces (often highly polished) and craggy 'lee-side' faces where bedrock fragments have been removed. Scoured rock surfaces are found on individual landforms at all scales from a few centimetres in size to valley sides, spurs and whole mountain tops many tens of metres in size and often in roche moutonnée form. Since it is now believed that the majority of such features develop most efficiently where the glacier base is at the pressure melting temperature, the distribution of landscapes dominated by scouring in north-east Scotland is informative. Areal scouring phenomena are either poorly developed or absent in higher parts of the Cairngorm plateau, but they become much more widespread on lower parts of the plateau and on adjacent hills and interfluves. This is particularly so on the flanks and interfluves of the Dee catchment area upstream from Aboyne and in the Spey catchment upstream from Grantown and in some of the main tributary valleys. Seawards from these localities, even along the high plateau of the eastern Grampians, the landscape loses its roughened appearance partly because the scale of scoured features is greatly reduced and partly because there is a thicker mantle of superficial deposits such as decomposed or broken bedrock, glacial drift and peat; this is particularly noticeable as the topography becomes progressively lower. Locally, however, as on the flanks of the Hill O'Fare (Fig. 3), patches of scoured and streamlined bedrock are prominent, while groups of small rock drumlins and shallow rock basins are present even on the low Kincardineshire plateau (120 m) only 3.5 km from the east coast. Highly polished bedrock occurs beneath glacial deposits along parts of the Moray coast near Banff and recent pipeline trenches a few kilometres west of Aberdeen have exposed scoured bedrock in places beneath thin till. It therefore appears that although glacial erosion has created obvious and large scale features of areal scouring in some western parts of the region there is clear evidence of distinct but perhaps more subtle and localised areal erosion even in the lower coastal districts. In a study of all available sections in the Aberdeen-Peterhead area D. K. Chester (pers. comm.) observed that distinct remnant tails of decomposed bedrock lie in the lee (eastern side) of bedrock knolls that have been scoured on all other sides. Although a detailed study has not been made of such phenomena it is apparent that the areas of heavy scouring predominate where basal flow lines of the ice sheet would have been channelled by the configuration of underlying topography (as in the upper valleys of the Dee and Spey) or else plastically compressed to flow more rapidly past obstacles such as the Hill O'Fare and other hilly ground rising above the coastal lowlands. Such a landscape is therefore one of selective areal scouring.

Features of linear erosion are essentially confined to the higher western parts of the region where pre-existing river valleys have been severely widened and, in places, deepened into trough forms (Sugden, 1968). Some terminate at prominent trough headwalls that slope steeply up to adjacent plateau surfaces rather than to corrie basins, the Einich, Avon and Slochd Mor features being particularly impressive examples (Fig. 4). Spectacular through troughs forming watershed breaches and partial breaches, such as the Lairig Ghru, Glen Derry and Loch Builg troughs also occur in these mountains. Apart from Glen Callater and Glen Muick and a few other less well developed troughs in the eastern Grampians the most prominent features of linear glacial erosion are confined to the Cairngorms. The characteristics and restricted distribution of troughs such as these are believed to relate to critical situations in which the basal layers of ice sheets flowing across and off high plateau country stream into and selectively exploit pre-existing valleys which are developed headwards in addition to being widened and deepened into trough forms (Sugden and John, 1976). Beyond the high Cairngorm and Grampian plateau in north-east Scotland the amount and rate of glacier discharge was apparently reasonably accommodated by the existing topographic geometry which was therefore modified in a much more restricted way.

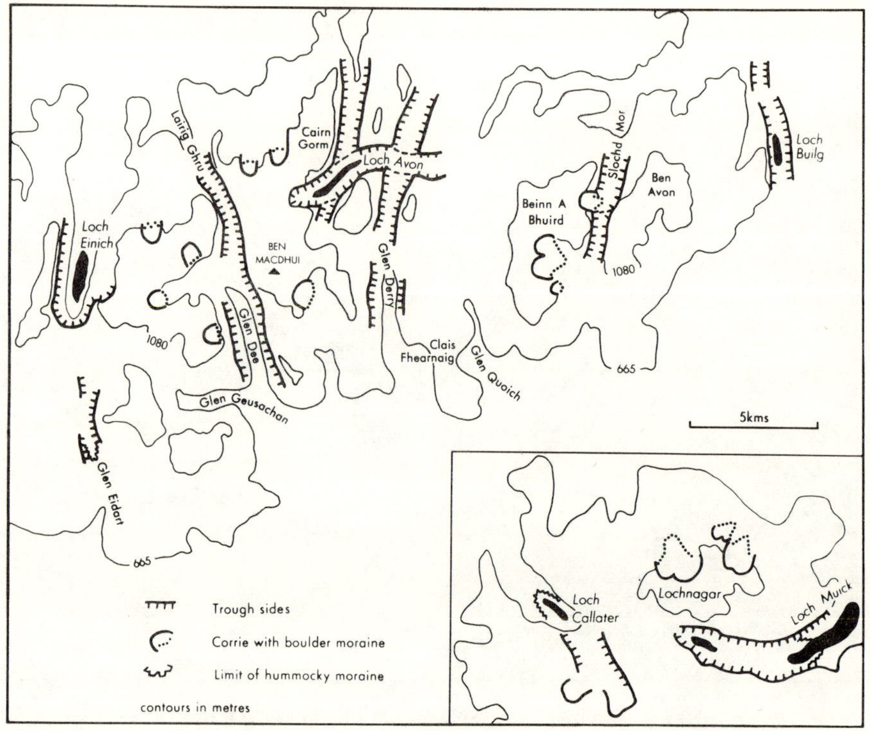

Fig. 4. Distribution of corries with boulder moraines, some hummocky moraines and some troughs in the Cairngorms and in the Lochnagar area (inset).

Impressive corrie forms are present on the highest plateaux and summits of the Cairngorm and eastern Grampian mountains but they are generally absent or only weakly developed outside these localities. In the Cairngorms some corries truncate the walls of larger ones, others are cut into the side of glacial troughs and many have clearly been modified by an over-riding ice sheet. It is therefore difficult to assign a specific age to any particular corrie since it appears that there may be several generations of them (Sugden, 1969). A number of corries are believed to have contained glaciers during the Loch Lomond Readvance (Clapperton and Sugden, 1975) and these possess particularly sharp back walls and plan profiles.

A number of corrie-like features with more subdued and vegetated walls are present in many parts of the eastern

Grampians and in other high hill masses within the region. Some are located at the heads of river valleys, as in the Ladder Hills, whereas many along the southern edge of the eastern Grampians overlooking Strathmore are independent features cut into the plateau. If all of these are true corries then they may have formed under glacial conditions earlier and possibly somewhat thermally different from those of the Late Devensian. It may be suggested, however, that the majority of Cairngorm and Grampian rock-walled corries were partly fashioned during conditions of marginal glaciation at some time in the Late Devensian and that most were inundated by the ice sheet of that age.

The overall characteristics and distribution of features of glacial erosion in north-east Scotland indicate that the last ice sheet was certainly temperate (wet)-based in places and effected much erosion. At localities such as the Cairngorm plateau and parts of the eastern Grampians, however, where widespread blockfields overlie soliflucted layers of decomposed bedrock fragments (with incorporated erratics), the ice sheet may have been cold-or-dry-based, thus achieving minimal erosion. The ice sheet would certainly have been thinner and slower moving over such areas. Answers to such problems may be obtained when it is possible to reconstruct spatial variations in the basal thermal regime of the last ice sheet over north-east Scotland.

FEATURES OF GLACIAL DEPOSITION

Deposits of glacial till are widespread in north-east Scotland, particularly over low-lying areas and on the floors and lower slopes of valleys in the higher hill massifs. They are absent or only thinly developed where features of glacial erosion are dominant and where glacial meltwater activity has incorporated them into the fluvio-glacial depositional system. Solifluction has redistributed substantial amounts of till downslope and there are probably widespread areas of higher ground where very little till was ever deposited.

The till is typically very stoney throughout the region and contains many blocks greater than one metre in size but, whereas it is commonly characterised by a clay-silt matrix in the coastal lowlands, sand and small pebbles dominate the matrix of tills in the higher western parts. This is partly due to the presence there of so much coarse grained granitic rock which breaks down into sandier material. Within the higher districts there appears to be only one till unit although in some places, such as the Quoich valley (Fig. 4), small and discontinuous lenses of water-laid sand and gravel are present within it. Contrasting with the apparent simplicity of the upland till a great deal of complexity was recognised in the so-called 'boulder clays' of the coastal lowlands by early workers such as Jamieson and Bremner. On the basis of colour differences, erratic content and interpolated stratigraphy between sites, four tills of supposedly different age were identified; a lower 'indigo' (sometimes shelly) till, a grey 'Highland' till, a red 'Strathmore' till and a brown 'Aberdeen Readvance' till (Synge, 1956). The four apparently different tills were never observed in the same section however, and the chronology applied to them was essentially based on stratigraphical interpolation over wide tracts of the region. Recent investigations by the writers and others of till exposed in pipeline trenches cut through the eastern part of Aberdeenshire have clarified the till sequence. It has become clear by following over 50 km of continuous section that the colour of the till is very closely related to that of underlying or adjacent bedrock. Since the Highland Series of metamorphic and igneous rocks and some Palaeozoic sedimentary rocks are extremely varied in type and colour it is not surprising that tills ranging in colour from indigo to bright red have developed over comparatively short distances. Detailed revision of the hard rock geology over parts of north-east Scotland has also cast doubt on the validity of some indicator erratic stones described in earlier literature. Not only is there a much greater diversity of rock types than previously known but also some of the indicator parent masses are either more widespread than previously suspected or have similar counterparts in other parts of the region (I. M. Munro, pers. comm.).

Recent exposures of till in the Aberdeen area indicate that three stratigraphical units may be identified (Murdoch, 1975). The first is typically a clay-rich till that is gleyed and grey-blue in colour. It has a strong fabric and contains stones typical of the immediate vicinity. The second unit is the most widespread. Containing a much lower proportion of clay-size particles this unit is generally sandier and normally has a much weaker fabric than that of the first. It commonly contains pockets and lenses of sand and gravel. The second unit always overlies the first or rests directly on bedrock wherever the first is absent. Where the two units occur together the upper appears to succeed the lower conformably, although a thin and impersistent layer of sand and gravel is sometimes present between them. The third till unit is a distinctive bright red-brown colour and is normally very clay rich. It is essentially an east coast phenomenon and seldom extends for more than 8 km inland (Fig. 5). Pebbles contained in this till unit have been derived not only from local metamorphic and intrusive rocks composing the Highland Series but also from sedimentary beds of Devonian, Permo-Triassic, Jurassic and Cretaceous ages, most of which outcrop in the adjacent North Sea basin. Analyses of clay particles in the third till unit (Glentworth

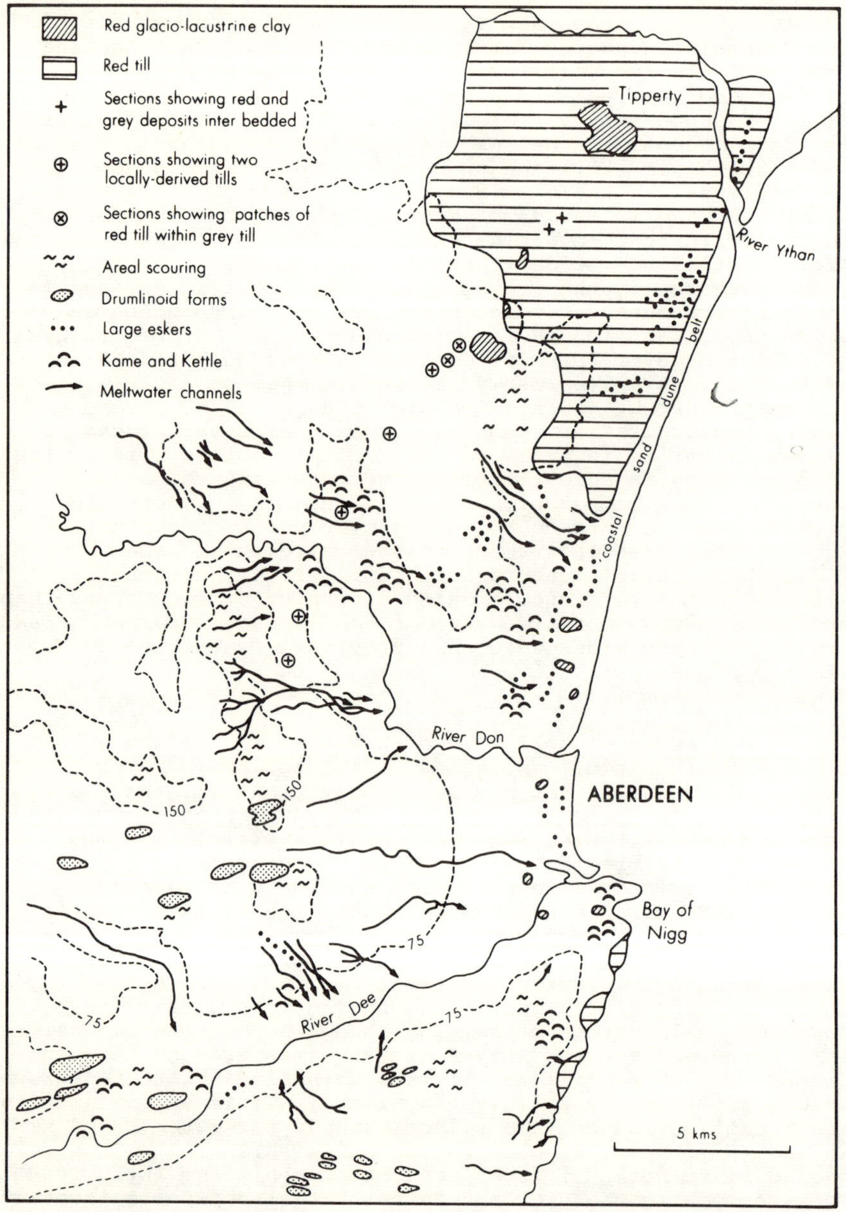

Fig. 5. Some glacial landforms and deposits in the Aberdeen area (after Clapperton and Sugden (1972) and Murdoch (1975)).

et al., 1964) indicated that the deposit is unlikely to have been derived from the red Devonian rocks of Strathmore (25 km to the south-west) as formerly believed. It therefore seems much more probable that the till was derived immediately off the present shoreline and was transported to the coastal lowlands by an ice sheet with an on-shore component of movement. Wherever the third unit is present with either of the other two it is always the uppermost, a relationship which is very well displayed near Aberdeen in the sea cliff section at the Bay of Nigg where a lower grey-brown till passes conformably upwards into a red-brown till. At several sites exposed a few kilometres inland the two units were observed to be occasionally interdigitated. Similar relationships may exist between the till units exposed in coastal Banffshire (Peacock, 1971c) but a rigorous comparative study has not yet been made.

Perhaps the most important conclusion to be drawn from recent work on the tills of north-east Scotland is that the units described can all be explained with reference to only one glaciation. The fabric characteristics of the first unit suggest that it is either a lodgement till or a basal melt-out till that has retained its fabric. The second unit is very similar in nature to top melt-out tills described from other glaciated regions (Boulton, 1972; Shaw, 1971) and may even be composed partly of flow till units. The sand and gravel lenses within the melt-out till and between the supraglacial and basal tills are to be expected since melt-water seeping from wasting ice constantly winnows particles of debris from the ice and deposits them in supraglacial ponds or in melt-stream channels and small lakes within the stagnant ice; these deposits are then often buried by melt-out or flow till. The third till unit is clearly contemporaneous with the second with which it achieved its stratigraphical relationships through fluctuating strengths and flow directions of ice streams emanating from different provenances (Fig. 6) - *i.e.* ice moving eastwards and north-eastwards from the Cairngorm-Grampian area in confluence with ice streaming from the southern Highlands moving north-eastwards along the North Sea coast and with ice streaming from the northern Highlands moving eastwards and south-eastwards along the Moray Firth coast; both of these coastal ice masses had an on-shore component of movement.

The three units of glacial till form not only a widespread mantle of varying thickness over the lower parts of the region but also compose distinct streamlined or moulded landforms. Although there are no true drumlin fields in north-east Scotland, and apparently very few individual till drumlins, substantial tracts of streamlined or drumlinoid topography composed partly of glacial drift occur in parts of the region. They form conspicuous elements of the landscape along the coastal lowlands of the Moray Firth and

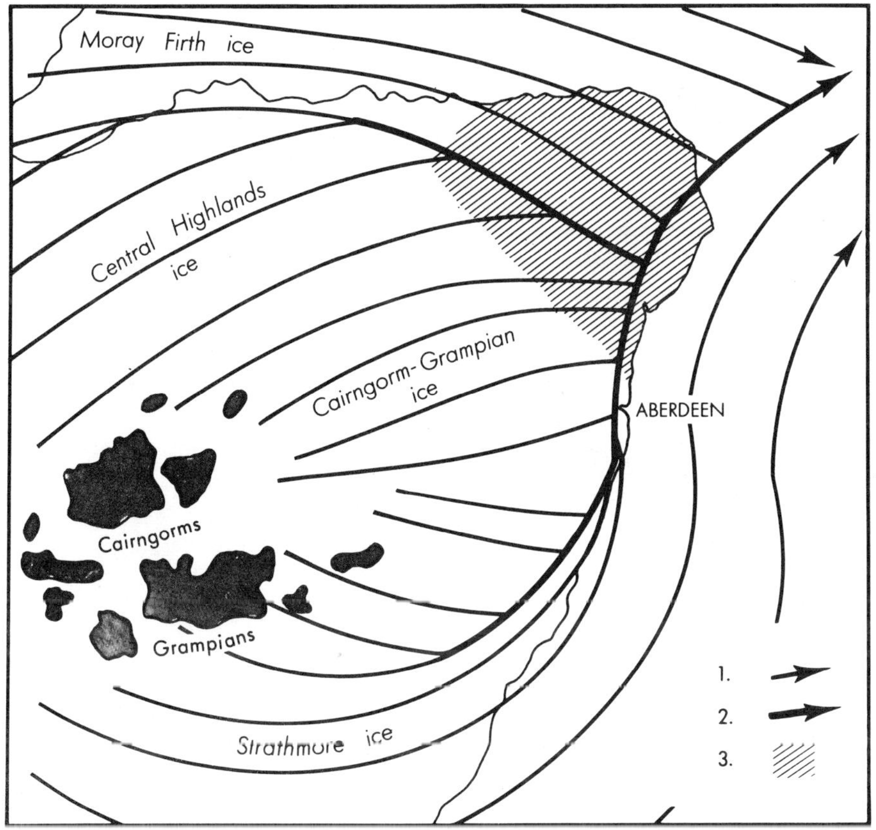

Fig. 6. 1, Generalised flow-lines for the Late Devensian ice sheet. 2, Ice confluence zone of locally fluctuating flow strength and direction through time during the Late Devensian. 3, Buchan area peripheral to ice dispersal centres; a zone of low glacial erosion and deposition, exposed earlier than the Lateglacial to periglacial conditions.

particularly on the Dee-Don interfluve west of Aberdeen where they are typically from 500 m - 2 km long and 15 - 40 m high. The few sections which have exposed underlying materials suggest that the majority of the drumlinoid forms may be composed mainly of streamlined bedrock with tails of till, although in places there is also a thin overall veneer of till. In both localities they have a strong west-east alignment which corresponds to the last direction of ice movement as deduced from other evidence, but local divergences often show the influence of large topographical units on basal flow lines; for example in the lee of the Hill O'Fare.

Since the only moraines deposited at the margins of formerly active glaciers in north-east Scotland are located in only a few of the Cairngorm and Grampian corries it may be suspected that the last ice sheet over the area terminated somewhere beyond the coasts. Moraines analogous to those deposited by the last ice sheets in northern Europe and North America are wholly absent. Composed mainly of large boulders and pebbles, the corrie moraines lack the clay and silt matrix of the lowland tills but have a distinctive morphology. In most corries prominent ridges 1 - 5 m high form clear lateral moraines that slope down the side walls before curving across the corrie floors as single or multiple end moraine complexes. Where ridges are absent the former glaciers are marked by a wilderness of large blocks thickly strewn over the corrie floor. The rather restricted distribution of the corrie moraines (Fig. 4) suggests that they were formed during a period of marginal glaciation when only small glaciers could be nourished by snow drift from adjacent plateau surfaces. Down valley from the corries there are very few morainic landforms composed of glacial till, but in the upper parts of some of the adjacent Grampian glens there are landform assemblages that have been termed 'hummocky moraine' (Sissons, 1972a). In Glen Callater and the three glens immediately to the west an apparent chaos of boulder-strewn hummocks and ridges extends for some 5 - 6 km from the valley heads before abruptly terminating, often at the apex of suites of terraced sediments that continue farther down valley. Detailed mapping shows that there is some morphological pattern in places. For example, a number of ridges are aligned along the lower slopes and floors of the valleys and some curve in towards the valley centres. The few available sections show that some of the hummocks and ridges are composed partly or wholly of coarse water-laid sand and gravel and that others consist partly or entirely of a rubbly melt-out till. Each assemblage as a whole tends to be morphologically and spatially distinct from those wholly of fluvioglacial origin farther down valley and throughout the lower parts of the region. The characteristics of the hummocky moraine imply an origin closely linked with short tongues of glacier ice that stagnated and wasted away *in situ* in the upper valleys of the Grampians; such phenomena are absent from much of the Cairngorms apart from Glen Eidart, Glen Geusachan, and possibly a few other localities. The origin and distribution of the hummocky moraine has been explained with reference to the readvance of glaciers during the Loch Lomond Stadial (Sissons, 1972a).

FEATURES OF GLACIAL MELTWATER ACTIVITY

Landforms created by glacial meltwater activity are widespread throughout north-east Scotland and include the channel forms associated with melt-stream erosion and a range of ice-contact forms relating to meltwater deposition in relatively inactive glacier ice. Since meltwater channel forms have previously been described from several parts of the region (Bremner, 1921; Synge, 1956; Sugden, 1970; Clapperton and Sugden, 1972, 1975) there is little need to repeat such descriptions here. Of much greater consequence is the interpretation not only of individual channels and channel systems but also of the entire spatial pattern and distribution of the phenomena over north-east Scotland in relation to the last ice sheet cover. Early explanations of the channels invoked former melt-streams flowing along ice-free hillsides at the glacier margin or from one ice-marginal lake to another and generally implied that water could not penetrate within or beneath the glaciers. Investigations at existing valley glaciers and ice sheets and developments in glaciological and glacio-hydrological theory over the last decade show that not only can water exist everywhere within and beneath glacier ice that is at the pressure melting temperature but also that it is an integral part of the glacier system (Sugden and John, 1976); as such it can flow in thin sheets and/or in discrete streams at the glacier bed, generally directed by pressure conditions towards the glacier margins. Detailed mapping of meltwater channels in parts of north-east Scotland has shown that morphological characteristics such as up and down longitudinal floor profiles and complex anastomosing networks are most satisfactorily explained with reference to subglacial flow as Sissons (1960, 1961a) advocated for many such features elsewhere in Scotland.

Meltwater channels are present throughout north-east Scotland from high altitudes in the Cairngorms almost to sea-level on the east and north coasts. Although a rigorous morphological study of the phenomena has not yet been made there is little apparent difference in sharpness of form or in the continuity of the systems that would imply an age difference between those located in the mountains and those in the coastal lowlands (other than those associated with the Loch Lomond Readvance glaciers by Sissons, 1974c). The majority are located in topographical situations that clearly imply some relationship between topography and former glacier hydrology. The channels are typically arranged in systems of various size and complexity. Some are dominated by one major channel which may be greater than 50 - 70 m in depth, such as the Clais Fhearnaig (Fig. 4) lying between Glen Lui and Glen Quoich 8 km west-north-west of Braemar; such channels normally have only small feeder channels located at their proximal end. Others, such as the system on the hillside above Glensaugh (near Fettercairn), that on the northern flanks of Morven between the Dee and Don valleys, and particularly those over much of Buchan possess a remarkable degree of connectivity with many tributaries, distributaries and isolated loops (Fig. 7). It is apparent that the large dominant channels are more common in narrow cols whereas the more complex channel systems are located in broad pre-existing valley heads and lowlands. This relationship between channel development and topography has been noted also in the Cheviot Hills (Clapperton, 1971) and occurs in many other parts of Britain. Sugden and John (1976) have recently summarised an explanation for such a relationship and stress that water flowing at the base of temperate ice would be directed by pressure conditions at the glacier bed to seek and become most firmly established in courses following routes of lowest pressure. Beneath an ice sheet overlying irregular topography (as over much of north-east Scotland) such courses would be in existing river valleys or in valley heads and cols across watersheds. Channel systems in these localities have been termed ice-directed (Clapperton and Sugden, 1972), and their significance is two-fold. First of all their existence suggests that the last ice sheet over the region was warm based at least during the later stages of its life and secondly they provide a clear indication of regional trends in the direction of ice flow since this broadly determines the alignment of pressure fields which control the direction of meltwater drainage at the glacier bed.

Groups of meltwater channels with a distinctly different alignment are commonly located on the lower hillslopes throughout the region. Since they are orientated either directly downslope or at a small angle to the contours, it is believed that they were formed when the ice sheet had stagnated and thinned to such an extent that free flow of meltwater could take place subglacially down the hillsides to the valley bottoms. In other words the strong hydraulic pressure gradient that had previously directed meltwater flow according to the regional ice flowlines no longer existed.

Landforms composed of fluvioglacial sands and gravels are also present throughout the region from high in the Cairngorms to the coasts. In general they show the variety of form and internal composition typically associated with such deposits in Scotland as a whole and which have been summarised by Sissons (1967a, 1976b). With regard to their significance in interpreting or reconstructing the last glaciation of north-east Scotland however, a consideration of the following points may be helpful. Like the meltwater channels the deposits seldom occur as isolated features. They are normally present as complex associations of eskers, kames, kame terraces and kettle holes. Concentrations of such phenomena are usually located on the lower slopes of

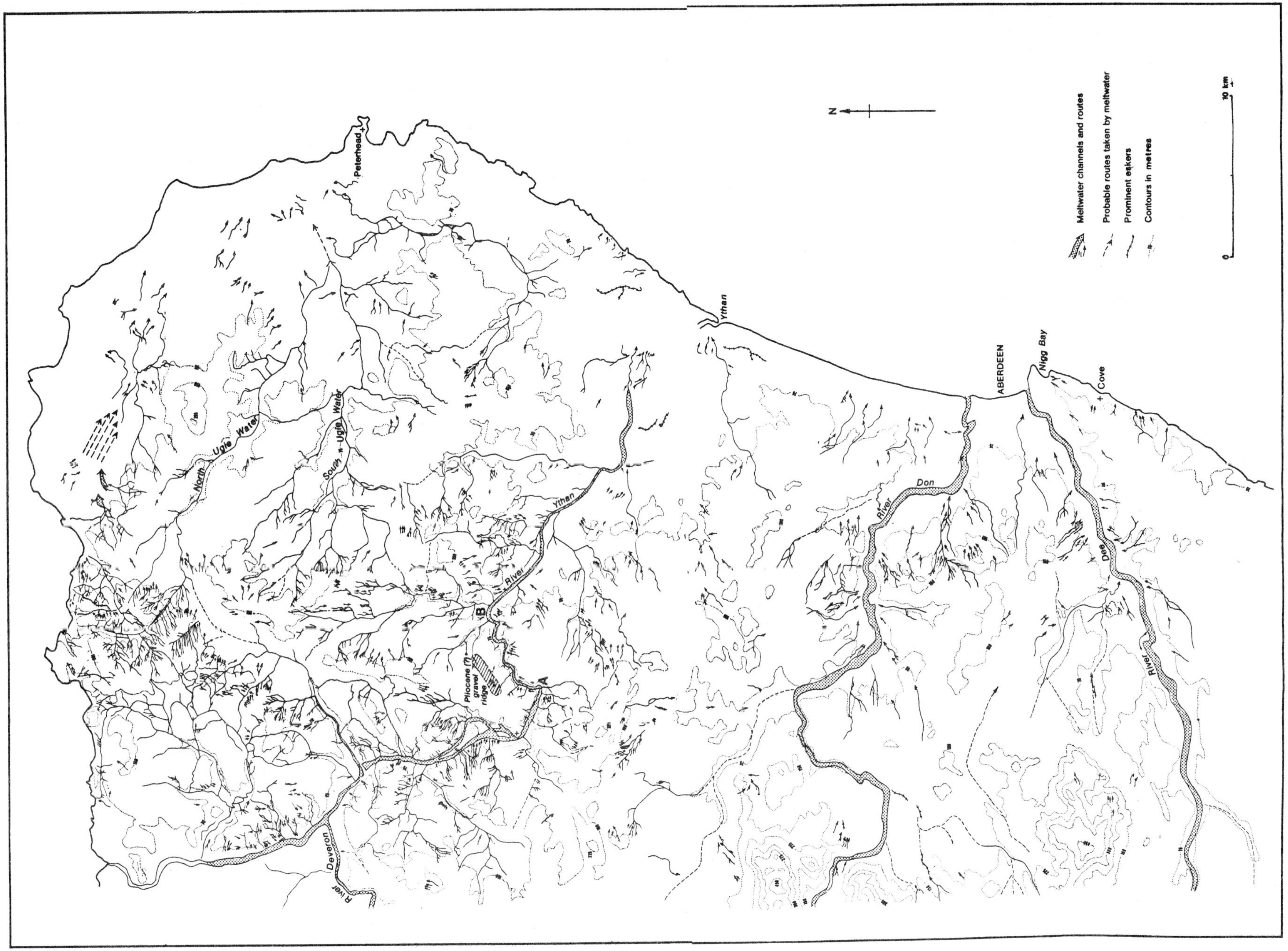

Fig. 7. Meltwater channels and associated features in part of north-east Scotland.

valleys in the mountains and aligned along major meltwater routes in the lowland basins and coastal districts. Those in the upland areas are normally more striking and sharply defined than those in low lying districts for various reasons; they have never been cultivated, they are composed of coarser materials, and since they were probably exposed later during deglaciation they have experienced less modification by weathering and slope processes. Very striking continuous eskers nevertheless do occur in some lowland localities, such as at Candy 8 km south-west of Stonehaven, at Kemnay 22 km west-north-west of Aberdeen and at Flemington 8 km south-west of Nairn to mention only three out of many well known examples.

The most impressive assemblages of glacial meltwater deposits are in the major valleys, lowland basins and coastal districts at sites in which large masses of glacier ice would have stagnated *in situ* during deglaciation of the region. In such localities interconnected eskers and esker-kame systems often show a similar alignment to that of adjacent ice-directed meltwater channels. As in the Cambus O'May-Dinnet Muir district of the Dee valley, in most of the Spey valley and along the Moray Firth coast from Inverness to Fraserburgh the fluvioglacial deposits are clearly an integral part of the meltwater drainage system which formerly functioned at the glacier bed. Indeed, the continuity of the features along the Moray Firth coastal lowland strongly implies a roughly contemporaneous origin for the entire assemblage and may discount previous opinions (Synge, 1956, Peacock *et al.*, 1968) that eskers and kames immediately east of Elgin represent a readvance or stadial phase of the so-called Moray Firth glacier.

Fluvioglacial complexes are sparse or absent from some high- and low-level basins such as those of the Cabrach and the Alford-Garioch area respectively and from parts of the central Buchan plateau. It is difficult to prove that such areas never functioned as major meltwater routes or as primary ice stagnation centres, but these explanations must be considered. In addition it may be significant that there is normally only a relatively thin till cover present over parts of these localities (*e.g.* one metre in the Alford basin and much less over much of Buchan) and since a substantial part of fluvioglacial deposits in general are probably derived from ice thickly laden with debris (Clapperton, 1971), it is not surprising that they are sparse in those localities.

The spatial continuity and morphological homogeneity of the meltwater features throughout the region suggest that they were created by the meltwater system of the last ice sheet which covered the whole of north-east Scotland. Their presence all over Buchan contradicts previous notions that the area remained ice-free during the Late Devensian glaciation (Charlesworth, 1955; Synge, 1956; Fitzpatrick, 1972). There may have been a degree of sequential development from the coasts inland as recession and stagnation proceeded but by and large the meltwater phenomena appear to relate to one continuous period of glaciation and deglaciation.

FEATURES OF GLACIOLACUSTRINE DEPOSITION

Deposits identified as glaciolacustrine in origin have long been recognised at various localities in the coastal districts of north-east Scotland. They consist mainly of horizontally bedded and little disturbed sequences of laminated clays and silts, sometimes with thin sand and pebbly layers, and are well known from sites such as that near Rothes in the lower Spey valley (Bremner, 1932), coastal Banffshire (Read, 1923) and along the east coast of Aberdeenshire (Bremner, 1915). The lacustrine sediments are generally the uppermost of any local stratigraphical sequence although small lenses of flow till may partially overlap them or be interbedded with upper layers (*e.g.* at Troup Head, Aberdeenshire). The colour of the sediments is normally similar to that of adjacent till and a most interesting observation made by Murdoch (1975) at the Tipperty locality north of Aberdeen (Fig. 5) is that the red-coloured clays and silts have been contaminated by material derived from grey till.

The most recent interpretations (*e.g.* Peacock, 1971c; Murdoch, 1975) conclude that the sediments are essentially glaciolacustrine rather than glaciomarine and that they were deposited in lakes which formed over low ground and coastal valleys that were first vacated by ice during deglaciation, particularly at localities where ice masses from different provenances were coalescent (Fig. 6). The abrupt seaward termination of the deposits supports the contention that glacier ice occupied the Moray Firth and western North Sea basin when the lakes were in existence. At some localities north of Aberdeen the lacustrine sediments have clearly been draped over undulating fluvioglacial deposits and must therefore have been last in the depositional sequence, possibly originating in ephemeral supraglacial ponds and lakes and subsequently superimposed onto underlying topography. Some may also have originated in ice marginal lakes during deglaciation. Depressions on the ice surface and marginal lake sites are most likely to have developed along the coalescent margins of the ice masses emanating from the Moray Firth, the western uplands and Strathmore (Fig. 6).

FEATURES OF PERIGLACIAL ACTIVITY

The presence of gravel wedges and involutions in fluvioglacial quarries is clear evidence of the former occurrence of permafrost in the Buchan lowlands following the dissolution of the last ice sheet. They imply that the contemporary climatic conditions must have been typified by mean annual temperatures no higher than about -1° C (Péwé, 1966; Sissons, 1976b). The largest and best developed gravel wedges appear to occur in fairly level terraces of fluvioglacial outwash - as near Capo (NO/628664) beside the river North Esk in Kincardineshire where six wedges more than 2.5 m in size have been observed, and at a locality (NJ/048484) 7.5 km west of Peterhead where the surface polygonal pattern shows clearly on aerial photographs because of vegetation contrasts. Since comparatively smaller wedges seem only to have developed where the fluvioglacial sediments are hummocky it is possible that morphological and drainage conditions rather than temperature and/or time variables have influenced the scale of development of the phenomena. The former presence of a seasonally active regolith over permafrost is also suggested by the pronounced vertical orientation of pebbles in the upper 2 - 3 m of the quartzite and flint deposit (so-called 'Pliocene'?) at Windy Hills (NJ/800400) near Fyvie in central Buchan and in other drift exposures in that area where the cryogenic disturbance appears to be stronger or at least more conspicuous than in drift sections in other parts of the region.

Strong pedological evidence in favour of two episodes of periglacial activity has recently been obtained by Fitzpatrick (1975) who identified a double discontinuity in some soil profiles in north-east Scotland. Changes in particle size distribution, stone concentration and orientation, and in consistency led him to conclude that the lower discontinuity originated when freeze-thaw cycles extended to a depth of at least two metres and that the upper discontinuity represents a later more gentle and probably final phase in freeze-thaw activity.

Since all of the cryogenic disturbances observed in the glacial drifts (where they are most numerous) of north-east Scotland are from the ground surface downwards it may be concluded that they relate to periglacial conditions following the last ice sheet cover. Although there is a lack of absolute dates on periglacial events in the region the following points are significant. Lateglacial stratigraphy from a number of lake sites in north-east Scotland shows the typical tripartite sequence of minerogenic sediments followed by organic muds and then minerogenic sediments of the Loch Lomond Readvance (Gunson, 1975). At a site near Tarves (NJ/833318) in Aberdeenshire 1.5 m of soliflucted sediment overlies peat radiocarbon dated to 12,200 ± 170 B.P., suggesting that some measure of periglacial activity occurred in the lowland area during the Loch Lomond Readvance, a view shared by Sissons (1976b) for other parts of Scotland. Such evidence implies that during the more prolonged period immediately following deglaciation there may have been even more scope for permafrost development and cryogenic disturbance, particularly at sites peripheral to the ice sheet centres which would have become ice free early during deglaciation (such as central Buchan). Thus the double discontinuity in the soil profiles may suggest an earlier phase of periglaciation associated with ice sheet decay.

The most impressive landforms developed by intense cryogenic activity occur on the upper parts of the Cairngorm and Grampian mountains where massive screes, boulder lobes, garlands and blockfields are common. Sissons has summarised his views and other evidence in support of a Loch Lomond Readvance age for most of the features and concluded that "The periglacial landscape of the mountains is essentially a fossil landscape" (Sissons, 1976b, p.110). The importance of Flandrian solifluction must not be underestimated, however, since radiocarbon dates of 4,880 ± 135 B.P. and 2,680 ± 120 B.P. have been obtained for organic material buried by solifluction lobes in the Cairngorms (R.M.G. O'Brien, pers. comm.).

FEATURES OF THE LATEGLACIAL

North-east Scotland has for long been an area of considerable interest and debate in interpreting events of the Lateglacial period. A number of authors have postulated readvances of glaciers down valleys such as those of the Dee, Don and Spey, down the glens of the eastern Grampians and in the Moray Firth. The supposed readvance episodes were termed the 'Aberdeen-Lammermuir Readvance' (Sissons, 1967a), equated with the early decay of the Late Devensian ice sheet, and the Aberdeen and Dinnet readvances (Synge, 1956) equated with Jessen-Godwin zones I and III of the Lateglacial pollen sequence respectively. Recent investigations have shown that supposed readvances of glaciers in Lateglacial times to the proposed type localities at Aberdeen and Dinnet in the Dee valley, to Bellabeg in the Don valley and Ballindalloch in the Spey (Fig. 3) are invalid and that clear evidence from kettle holes at the base of the eastern Cairngorms shows that the Dee and Don valleys had become ice free and remained ice free from the early Lateglacial (Clapperton *et al.*, 1975); similar results have been obtained in the upper Spey valley (Young, 1974). The only readvance of glaciers in north-east Scotland since the Late Devensian occurred during the Loch Lomond Stadial and produced only small corrie glaciers in parts of the Cairngorms (where some corries apparently even remained ice free) and on Lochnagar and possibly small plateau ice caps with short outlet glaciers in parts of the central Grampians (Sissons, 1974b). A relatively abrupt end to the Loch Lomond Readvance is suggested by the morphology of the so-called hummocky moraines which, as already suggested, appear to have developed by the stagnation *in situ* of short tongues of valley ice emanating from plateau ice caps.

Vegetation and soil developments in north-east Scotland during the Lateglacial and early Flandrian have been described in detail by Vasari and Vasari (1968), Gunson (1975) and Vasari (this volume).

CHRONOLOGY AND CONCLUSIONS

The spatial continuity of the pattern of glacial landforms in north-east Scotland from the Lateglacial moraines of the mountains to the coasts strongly indicates that the entire region lay beneath the last ice sheet to cover Scotland. It is also clear from the landforms and the sediments that the ice sheet was wet-based (*i.e.* temperate) during at least the later stages of its existence. There is considerable difficulty in applying a finite age to this ice sheet stage however, because of the paucity of incorporated or contiguous organic deposits. Only three such deposits have been found so far and two of these are of doubtful significance. For example, in a pit excavated in the glaciolacustrine clays at Cruden Bay (NK/089370) a black-brown layer of apparently organic mud 2 - 8 cm thick is found beneath 4 m of laminated clays, silts and sands with inclusions of flow till, much of which shows some sign of slumping and settling. Similar material is present within fine fluvioglacial sediments near Mintlaw (NK/000482) 16 km to the north-west. It is possible from the stratigraphy at both sites that the organic layer is a derived deposit. A sample from Mintlaw gave an infinite radiocarbon age in excess of 40,180 B.P. (SRR-595) and one from Cruden Bay an age of 36,718 +1,030, -910 B.P. (SRR-596). It is believed that the latter is also infinite but gave a finite age because of slight contamination (D. D. Harkness, pers. comm.). Until a satisfactory explanation can be obtained for the origin of these carbonaceous deposits, which may be lacking in pollen (A. R. Gunson, pers. comm.) it is not possible to fully assess their significance but it could be proposed that they have been derived by meltwater from Middle Devensian organic material incorporated by an ice sheet of Late Devensian age.

Of perhaps greater significance is the deposit at Teindland in Morayshire interpreted by Fitzpatrick (1965) as a fossil podsol buried beneath soliflucted till and outwash debris. The organic content gave a radiocarbon age of 28,100 +480, -450 B.P. Recent palynological analysis of the Teindland profile suggests that the radiocarbon-dated organic bands incorporate pollen of an interstadial nature, while pollen at the base of the profile is of a more thermophilous character and is possibly of interglacial origin (perhaps chronologically to be related to the Ipswichian of England) (Edwards, Caseldine and Chester, in preparation). Fabric analyses of the overlying deposit indicate that it is an *in situ* till with fabric characteristics in accordance with the known direction of ice flow over the area. East and south-east of the site there is no stratigraphical or morphological discontinuity to suggest that the ice sheet which deposited the Teindland till did not also cover the rest of north-east Scotland. The Teindland evidence therefore confirms the occurrence of a Middle Devensian interstadial period in north-east Scotland, probably corresponding in age and nature with deposits from Lewis (Weymarn and Edwards, 1973). This interstadial period was followed sometime after 28,000 B.P. by the growth of an ice sheet that covered the entire region and left the landforms and deposits described in earlier sections. The limits and possibly the terminal moraines of this ice sheet probably lie somewhere in the North Sea basin. Dated evidence from other parts of Scotland and from England suggest that this Late Devensian ice sheet reached its maximum extent by approximately 17,000 B.P. (Sissons, 1974c).

The detailed investigation of glacial landforms and stratigraphy at key localities between Aberdeen and the Cairngorms indicates that the decay of the Late Devensian ice sheet proceeded without any major readvances until the entire area of north-east Scotland had become ice free. Sissons (1974c) has concluded that the whole of Scotland was free from glaciers by 12,500 B.P. The massive accumulations of ice-contact meltwater deposits in many parts of the region and deep into the heart of the Cairngorm and Grampian mountains indicate that the ice sheet underwent comparatively rapid downwastage that led to its widespread stagnation in many lowland sites. Sites peripheral to the confluent ice masses which composed the Late Devensian ice sheet (Fig. 6) would have become ice free early in the deglaciation. Such localities are believed to have been central, northern and eastern Buchan where glacial lakes formed either in depressions on the thinning ice sheet surface or against freshly exposed land surfaces. The relative scarcity of ice-contact fluvioglacial forms in the same areas may be partly due to the early recession of the ice. Partition of the ice masses over Buchan is likely to have occurred by about 16,000 B.P. thereby exposing the district to at least 2,000 years of probable periglacial conditions before the gradual establishment of a closer vegetation cover by the Lateglacial Interstadial. The return to glacial and periglacial conditions during the Loch Lomond Readvance, is now well established from morphological, stratigraphical and palynological evidence. Although western Scotland supported a large ice cap, a somewhat irregular pattern of small plateau and corrie glaciers appears to have developed in north-east Scotland because of the relatively low precipitation and local variations in amounts of snow accumulation. Thus whereas plateau ice caps with short outlet glaciers may have developed over parts of the Grampian Mountains, only a few corries in the considerably higher central and eastern Cairngorms contained glaciers. Sissons and Sutherland (1976) have partly explained such a distribution by arguing that the main snow-bearing weather during the Loch Lomond Stadial approached eastern Scotland from southerly directions, thereby creating a rain-shadow effect for the Cairngorms. Although it seems certain that the corrie moraines were formed by the glaciers of this stage there is an opinion that until some dateable organic material can be found beneath or incorporated in the so-called hummocky moraine of the plateau ice-cap glaciers some doubt must remain concerning glaciological and climatic reconstructions for the Loch Lomond Stadial (Sugden and Clapperton, 1975).

ACKNOWLEDGEMENTS

We are most grateful to the Carnegie Trust for financial support in geomorphological projects undertaken in north-east Scotland and to many colleagues for field assistance and countless snippets of ad hoc information about the region over the years. We are particularly grateful to D. K. Chester, A. M. D. Gemmell, A. R. Gunson, F. McLean and W. Murdoch for details of pipeline trench observations. Dr. Gemmell also kindly provided helpful comments on the manuscript.

Lateglacial Raised Shorelines and Deglaciation in the Earn-Tay Area

R. A. CULLINGFORD (University of Exeter)

ABSTRACT

Analysis of altitude data obtained by accurate levelling at frequent intervals along raised shoreline and fluvioglacial terrace fragments permits the recognition of five Lateglacial raised shorelines sloping down towards E13°S at gradients that diminish with decreasing age and altitude from 0.61 to 0.34 m/km. There is evidence of several additional stages in the Lateglacial shoreline displacement on which the data are too sparse to permit the calculation of meaningful gradients. The lowest shoreline for which gradients were calculated descends eastwards beneath the carse deposits associated with the Main Postglacial Shoreline.

The highest and most conspicuous feature, the Main Perth Shoreline, was formed whilst the decaying ice-sheet margin lay near Dunning and Perth, and Shorelines LP-1 to LP-4 were formed when the ice-margin lay not far from that position. The present elevation of the shorelines is explained in terms of an interplay between glacio-isostatic recovery, which was rapid during ice-wastage, and eustatic changes of sea-level.

Consideration is given to the character of the raised beach sediments, and to the dissection of the raised estuarine flats.

INTRODUCTION

The purpose of this paper is to put forward evidence relating to the displacement of shorelines that were formed during the Late Devensian deglaciation of an area comprising the south-coastal fringe of the Firth of Tay west of Balmerino, the Earn valley below Aberuthven, the Tay valley below Stanley, and the Carse of Gowrie (Fig. 1). The Earn-Tay lowlands occupy a basin let down between two faults in the centre of the Tay anticline, and are underlain by Upper Old Red Sandstone sediments mantled with a variable thickness of drift, the lowest recorded rockhead altitudes being -62 m O.D. at Perth, and below -65 m on the line of the Tay Road Bridge (McManus, 1966, 1972). The basin is bounded on either side by abruptly-rising hills composed mainly of Lower Old Red Sandstone lavas - the Sidlaws to the north-west, and the Ochils to the south-east. From a width of 3-4 km near Bridge of Earn, the basin broadens north-eastwards to a maximal width of 9 km before narrowing sharply at its eastern end, where it is only 2 km wide between Dundee and Newport. The present and former Earn-Tay estuaries have therefore been greatly sheltered from wave attack, and both during and since deglaciation the basin has been the scene of abundant, sheltered estuarine deposition, with copious sediment supplied by two large rivers and their fluvioglacial forerunners.

METHODS OF INVESTIGATION

The methods used in this investigation have mostly been described and justified elsewhere (*e.g.* Sissons and Smith, 1965; Cullingford and Smith, 1966; Sissons, 1967a; Cullingford, 1972; Gray, 1975a), so only a brief summary is given here, except for those aspects that have perhaps been inadequately explained before.

Field Methods

Morphology and stratigraphy. Field data collection began with the unselective mapping of all identifiable marine, fluvial and fluvioglacial features below the 61 m contour.

This arbitrary upper limit was chosen because evidence relating directly to Late Devensian shoreline displacement peters out before this altitude is reached, and mapping extended above this level only when necessary to complete the mapping of fluvioglacial landforms or landform assemblages. Mapping of deposits from surface materials, available exposures, and information from shallow augering was carried out simultaneously with the morphological mapping, the principal aims of which were first to plot the position and extent of all steps, regardless of origin, and to record an assessment of their suitability for measurement, and secondly, to map the relationships of steps to each other and to other features. In describing the steps, particular attention was paid to the degree of flatness and the nature and amount of dissection of the surface, the sharpness of the break of slope between step and backslope, and the height and steepness of the latter. All mapping was carried out at a scale of 1:10,560, this being the smallest available scale at which adequate morphological detail could be shown, and sufficient accuracy of location obtained. The description of deposits in the field was entirely by visual judgment, although several laboratory analyses of particle size have subsequently been carried out.

The stratigraphic information obtained from the field mapping and shallow augering was supplemented by the use of more than 200 commercial borehole records obtained for engineering (mostly road-building) purposes.

Measurement of altitude. The method of altitude measurement used throughout this study was accurate instrumental levelling (using Ordnance Survey bench marks tied to the second geodetic levelling) to establish the present altitude above Ordnance Datum (Newlyn) of each terrace fragment at frequent (50 - 80 m) intervals along its length. All traverses were closed, and most of the bench marks used were checked against at least one other bench mark to establish whether road settlement or the rebuilding of walls had altered their altitude since the Ordnance Survey levelling. All 708 altitudes used in this study are accurate to within 0.1 m, and the vast majority of them to within 0.05 m or less.

The advantages of this method of terrace height determination, as compared with the use of a barometer or a hand level, have been dealt with in some detail elsewhere (*e.g.* Sissons, 1963a, 1967a; Cullingford and Smith, 1966; Kirby, 1969; Cullingford, 1972; Gray, 1975a), and need not be reiterated here.

Analysis of the Altitude Data

Shoreline diagrams. The first stage in the analysis of the altitude data was to plot the measurements on shoreline diagrams, of the height-distance or 'equidistant' type, which were used in preference to shoreline relation diagrams. Recent users of the latter for Scotland include Synge and Stephens (1966), Andrews (1970) and Gemmell (1973). Because of the magnitude of the underlying assumptions, however, the use of these diagrams has been under increasing attack (*e.g.* Marthinussen, 1960; Andersen, 1965; Donner, 1965; Sissons, 1967b, 1972b; Gray, 1975b) and the writer believes that the type of displacement implied by the use of relation diagrams requires to be demonstrated rather than assumed. Furthermore, they have the disadvantage that any errors in identifying or measuring a synchronous reference level are transmitted to the rest of the data. The use of height-distance diagrams involves no unwarranted assumptions about the nature of deformation geometry, but poses problems that include the following:

(i) A true representation of the two-dimensional form of the raised shorelines depends on the vertical plane of projection being normal to the isobases, and major deviations from this alignment cause distortions that may lead to miscorrelations, especially where the vertical separation of successive shorelines is small.

(ii) Isobase curvature within the area of field data also causes distortions, because the plane of projection can only be truly orthogonal to the isobases within a limited area.

The first of these problems was eased greatly by the presence in the Tay-Forth region of two widespread, conspicuous raised shorelines that stand out clearly on diagrams plotted in a wide range of planes of projection, and for which isobases could therefore be drawn with confidence. Trend-surface analysis was used to construct isobases for both the Lateglacial Main Perth Shoreline (Smith *et al.*, 1969) and the Main Postglacial Shoreline (Cullingford, 1972), and these isobases, together with those determined for the shorelines in eastern Fife (EF-1 to EF-6; Cullingford and Smith, 1966; Cullingford, 1972), enabled the selection of 4 planes of projection that could be expected to cover all reasonable possibilities concerning the most appropriate plane for the Earn-Tay Lateglacial data: W7°N-E7°S, W13°N-E13°S, W18°N-E18°S and W37°N-E37°S. The first is the direction in which the Main Perth Shoreline slopes in the Earn-Tay-East Fife area; the second is the local direction of slope of the Main Perth feature in the Earn-Tay lowlands; the third is the most appropriate plane for the East Fife shorelines, and is very close to the regional direction of slope (down towards E17°S) of the Main Perth Shoreline in the Tay-Forth region; and the last is orthogonal to the trend of the Main Postglacial isobases in the Earn-Tay area. The second of these planes (W13°N-E13°S) might be expected to be the most appropriate for Lateglacial data in this area, but it was thought advisable to allow for changing isobase trends by considering other possibilities. Shoreline diagrams in the above four planes appear in Figs 3 and 4. The distances of the measured points in each plane were computed from the National Grid co-ordinates by trigonometry.

The problem of isobase curvature is not likely to lead to significantly large errors within the relative small area studied.

Correlation of shoreline fragments. Shorelines are usually identified on the basis of marked alignments of points on shoreline diagrams, and they are usually depicted as straight lines, despite the long-established curvilinear nature of shoreline deformation (*e.g.* Goldthwait, 1908; Robinson, 1908; Ramsay, 1924; Broecker, 1966). Since the Earn-Tay area is too small for any marked curvilinearity to be apparent (Cullingford, 1972), the recognition of straight alignments is a reasonable approach to the problem of correlation provided that account is taken of the following constraints imposed by the field evidence:

(i) The staircase constraint: Raised beaches that occur one above the other in a staircase cannot be contemporaneous with each other. As shown in Fig. 1, both separate and overlapping staircases are frequent, and impose a valuable limit to the number of possible correlations.

(ii) The continuity constraint: Morphological continuity is often greater than the distribution of measured altitudes might suggest, some measured terrace fragments being connected by a clear shoreline and/or terrace that for some reason could not be measured.

(iii) The ice-margin constraint: A raised shoreline remnant cannot be correlated with features that lie within an area known to have been contemporaneously ice-covered.

Calculation of shoreline gradients and isobase trends. The gradients of the raised shorelines identified from the shoreline diagrams in the four different planes were calculated by linear regression analysis, and isobases for certain shorelines were constructed by fitting trend surfaces to the altitude and grid co-ordinate data. These techniques were used purely as an objective way of calculating shoreline gradients and constructing best-fit isobases. The correlation coefficients and levels of explanation (Tables 4 and 5) express the closeness of fit, and cannot be used reliably to assess the validity of the correlations, which are based on visual inspection of shoreline diagrams in the light of the field constraints.

Since the object is to correlate shoreline fragments rather than individual altitudes, and since the altitudes on a single fragment are automatically correlated with each other, Gray (1972b, 1975a) has advocated the use of geographic-mean samples (*i.e.* one mean value per fragment) rather than total sampled populations in linear regression and trend-surface analysis, thus also partly alleviating the undesirable effects of spatial clustering on measures of significance. This is contrary to the present author's previous practice (Cullingford and Smith, 1966; Smith *et al.*, 1969; Cullingford, 1972). One problem of using fragment means or grouped data in the Earn-Tay area is that the data consists of a mixture of short and long fragments,and, if the altitude of each fragment is expressed only by a single mean value, the data on the short fragments in effect are emphasized at the expense of the more abundant (and often more reliable) data on the long fragments. There is also a serious loss of relevant information concerning gradient from the long fragments. These problems can partly be alleviated by dividing each long fragment into two or more 'sub-fragments', but this then introduces arbitrary decisions concerning the optimal length of fragment, and begins to increase again the clustering one has been attempting to reduce.

As far as the calculation of gradients by linear regression analysis is concerned, the writer is therefore unconvinced of the geomorphological advantage to be gained by using grouped data in preference to ungrouped data in an area such as the Earn-Tay lowlands, though it is probably to be preferred in areas where the shoreline remnants are much more highly fragmented. With regard to trend-surface analysis, the use of grouped data results in more meaningful measures of significance that enable a more objective decision to be reached concerning the highest appropriate order of trend surface, as pointed out by Gray (1974a). For this latter reason, and for ease of comparison with studies that have used grouped data, analyses were carried out for the latter as well as for the ungrouped data, and regression details for both are given below (Tables 3 - 5). For the analyses using fragment means, measured fragments more than 0.75 km long were subdivided into two 'sub-fragments', and those over 1.5 km long into three.

LATEGLACIAL RAISED SHORELINES AND BEACHES

Morphology, Altitude and Composition

The locations of the Lateglacial shorelines and the associated raised beaches and estuarine flats are shown in Fig. 1, and the altitude data for shoreline fragments along the south coast of the Firth of Tay and in lower Strathearn are summarized in Table 1, and for fragments in the lower Tay valley and the Carse of Gowrie in Table 2. The height

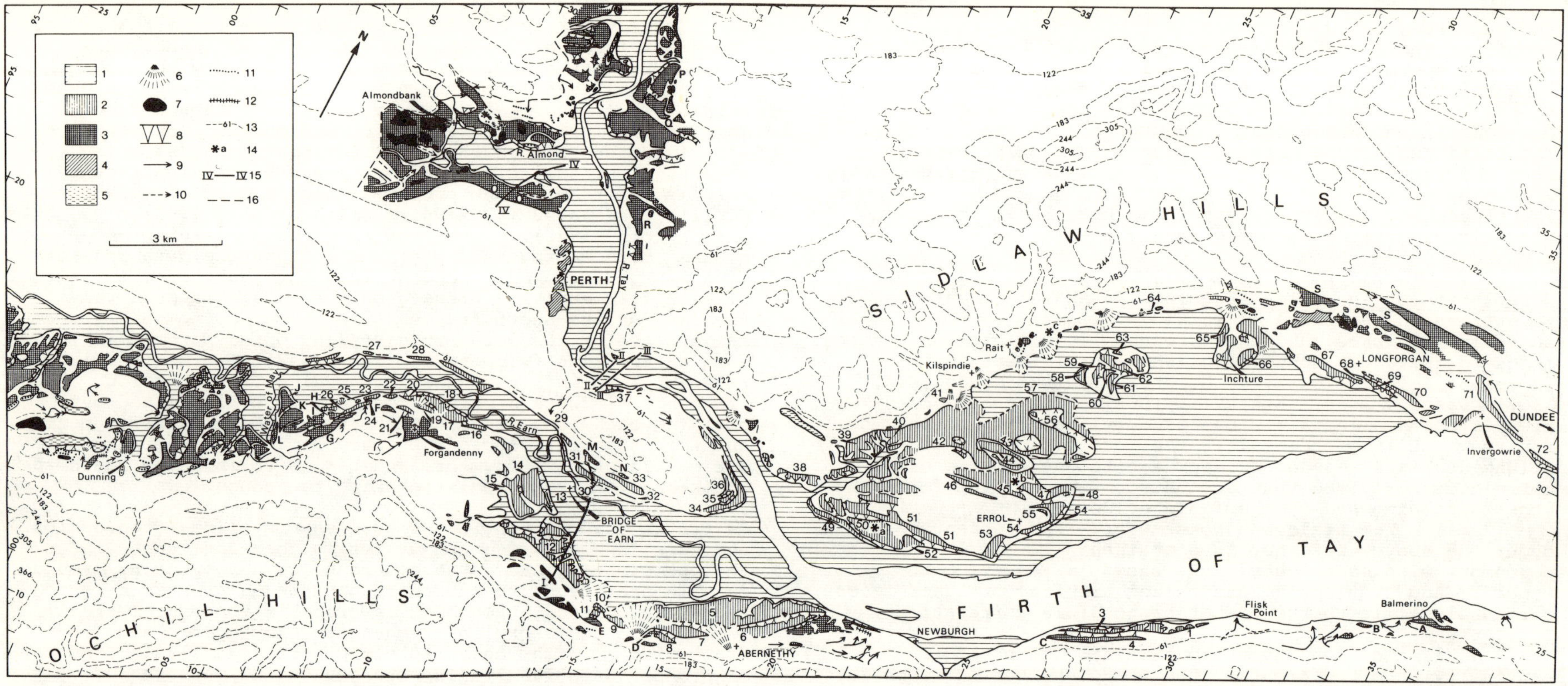

Fig. 1. Lateglacial raised beaches and related features in the Earn-Tay area. 1, Flandrian flats. 2, Lateglacial raised beaches. 3, fluvioglacial terraces. 4, steps of uncertain origin. 5, lacustrine flats. 6, alluvial fan. 7, kettle hole. 8, steep bluff (shown only where other detail not obscured). 9, meltwater channel. 10, linear depression. 11, round-crested ridge of sand and gravel. 12, sharp-crested ridge of sand and gravel. 13, contour (altitude in m O.D.). 14, pit referred to in text. 15, line of section through boreholes. 16, limit of detailed mapping in Almond and Tay valleys.

values obtained on some fragments probably depart considerably from 'true' shoreline altitudes owing to severe dissection of the raised beach surface, or to the presence of a gully or dry valley along the former shoreline. For example, the latter situation is estimated to have caused altitudes, taken well out from the former shoreline on fragments (14) and (15), to be about 0.6 m lower than 'true' shoreline altitudes. Such fragments have been excluded from the altitude analyses, and they are identified in both text and tables by the use of parentheses.

Along the south coast of the Firth of Tay steps of any sort are rare, and those that do occur are usually the expression of outcropping Old Red lava beds. Well developed raised beaches and estuarine flats occur only to the west of Flisk Point, and become extensive in lower Strathearn, west of Newburgh. North of the Firth of Tay, the Carse of Gowrie is surfaced largely by Lateglacial and Flandrian raised estuarine flats through which protrude 'islands' of Old Red sandstone veneered with till and fringed by Lateglacial raised beaches. The latter are also well developed on the slopes backing the carselands between Inchture and Dundee but, by contrast, the lower slopes of the Sidlaws between Kilspindie and Inchture bear only scant evidence of Lateglacial shorelines. The raised beach fragments occur mostly in staircases of between two and four which, together with

TABLE 1 Lateglacial raised shoreline fragments 1 to 36

Fragment	Altitude (m O.D.) range	Altitude (m O.D.) mean	No. of points	Length km
(1)	20.0-20.2	20.1	2	0.05
2	20.5-21.6	21.1	9	0.30
+ 3	21.2-22.2	21.6	14	1.00
+ 4	26.3-28.3	27.3	14	1.05
* 5	12.1-13.5	12.8	40	1.95
+ 6	25.1-25.9	25.6	17	1.15
7	14.3-14.8	14.6	7	0.30
8	26.7-27.8	27.1	13	0.55
9	16.0-16.6	16.3	6	0.55
10	16.3-16.9	16.6	6	0.20
11	22.2-22.6	22.4	2	0.05
+12	16.0-17.2	16.7	15	1.35
13	11.0-11.3	11.2	6	0.25
(14)	13.7-14.2	14.0	10	0.60
(15)	16.5-16.8	16.7	3	0.15
16	15.8-16.1	16.0	2	0.05
17	14.7-14.9	14.8	3	0.10
18	14.6-15.0	14.9	4	0.20
19	17.9-18.3	18.1	5	0.25
(20)	13.0-13.2	13.1	2	0.10
21	30.9-31.7	31.3	5	0.35
22	14.5-14.8	14.6	5	0.20
23	15.5-16.0	15.8	4	0.20
(24)	17.3-17.6	17.4	2	0.05
25	20.9-21.0	21.0	3	0.10
26	31.6-32.3	31.8	8	0.65
(27)	17.6-18.7	18.2	4	0.40
28	18.4-19.0	18.7	4	0.15
29	11.2-11.6	11.4	3	0.15
(30)	16.1-16.8	16.4	4	0.15
31	28.8-29.6	29.1	7	0.30
32	13.7-14.1	13.9	5	0.30
33	27.6-29.1	28.4	10	0.60
+34	11.1-12.2	11.6	19	1.35
35	15.2-16.0	15.6	6	0.35
36	26.6-27.1	26.8	9	0.55

\+ denotes fragments divided into two for grouped data
* denotes fragments divided into three for grouped data
() parentheses explained in text

the continuity of some of them over relatively great distances, greatly eases the problems of correlation.

The composition of the raised beaches and estuarine flats varies from sand and gravel to silty clay. An exposure (NO/1892 1753) showed brown, laminated, silty fine sand down to at least 3 m beneath the raised beach surface of fragment 5. Fragment 12 is crossed by the line of a future motorway, bores along which revealed a composition of brown, laminated clay near the shoreline, changing to fine sand with occasional fine gravel near the front, with the clay still present at depth (Fig. 2). A sandpit on (14) showed up to 3 m of fine to medium micaceous sand with scattered, well rounded small stones. An approximately similar surface composition is evident on all of the other fragments in lower Strathearn below about 20 m O.D. (Table 1) that have not already been mentioned, apart from (30), whose composition is more clayey. Fragments above 20 m are generally composed of coarser sand and gravel.

In the Carse of Gowrie an extensive area of Lateglacial estuarine flats includes fragments (39), 40, 41 to 45, 47, (48), 56 and 57. Exposures and augering revealed a surface composition of yellowish-brown or grey-brown, micaceous, silty fine sand in the south-western parts of the flats, becoming finer towards the east and north-east, where laminated, silty clay and silty fine sand predominate. Occasional small stones occur throughout the area. The records of 27 site investigation bores sunk along a road crossing these flats show depths of between 4 and 12 m of such deposits, in most cases described as alternating clayey and fine sandy material. Four bores sunk in a valley crossing the flats revealed till beneath the marine deposits, bedrock being reached in 3 bores, at 3.2 - 7.4 m O.D.

The laminated clay is known to be fossiliferous to the north of Errol, claypit exposures having been examined by Brown (1867) and Davidson (1932), who both described an arctic shell fauna. The section examined by Davidson (NO/2365 2340) revealed:

3. Yellowish brown sandy clay 2.5 - 3.0 m thick
2. Fine blue clay, coarsening upwards 1.5 - 2.1 m thick
1. Fine red clay over 1.2 m thick

and the most recently excavated pit (*b, Fig. 1) reveals a similar sequence. Deposits 2 and 3 are laminated, and consist of repetitions of grey clay (8 - 13 cm thick), red clay (2 - 3 cm) and sand (up to 2 cm). The basal red clay is also laminated, and rests on a layer of gravel, cobbles and boulders. Stones and boulders, often striated, are also scattered through all layers of the clay. The rich arctic estuarine fauna, which is *in situ* and is most abundant in the lower layers, "includes *Musculus* spp., *Pecten groenlandica*, *Astarte borealis*, *Buccinum groenlandicum*, *Saxicava* spp., over 20 species of ostracods and 4 species of Foraminifera"

TABLE 2 Lateglacial raised shoreline fragments 37 to 72

Fragment	Altitude (m O.D.) range	mean	No. of points	Length km
(37)	28.1-28.2	28.2	2	0.05
38	16.2-16.8	16.5	8	0.45
(39)	19.4-20.1	19.6	4	0.50
+ 40	14.6-16.1	15.2	27	1.50
41	21.2-22.0	21.7	3	0.10
+ 42	13.4-14.1	13.7	16	0.80
43	11.9-12.6	12.2	5	0.35
44	14.0-14.6	14.2	4	0.25
+ 45	16.2-16.5	16.3	5	0.95
(46)	24.2-24.3	24.2	2	0.25
+ 47	11.2-11.9	11.6	11	0.90
(48)	13.1-13.2	13.1	3	0.10
+ 49	10.9-11.2	11.1	9	1.20
50	20.5-20.6	20.5	5	0.15
* 51	24.5-25.6	25.1	32	2.05
52	11.6-12.1	11.8	9	0.40
53	23.8-24.4	24.0	7	0.30
+ 54	23.0-24.7	23.8	11	1.30
(55)	27.1-27.5	27.2	3	0.10
56	11.8-12.4	12.0	10	0.80
57	12.2-12.6	12.4	9	0.65
58	11.1-11.2	11.1	3	0.15
59	20.6-20.9	20.8	4	0.15
60	15.6-16.0	15.8	4	0.25
61	12.5-12.9	12.6	6	0.35
62	23.4-24.0	23.7	7	0.35
63	14.0-14.4	14.2	3	0.10
64	11.8-12.2	12.1	5	0.20
65	12.8-12.9	12.9	5	0.20
66	12.5-12.8	12.6	5	0.30
+ 67	21.2-22.3	21.7	15	0.95
68	25.2-25.7	25.4	3	0.10
+ 69	27.3-28.0	27.6	11	0.80
70	31.6-32.1	31.8	8	0.35
+ 71	18.8-19.8	19.3	14	1.55
(72)	19.6-19.9	19.7	3	0.15

+ * ( ) Symbols as in Table 1

(MacGregor, 1968, p.105). Bones of the common seal *Phoca vitulina* have also been found (Davidson, 1935). A more temperate fauna including *Mytilus* and *Ostrea* has been found in the upper part of the clay, but the latter is unlikely to be carse clay as assumed by MacGregor, for apart from its great dissimilarity from known carse deposits, its top lies at an altitude of 11.8 m, almost 2 m higher than the highest carse flats in the vicinity.

The surface of fragment 50 has been extensively pitted by clay workings, including one pit (*a, Fig. 1) that is still in use. The clay, about 3 m thick, shows similar lamination and banding to that at Inchcoonans (*b), and contains pulpy shell masses that cannot be separated from the clay. It also contains rootlets near the bottom, but these may be recent intrusions from a belt of trees that once grew on the site. The floor of the pit is strewn with erratic boulders of varied lithology, up to 1 m across and sometimes striated. Some of these are *in situ*, belonging to the gravel and boulder layer beneath the clay, but many have come from the clay itself.

Two raised beach fragments, both of sand and gravel, merge with fluvioglacial terraces. Fragments 4 and C (Fig. 1 and Table 6) are components of one 2.4 km-long terrace whose back-feature slopes down eastwards from 33.3 - 33.5 m to 26.3 - 27.1 m at an overall gradient of about 3.3 m/km. This would seem to suggest a fluvioglacial origin, but there is a marked change of gradient about half way along the feature, the gradient of the western half (C) being 5.8 m/km, but of the eastern half (4) only 1.2 m/km. It is probable that the eastern part is of marine or glaciomarine origin. Raised beach fragment 26, backed by the westernmost and highest measured shoreline fragment in lower Strathearn, similarly merges westwards with fluvioglacial terrace fragment H (Fig.1 and Table 6).

Stratigraphy

The stratigraphic relationships of the Lateglacial estuarine sediments are revealed in three sets of commercial borehole records obtained for road- and bridge-building purposes (sections I, II and III, Figs. 1 and 2). The stratigraphy revealed in section I may be summarized thus:

viii. Main Postglacial and later estuarine deposits
vii. Sub-carse peat
vi. Buried terrace deposits (estuarine)
v. Gravel and coarse sand layer
iv. Lateglacial marine deposits
iii. Fluvioglacial sand and gravel
ii. Till
i. Bedrock

Sections II and III show a similar sequence except that elements vi and vii are absent from section II, and element ii from section III. It is assumed that the sand and gravel that in some bores occurs immediately above the till, and in a lower stratigraphic position than the sand-and-gravel layer, is likely to be of fluvioglacial origin.

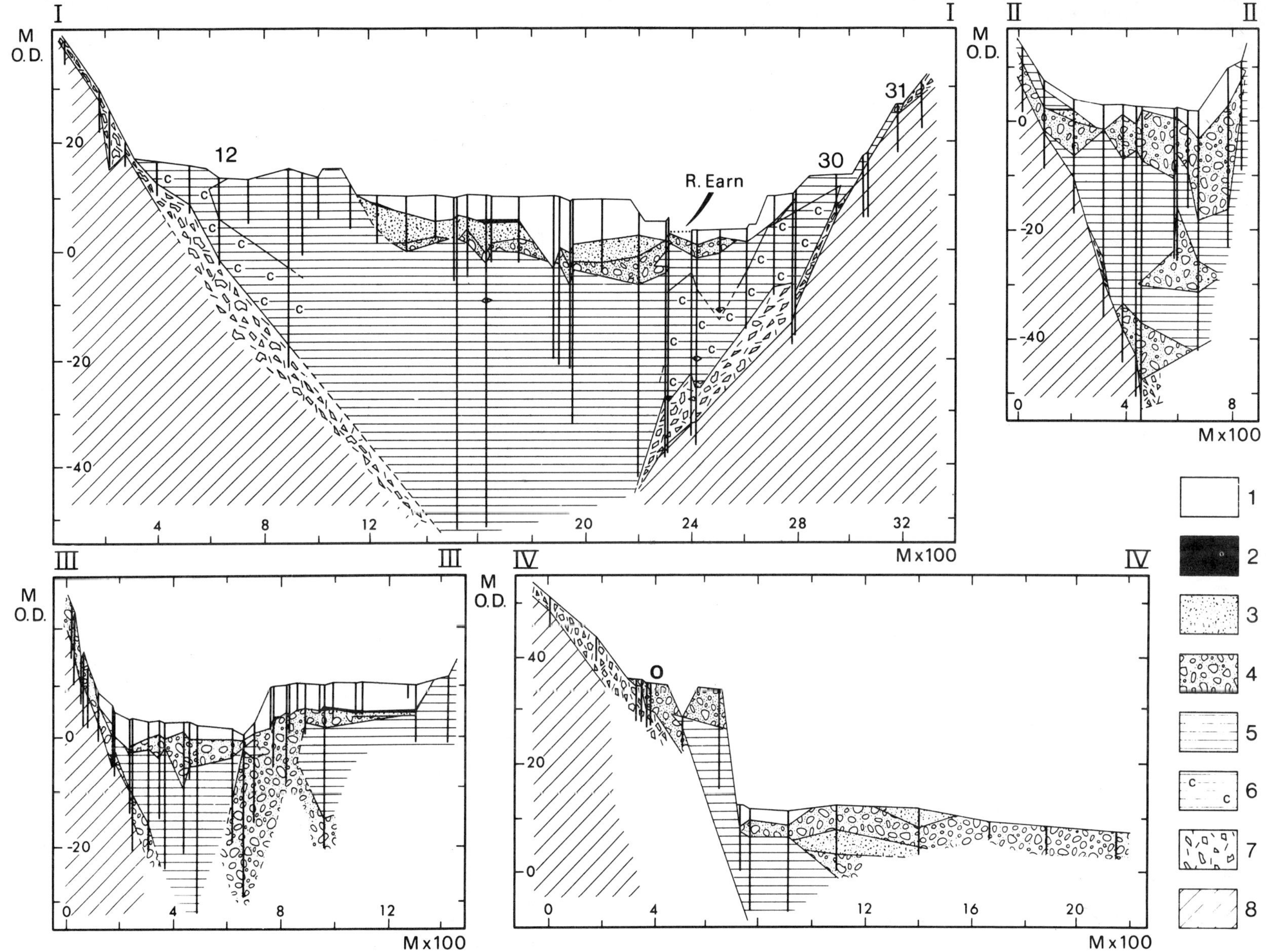

Fig. 2. Sections through boreholes. 1, Carse and other Flandrian estuarine deposits. 2, peat. 3, fine or medium sand. 4, sand and gravel. 5, Lateglacial estuarine deposits (mostly laminated clays, silts and sands). 6, plastic clay facies of 5. 7, till. 8, bedrock. Each heavy vertical bar represents a borehole. Terrace designations (12, O, *etc.*) relate to Fig. 1 and Tables 1, 2 and 6.

The Lateglacial estuarine deposits revealed in these bores are a mixture of sands, silts and clays, often laminated, and containing scattered stones and boulders. In sections II and III, larger inclusions of sand and gravel occur in these materials. In section I, by contrast, they sometimes consist of a very distinctive laminated, plastic clay, chiefly near the flanks of the valley (Fig. 2).

The lowest altitude at which the Lateglacial estuarine materials have been proved is -51 m O.D. in section I, -45 m in bores in Perth (McManus, 1966; Cullingford, 1972), and -42 m in section II.

The Shoreline Sequence

Inspection of the shoreline diagrams (Figs. 3 and 4) in the light of the field constraints allows the recognition of certain alignments that might represent former shorelines. The field constraints are set out below, using the following conventions:

-))4 : fragment 4 was formed adjacent to glacier ice;
- 2-3 : fragments 2 and 3 are morphologically continuous;
- $\frac{6}{5}$: fragment 6 occurs above 5 in a staircase;
- 5,7 : fragments 5 and 7 are isolated from each other.

Field constraints, south side of Firth of Tay and lower Strathearn:

```
  ))4     6   8    11    12,(15)  16,19    21     ))26
 -----   --  --  ------  -------  -----   ----  ---------
(1)-2-3 , 5 , 7 , 9-10 ,  (14)   , 17-18 , 22  , (24),25,
              --          -----           ----  ---------
              5            13             (20)     23

                  31,33   36
                  -----   --
(27), 28, 29,     (30)  , 35
                  -----   --
                   32     34
```

Field constraints, lower Tay valley and Carse of Gowrie:

```
      (39), 41    44           (46)                (55)
      --------    --           ----                ----
37,      40     , 43-56-57 ,    45  ,    51-53-54 -------- ,
         --                     --       --
         42                     47         50     45,(48)
                                      --------    --------
                                        49,52        47

62
--
59                              70
--                              --
-----     62                    69
58,60  ,  --  , 64,65,66,       --  , 71, (72)
   --     63                    68
   61                           --
                                67
```

The Main Perth Shoreline. By far the most conspicuous Lateglacial raised shoreline in the Earn-Tay lowlands, in terms of both field morphology and its representation on shoreline diagrams (Figs. 3 and 4), is the Main Perth Shoreline. It stands out on diagrams covering a wide range of planes of projection as a striking alignment of 178 heights, comprising the following fragments (Fig. 1 and Tables 1 and 2): 2, 3, 6, 8, 21, 26, 31, 33, 36, 51, 53, 54, 62, 67 and 71. It also includes fragments (1), (37) and (46), on which the altitude measurements were deemed to be of doubtful reliability, resulting in their exclusion from the analysis. It thus includes some of the longest and most extensive raised beach fragments, and in terms of morphology forms the highest recognizable Lateglacial marine feature west of Errol and Newburgh.

The shoreline slopes down from about 32 m west of Fordgandenny in the Earn valley, to 19-20 m near Invergowrie, west of Dundee, its gradient in the Earn-Tay lowlands being 0.47, 0.51 and 0.54 m/km in the W7°N-E7°S, W13°N-E13°S and W18°N-E18°S planes respectively (Table 4). The linear trend surface for the same data (Fig. 5A and Table 5) slopes down towards E13°S at a gradient of 0.51 m/km.

The Main Perth Shoreline has also been identified as the most conspicuous Lateglacial shoreline in the vicinity of the western part of the Firth of Forth (Sissons and Smith, 1965), where its gradient in the W18°N-E18°S plane, 0.44 m/km, is similar to that in the Earn-Tay lowlands. Isobases for this shoreline in the Tay-Forth region have been presented and discussed in a previous paper (Smith, *et al.*, 1969).

Lower Perth raised shorelines. There is abundant evidence in the Earn-Tay lowlands of Lateglacial raised shorelines that are lower and younger than the Main Perth Shoreline, but the recognition of contemporaneous features is not as straightforward as in the case of the Main Perth feature because successive shorelines are relatively closely spaced in altitude, and the approximately SW-NE alignment of the Earn-Tay lowlands is somewhat oblique to the general trend of the isobases. This means that small changes in the bearing of the plane of projection can result in large changes in the relative positions of the measurements on shoreline diagrams. This caused little difficulty in the case of the Main Perth Shoreline because of the clarity and continuity of this feature over long distances, but the Lower Perth fragments, although sometimes extensive, are in general more fragmented, and diagrams in different planes suggest different correlations. For this reason, 4 planes of projection were used as mentioned earlier: W7°N-E7°S (Fig. 3), W13°N-E13°S (Fig. 3), W18°N-E18°S (Fig. 4), and W37°N-E37°S (Fig. 4).

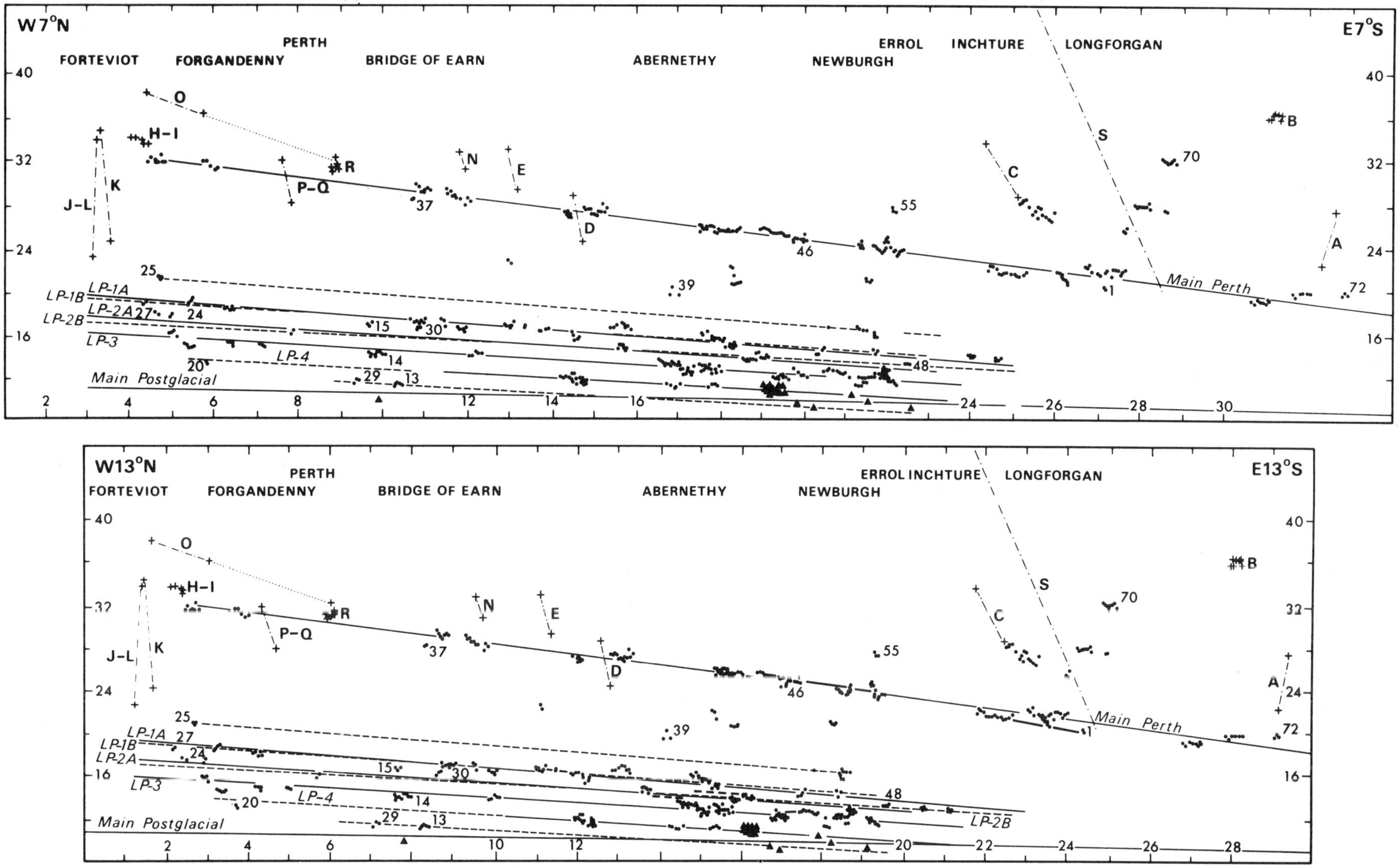

Fig. 3. Shoreline diagrams in the W7°N-E7°S and W13°N-E13°S planes of projection. Altitudes on visible raised shorelines are represented by dots, those on buried raised shorelines by triangles, and those on fluvioglacial terraces by crosses. Correlation coefficients and gradients for best-fit lines through shoreline altitudes are listed in Table 4. Numbers and letters relate to Fig. 1 and Tables 1, 2, 3 and 6. Parentheses explained in text. The apparently anomalous position of buried shoreline measurements above the best-fit line for the Main Postglacial Shoreline results from the higher-than-average elevation of the latter in an embayment.

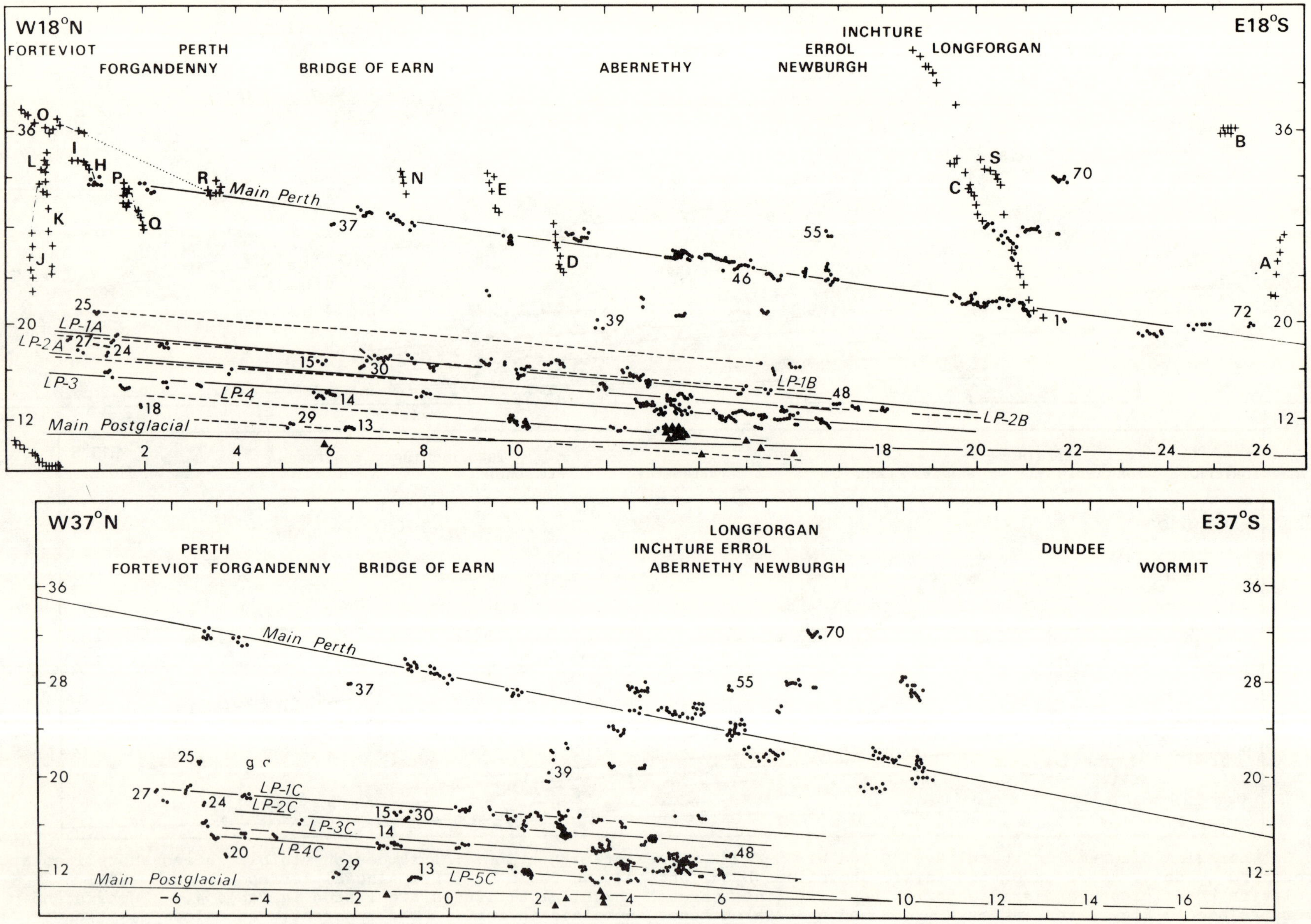

Fig. 4. Shoreline diagrams in the W18°N-E18°S and W37°N-E37°S planes of projection. Symbols and conventions as for Fig. 3.

TABLE 3 Component fragments of postulated Lateglacial shorelines

Scheme	Shoreline	Component fragments	NF*	N†
-	Main Perth	2,3,6,8,21,26,31,33,36,51,53,54,62,67,71	22	178
A	LP-1A	9,10,12,19,28,35,38,40,44,63,65,66	14	94
	LP-2A	7,16,42,61	5	31
	LP-3	5,17,18,23,32,43,47,52,56,57,58,64	15	108
	LP-4	34,49, Kilspindie-Rait 'Main Postglacial'	5	40
B	LP-1B	Same as Scheme A, excluding 65 and 66	12	84
	LP-2B	Same as Scheme A, including 65 and 66	7	41
	LP-3	Same as Scheme A	15	108
	LP-4	Same as Scheme A	5	40
C	LP-1C	9,10,12,19,28,38,60	8	48
	LP-2C	7,16,35,40,44	6	46
	LP-3C	5,17,18,23,32,42,65,66	11	82
	LP-4C	22,43,47,52,56,57,61,64	9	60
	LP-5C	34,49,58	5	31

* NF is number of fragments in grouped data
† N is number of altitudes in ungrouped data

For the correlation of the Lower Perth fragments, only 3 schemes may be proposed that both obey the field constraints and produce close alignments on shoreline diagrams. Two of the schemes differ only in respect of 2 fragments. The shorelines of the 3 schemes have constituent fragments as listed in Table 3, and the gradients and correlation coefficients of the best-fit lines in all 4 planes of projection, for both grouped and ungrouped data, are listed in Table 4.

Schemes A and B clearly produce close alignments in the W7°N-E7°S, W13°N-E13°S, and W18°N-E18°S planes, and since they differ only in respect of 2 fragments, there is little to choose between them. The best-fit lines of both schemes give relatively poor fits in the W37°N-E37°S plane. Scheme A involves slightly higher gradients than Scheme B for LP-1 and LP-2 and shows a more regular decrease in gradient of successive shorelines, although the comparative paucity of data on LP-2 and LP-4 probably makes this an insufficient basis for selecting it as the better scheme. Scheme C is markedly inferior to A and B in respect of the alignment of possible shorelines, and is tenable only in the W37°N-E37°S plane. In the other 3 planes, which cover the most likely range of orientations of the orthogonal plane as suggested by other Lateglacial evidence, Scheme C produces alignments that are mostly greatly inferior to those produced by other schemes in that the best-fit lines in general fit the data less closely than is usual with this type of shoreline data. Moreover, even in the W37°N-E37°S plane, 2 of the 5 best-fit lines, including a quarter of the number of fragments, are poor fits for this type of data, and the change in gradient of successive shorelines is somewhat erratic.

It is therefore concluded that the Lower Perth data are best explained by Schemes A and B. The correlation coefficients of the best-fit lines in the different planes (Table 4) suggest that the bearing of the most appropriate plane of projection is best approximated by the W7°N-E7°S and W13°N-E13°S planes. This is confirmed by the isobases calculated for Shorelines LP-1 and LP-3, on which the data are most abundant. The directions of slope and the gradients of the

TABLE 4 Gradients and correlation coefficients of best-fit lines in different planes of projection

Shoreline	Ungrouped data								Grouped data							
	E7°S		E13°S		E18°S		E37°S		E7°S		E13°S		E18°S		E37°S	
	m/km	r	m/km	r	m/km	r	m/km	r	m/km	r	m/km	r	m/km	r	m/km	r
Main Perth	0.47	-0.99	0.51	-0.99	0.54	-0.99	0.75	-0.94	0.47	-0.99	0.51	-0.99	0.54	-0.99	0.75	-0.99
LP-1A	0.29	-0.96	0.32	-0.96	0.35	-0.95	0.50	-0.84	0.29	-0.98	0.32	-0.98	0.35	-0.97	0.49	-0.87
LP-2A	0.24	-0.95	0.27	-0.93	0.29	-0.91	0.27	-0.55	0.23	-0.98	0.24	-0.97	0.26	-0.95	0.31	-0.79
LP-3	0.22	-0.94	0.24	-0.94	0.25	-0.94	0.30	-0.83	0.23	-0.97	0.24	-0.97	0.25	-0.97	0.33	-0.91
LP-4	0.21	-0.87	0.24	-0.87	0.28	-0.87	0.34	-0.53	0.20	-0.99	0.23	-0.98	0.25	-0.96	0.24	-0.58
LP-1B	0.26	-0.94	0.28	-0.94	0.31	-0.94	0.40	-0.83	0.26	-0.97	0.29	-0.97	0.31	-0.97	0.40	-0.88
LP-2B	0.21	-0.95	0.24	-0.94	0.26	-0.93	0.34	-0.64	0.21	-0.98	0.23	-0.98	0.25	-0.97	0.35	-0.83
LP-1C	0.17	-0.86	0.19	-0.89	0.20	-0.91	0.27	-0.95	0.17	-0.91	0.19	-0.93	0.20	-0.95	0.28	-0.99
LP-2C	0.10	-0.48	0.12	-0.52	0.15	-0.56	0.24	-0.67	0.11	-0.73	0.12	-0.76	0.14	-0.78	0.20	-0.82
LP-3C	0.14	-0.78	0.16	-0.81	0.17	-0.84	0.25	-0.92	0.14	-0.89	0.15	-0.91	0.17	-0.92	0.25	-0.96
LP-4C	0.16	-0.87	0.17	-0.89	0.19	-0.90	0.27	-0.94	0.16	-0.90	0.17	-0.92	0.19	-0.93	0.27	-0.96
LP-5C	0.11	-0.62	0.13	-0.65	0.14	-0.67	0.30	-0.76	0.08	-0.75	0.10	-0.78	0.11	-0.81	0.27	-0.96

TABLE 5 Linear trend surface data: Main Perth, LP-1 and LP-3 shorelines

Shoreline	Ungrouped data				Grouped data			
	N	Direction of slope down	Gradient m/km	% RSS	NF	Direction of slope down	Gradient m/km	% RSS
Main Perth	178	E13°S	0.51	98.24	22	E14°S	0.51	98.80
LP-1A	94	E 7°S	0.29	91.76	14	E 6°S	0.29	95.49
LP-1B	84	E12°S	0.28	88.15	12	E12°S	0.28	94.13
LP-3	108	E10°S	0.23	89.23	15	E 9°S	0.23	94.38

N and NF as in Table 3

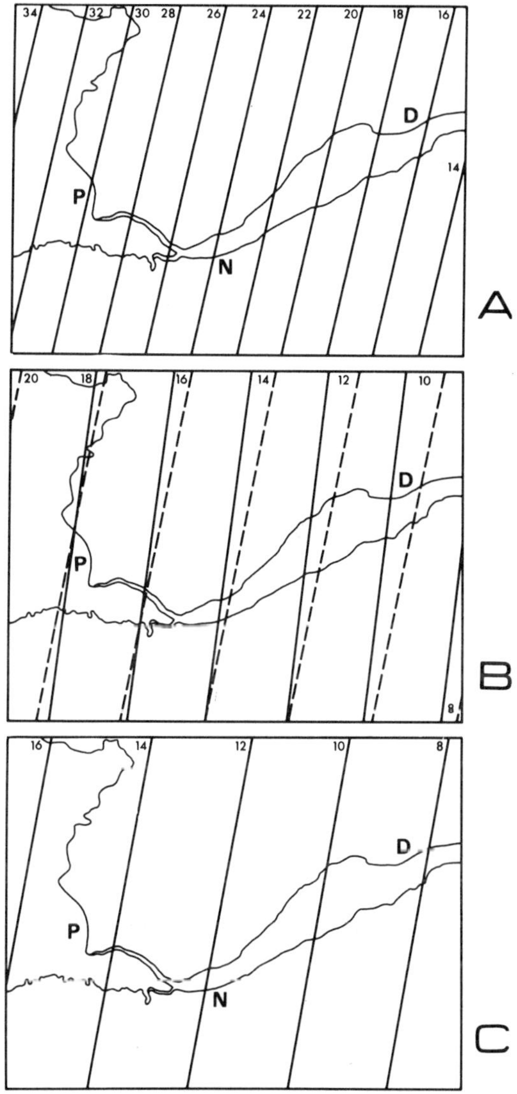

Fig. 5. Isobases in the Earn-Tay area for the Main Perth Shoreline (A), Shorelines LP-1A (continuous lines) and LP-1B (dashed lines) (B), and Shoreline LP-3 (C), as determined by linear trend-surface analysis. Contour values in m O.D. Regression details are summarized in Table 5. P. Perth; N. Newburgh; D. Dundee.

linear trend surfaces for these and the Main Perth Shoreline are listed in Table 5, and the isobases are depicted in Fig. 5.

The lowest major shoreline, LP-4, includes a set of Main Postglacial Shoreline measurements between Kilspindie and Rait (Figs. 1, 3 and 4) obtained where the veneer of Flandrian carse deposits is so thin over the buried raised estuarine materials that the two deposits are mingled at the surface, so that the measurements may be regarded as also applicable to the buried Lateglacial feature. If the best-fit line for LP-4 is extended towards E7°S, E13°S, or E18°S (Figs. 3 and 4), it passes close to the altitudes of at least 3 buried terrace fragments in the Carse of Gowrie, suggesting that shoreline LP-4 continues eastwards from Kilspindie-Rait as a buried feature (Cullingford, 1972).

Below LP-4, a later phase of shoreline displacement is represented by fragments 13 and 29, and it is reasonable to assume that this feature too continues eastwards below the carse, perhaps being represented by the other 3 buried terraces in the Carse of Gowrie shown on the shoreline diagrams. A still later stage in the Lower Perth sequence may be represented by a buried terrace at 9.8 m in lower Strathearn (Figs. 3 and 4). Many more bores will be necessary, however, before sufficient reliable buried shoreline heights are available to allow a reliable correlation of buried shorelines, and of visible with buried shorelines.

Within the altitudinal belt occupied by Shorelines LP-1 to LP-4 there are 8 fragments that have not so far been allocated. Seven of these fragments (14), (15), (20), (24), (27), (30) and (48) have been excluded from the calculations for reasons stated above, but they all lie close to the best-fit line of one or other of the shorelines postulated. The measurements on fragments (14) and (15), which were stated earlier to be up to 0.6 m below 'true' shoreline altitudes, accord very well with Shorelines LP-3 and LP-1 respectively if this is taken into account. Fragment (30) lies close to LP-1, (24) and (27) to LP-2, and (20) may be part of LP-4. No reservations were expressed about the measurements of the only fragment still unaccounted for, 22, which lies between the best-fit lines of LP-3 and LP-4.

In the altitudinal zone between LP-1 and the Main Perth Shoreline there are 8 fragments, of which the lowest 3 (25, 45 and 60) are about 2-2½ m above LP-1, and probably represent a slightly earlier shoreline. The remaining 5 fragments (11, 39, 41, 50 and 59) are the only ones in a rather wide (7-10 m) altitudinal band between the Main Perth Shoreline and the features already discussed. In the Forth area, between Polmont and Stirling, a well-developed raised shoreline occurs about 2-2½ m below the Main Perth feature, and in one locality a third shoreline occurs, about 8-9 m below the Main Perth, but evidence of lower features relating to the immediate post-Main Perth period of displacement is lacking (Sissons and Smith, 1965). It therefore appears that the evidence in the Forth and Tay areas is complementary rather than confirmatory, evidence of the higher post-Main Perth features being largely confined to the Forth, and of the lower features entirely to the Tay.

Other raised shoreline remnants. Evidence in the Earn-Tay lowlands of possible raised shorelines higher than, and predating the Main Perth Shoreline consists of only 5 fragments: 4, (55), 68, 69 and 70. With such scanty information it is not certain that all of these features are in fact raised beach fragments, particularly as 2 of them (55 and 68) are very small features each measured at only 3 points. Fragment 70, at 31.6-32.1 m, is so much higher than any other shoreline feature east of Forgandenny that its marine origin might be questioned on that basis alone: it could, for example, be a structural bench veneered with drift. It is the highest member of a staircase of 4 terraces, the lowest of which, 67, is part of the Main Perth Shoreline. The other members, 68 and 69, occur in a similar altitudinal belt to that occupied by 4 and (55) (Figs. 3 and 4).

FLUVIOGLACIAL FEATURES

Morphology and Altitude

Fluvioglacial terraces of great extent dominate the landscape of the Earn valley west of Forgandenny and the Tay and Almond valleys north of Perth, but are relatively sparse elsewhere in the area of study (Fig. 1). The altitude data obtained on fluvioglacial terrace fragments are summarized in Table 6, and are plotted on the W18°N-E18°S shoreline diagram (Fig. 4). They are represented on the W7°N-E7°S and W13°N-E13°S diagrams by intermittent lines connecting the measurements at either end of each terrace, except for three features (B, H-I, and R), for which the individual heights are plotted in all three diagrams. B and R show no marked gradient along their length, and are thought to be fluvioglacial on the basis of composition and their relationship to adjacent features of known fluvioglacial origin. B is too short for the height evidence to be conclusive, and the lack of gradient along the 350 m length of R may be at least partly the result of its location between two side valleys that contributed sediment to it. Terrace H, the steeply-sloping westward continuation of raised beach fragment 26, is separated by a composite gully from I, the surface of which is so severely gullied that it could only be measured at its extreme eastern end.

TABLE 6 Fluvioglacial terrace fragments

Fragment	Altitude range (m O.D.)	No. of points	Length (km)
A	22.3 - 27.2	7	0.30
B	35.7 - 36.1	7	0.25
C	28.5 - 33.5	11	0.80
D	24.4 - 28.4	8	0.35
E	29.3 - 32.7	7	0.30
F	40.6 - 41.9	5	0.30
G	42.0 - 43.3	7	0.95
H	33.2 - 33.6	3	0.20
I	33.8	2	0.10
J	22.8 - 27.6	6	0.45
K	24.3 - 34.4	11	0.70
L	31.9 - 33.8	3	0.20
(M)	35.7 - 37.5	3	0.10
N	30.8 - 32.6	4	0.15
O	36.1 - 38.0	11	1.45
(P)	30.8 - 32.0	6	0.35
Q	28.0 - 29.6	4	0.20
R	29.9 - 31.8	8	0.65
S	20.6 - 49.9	36	3.70

( ) as in Table 1

Two terraces, A and J-L, have an apparent westward slope on the shoreline diagrams. In the case of J-L the effect is spurious, and results from the projection into the planes used of a feature that slopes approximately south - north. Fan A, however, is unique amongst measured fluvioglacial terraces in the Earn-Tay area in sloping markedly from east to west, against the regional trend of both direction of former ice movement and of meltwater flow.

West of the Water of May the Earn valley terraces take the form of two great outwash fans that have been severely dissected by both proglacial meltwater rivers and later gullies. The complex fan between the Water of May and Dunning is the more dissected of the two, and includes terraces at several different levels, the main level, overlooked by higher terrace remnants, having an altitude of about 38-40 m at its front and over 61 m at its apex, at Dunning (O.S. contour and spot heights), where it is overlooked by several large, steep-sided mounds and ridges of coarse fluvioglacial material. These kames, up to 20 m high, clearly predate the main terrace, although they might be contemporaneous with the highest remnants of the fan complex. Immediately west of Dunning, however, are eskers, kames, kettles, and dead-ice hollows at a considerably lower altitude than that attained by the main terrace, and it therefore seems probable that glacier ice existed here during the deposition of most of the fan complex.

The second outwash fan complex, at the extreme western edge of Fig. 1, probably postdates the Dunning complex, and was probably deposited while the wasting ice margin lay near Auchterarder (off the area of Fig. 1).

Extensive fluvioglacial terraces occur north and west of Perth, and possibly in Perth itself, for it is likely that at least parts of the dissected terraced area depicted on the map as of uncertain origin are fluvioglacial (Fig. 1). In the eastern part of the Crieff-Methven depression, which the River Almond enters at Almondbank, the Flandrian flats are flanked on both sides by an extensive spread of fluvioglacial material, the flat surface of which is deeply scored by valleys and gullies, and pitted with several kettles. West of Almondbank the depression is almost completely choked by outwash, and the sharp-lipped kettles reach large dimensions, some being 12-20 m deep and up to 800 m across. Measurements along the portion of this terrace labelled O (Fig. 1 and Table 6), together with relevant Ordnance Survey spot heights and the height information obtained by Paterson (1974), permit the recognition of a fan with its apex near Almondbank at over 47 m O.D., sloping down towards the southwest, south and southeast. The thickness and coarseness of the gravel and sand diminish in the same directions (Paterson, 1974).

Outwash terraces occur on both sides of the Tay valley north of Perth, varying from extensive, little-dissected spreads to features so greatly dissected that only isolated flat-topped knolls remain. At least three terrace levels are represented in several staircases that continue beyond the edge of Fig. 1 (Cullingford, 1972), and into the area being currently investigated by E. V. Insch. The intense dissection responsible for isolating the mesa-like remnants around Luncarty, opposite fragment P (Fig. 1) was probably due largely to the cutting of proglacial meltwater channels, some of which descend to the level of the Flandrian flats.

Stratigraphy

The coarse outwash materials discussed above lie over laminated silts and clays, as seen in several exposures in the Earn, Tay and Almond valleys (Simpson, 1933; Cullingford, 1972; Paterson, 1974), and in the records of commercial boreholes (section IV, Figs. 1 and 2). These laminated sediments are fossiliferous in the lower Almond valley, where the marine fauna includes *Portlandia arctica*, foraminifera and ostracoda (Paterson, 1974), and they rest upon till. The junction of the marine sediments with the overlying gravel lies at about 27 m O.D.

The laminated deposits in section IV (Fig. 2) also continue beneath a lower sand-and-gravel layer underlying the Flandrian deposits, a relationship that has already been noted in sections I - III. A gravel layer in a similar stratigraphic position in the Grangemouth area of the Forth valley, identified on the basis of many hundreds of borehole records, has been ascribed to the widespread marine planation of Lateglacial marine deposits, till and bedrock (Sissons, 1969), and together with buried rock platform remnants farther east, has been related to the Main Lateglacial Shoreline (Sissons, 1974d, 1976a). It is tempting to postulate a similar origin for the Earn-Tay buried sand-and-gravel layer, but apart from the difficulty of visualizing large-scale marine erosion in the confines of lower Strathearn and the lower Tay valley, there are certain differences that imply a different origin. First, there is no evidence of planated surfaces of rock or of till, and at most only limited planation of marine clay, associated with the sand-and-gravel layer or any other deposit. On the contrary, the sand and gravel is of extremely variable thickness (0.3-14 m) and rests upon a highly irregular surface that is more suggestive of fluvial or fluvioglacial channelling of the underlying deposits than of planation (Fig. 2). Secondly, the surface of the sand-and-gravel layer in section III (Fig. 2) lies at two distinct levels, suggesting terraces at about 5 m and 0-1 m, the surface of the lower one bearing irregularities that may represent former stream channels.

It is therefore likely that the Earn-Tay sand-and-gravel layer, probably including that in the lower Almond valley (section IV, Fig. 2) whose surface lies at 7½-12½ m O.D. and is only partly buried, is of fluvial or fluvioglacial origin, and was deposited when local relative sea-level was 5 m or less in the case of the higher terrace, and at or below O.D. in the case of the lower. It remains problematic whether the gravel layer is outwash graded to a later stage in the Lower Perth phase of shoreline displacement than those identified above, or to the so far unidentified Main Lateglacial marine feature in the Tay area.

The Pattern of Deglaciation

On the basis of the stratigraphic sequence described above, Simpson (1933) concluded that two periods of ice advance were separated by a period during which estuarine conditions prevailed up the Earn valley as far as Crieff, and up the Tay and Almond valleys to near Luncarty and Almondbank respectively. Simpson termed the later ice advance, responsible for the high-level outwash gravels over the laminated deposits, the Perth Readvance, the limit of which was mapped over much of central Scotland by Sissons (1963b, 1964), and tentatively dated at about 13,000 - 13,500 B.P. (Sissons, 1967c). In the Earn-Tay area Sissons' limit, largely confirmed by the writer (Cullingford, 1972), crosses the Earn valley near Dunning and is marked by the sharp downvalley limit of large kettles in the Almond valley (Fig. 1).

The reality of the Perth Readvance has been challenged by Francis *et al.* (1970) in the Forth area, and by Paterson (1974) in the Perth district. The graded units in the laminated sediments near Almondbank, assumed to be varves by Simpson, are interpreted by Paterson as a result of the repeated discharge of sediment down the front of a marine delta advancing down the Tay valley and into the lower Almond valley, to be capped by outwash gravel reflecting progressive delta extension, and graded to the Main Perth Shoreline.

A crucial test of whether glacier ice overrode the marine sediments would be to determine whether the latter pass beneath the large kettles (which are much too big to result from ice rafted by fluvioglacial rivers), as required by the readvance hypothesis, or whether the marine sediments were deposited amongst detached masses of stagnant ice and are therefore absent beneath the kettles, as assumed by Paterson. The writer has proved clay beneath two of the smaller kettles east of Almondbank, but attempts to sink bores in the large kettles have so far failed.

In view of the two alternative explanations of features in the type locality of the Perth Readvance, and taking into account doubts concerning the validity of some other arguments used to support it over a wider area, it is necessary to conclude that "there is at present no satisfactory evidence in Scotland that the decay of the last ice-sheet was interrupted by a significant readvance" (Sissons, 1974c, p.314). The ice-limits specified above may therefore mark either a stage or a halt in the continued westward recession of the wasting ice sheet that reached its maximal extent 17,000 - 18,000 years ago, and had virtually or even entirely disappeared by about 13,000 - 12,500 B.P. (Sissons, 1974c, 1976b).

RELATION OF THE RAISED SHORELINES TO DEGLACIATION

The westward dissipation of the ice sheet from northern Fife to the vicinity of Perth and Dunning is recorded by only fragmentary evidence of fluvioglacial features and associated raised shorelines, including terraces 4-C and S (Figs. 1, 3 and 4). The former, the shoreline part of which lies at about 27 m (Table 1), was formed while ice lay to the west of Newburgh, where the hillslopes are scored by a meltwater channel system down to about 30 m O.D. Fluvioglacial terrace S, however, implies contemporaneity of wasting ice in the Benvie depression north of Inchture and Longforgan with a much lower relative sea-level, for in 3½ km it slopes down from 50 m to 21 m, close to the altitude (19 - 20 m) of the nearest remnant of the Main Perth Shoreline (71). Unfortunately, dissection and subsequent fluvial deposition have obscured the relationship between terraces S and 71, so it is not clear whether the Main Perth Shoreline is contemporaneous with S, or has transgressed the lower end of it. The former possibility would require the survival of stagnant ice over the site of the kames and kettles north of Inchture until the main edge of the ice sheet had receded to near Perth and Dunning, and the remnant ice mass would have had to be large enough to supply the copious outwash sand and gravel that now choke the Benvie depression.

The Main Perth Shoreline was formed while the ice margin lay near Dunning and Perth, as shown by the merging of raised beach fragment 26 with outwash deposited when wasting ice lay west of Dunning, and by the pitted outwash fan in the lower Almond valley that declines downvalley to a level close to that at which the Main Perth Shoreline would lie if it were identifiable and measurable at that point (the town of Perth obscures the relationship between the shoreline and outwash features).

In both the Tay and Earn valleys, outwash terraces younger than those formed contemporaneously with the Main Perth Shoreline descend well below the level of the latter. In the Tay valley, terrace P-Q descends 3-4 m below the level given by extrapolation of the Main Perth Shoreline; and in Strathearn, terraces K and J-L descend to altitudes about 8 m and 9 m respectively below that of the Main Perth Shoreline in the vicinity, and show no signs of levelling out at those altitudes. Terrace J-L continues down almost to the level of the shoreline, represented by fragment 25, that occurs immediately above Shoreline LP-1 (Table 1 and Figs. 3 and 4). These outwash terraces too were formed when the ice margin still lay near the site of Dunning, so it is evident that a considerable amount of negative shoreline displacement occurred whilst decaying ice still lay at or near the limits specified above. This is further emphasized, especially in the Tay valley, by the descent of some proglacial meltwater channels trenching the outwash to the level of the Flandrian flats at about 15 m O.D., considerably below the level given by Shoreline LP-3.

It is therefore probable that at least the Main Perth and LP-1 to LP-4 shorelines were formed while the decaying edge of the ice sheet receded only a short distance, and therefore probably during a relatively short period of time, for there is evidence that the climatic amelioration responsible for the later stages of ice-sheet decay was rapid after about 13,500 B.P. (Sissons, 1974c, 1976b). These shorelines slope down in directions between east and east-southeast, and show a consistent decrease in gradient with decreasing age and altitude when plotted in the most appropriate planes, from 0.47 to 0.21 m/km and from 0.51 to 0.24 m/km in the W7°N-E7°S and W13°N-E13°S planes respectively.

The most likely explanation of the above relationships is that the shorelines have been uplifted and deformed by differential isostatic recovery consequent upon deglaciation, and despite the lack of relevant absolute dates the implication is that this recovery was immediate and rapid. While the ice lay near Dunning and Perth there is evidence that relative sea-level fell by more than 9 m in Strathearn (the altitude difference between the Main Perth Shoreline and the lower end of terrace J-L), and by at least 17 m in the Tay valley (as shown by dissection by proglacial channels), and since world sea-level was probably rising at the same time in response to widespread deglaciation, the actual land uplift was greater than this. The total uplift since the Main Perth Shoreline was formed may be estimated by adding the present altitude of the shoreline to the net amount by which world sea-level has risen since that date, and subtracting about 2 m to allow for the fact that the shoreline measured relates to high water mark. The age of the Main Perth Shoreline is not yet known, but if it is assumed to have been formed about 13,500 years ago, when the rising world sea-level is estimated to have been 60-65 m lower than it is now (Shepard, 1963; Curray, 1965; Mörner, 1969), then the total uplift since then is about 90-95 m near Forgandenny, and 76-81 m at Dundee.

The formation of the raised shorelines is best viewed in terms of an interplay between isostatic uplift and eustatic rise of sea-level, major shorelines having been formed when the latter kept pace with the former. The state of balance need not have been maintained for long in view of the rapidity of sedimentation that may be postulated. Indications that the Main Perth Shoreline may have been formed at the limit of a transgression include the burial of an alluvial fan beneath Lateglacial marine sediments (see below), and the possible truncation by the shoreline of the lower end of terrace S.

GENERAL CHARACTERISTICS OF THE RAISED BEACHES

The Sediments

The variable composition of the raised beaches is partly related to distance from the former ice margin. For example, the Main Perth Raised Beach is composed of coarse sand and gravel,where it merges with outwash west of Forgandenny, sand and finer gravel lower down Strathearn, and fine sand, silt and clay in the Carse of Gowrie. This gradation is locally interrupted where incoming streams contributed coarser materials (*e.g.* fragment 8). The surface composition of the Lower Perth beaches is generally finer than that of the Main Perth in lower Strathearn.

The fine-grained, laminated glacioestuarine sediments of the Carse of Gowrie contain scattered stones and boulders whose deposition is difficult to explain other than in terms of ice-rafting (Jamieson, 1865; Sissons, 1967a), a process that was observed in Greenland and eloquently described by Brown (1870). The clays lie at different levels, but cannot be related simply to different raised beaches because estuarine deposits were accumulating throughout the period considered in this paper, and only the surface layers of material beneath a Lateglacial flat relate directly to the flat and shoreline. These upper layers are usually more sandy than the material beneath, which is consistent with deposition in shallower water.

Relation to Alluvial Fans

The numerous alluvial fan complexes in the Earn-Tay area include fan remnants of several ages, from high-level features predating the raised beaches to low-level ones that would still be active but for artificial containment of the streams. Contemporaneity of fan deposition with the Lower Perth shoreline sequence is suggested by the apparent merging of parts of the Farg fan complex with beach fragments 5, 7 and 9 (Fig. 1). Fan deposition related to a base-level lower than that represented by the Main Perth Shoreline, but pre-dating the latter, is demonstrated by a former gravel pit (*c, Fig. 1), which showed slightly contorted fan material descending beneath laminated clay at about 24-25 m O.D.

Dissection

A universal characteristic of the surface form of the raised beaches and fluvioglacial terraces is their dissection by numerous gullies and valleys of varying size. Evidence that gullies and valleys of different ages are present includes the following:

(i) Older and higher terraces are generally more dissected than lower and younger ones, with some exceptions reflecting the different susceptibility to erosion of different materials.

(ii) Some terraces are dissected both by deep gullies and valleys that descend low enough to have been infilled by later estuarine sediments, in some cases including Flandrian carse deposits and sub-carse peat (Sissons *et al.*, 1965), and by shallow ones that indent the front of the terrace and terminate at their downstream end at the apex of a fan that rests on the later terrace. These two types of feature are clearly of different ages.

It is therefore evident that dissection of Lateglacial raised beaches has continued throughout Lateglacial and well into Flandrian times. Some gullies may have originated contemporaneously with the beaches into which they are cut, as tidal creeks or channels on the sandflats or mudflats, and some minor gullying almost certainly still occurs on the steep frontal bluffs of some terraces. However, the large number of gullies and valley systems that descend below the main carse level implies one or more periods of particularly intense dissection postdating the lowest visible Lower Perth raised beaches and predating the sub-carse peat (Sissons *et al.*, 1965). Evidence in the mid-Forth area shows that at least some of the valleys are infilled by Main Buried Beach deposits, and were therefore largely or entirely formed in Lateglacial times (Sissons, 1969).

It is possible that they were initiated in Lateglacial times when relative sea-level was high and groundwater consequently higher than now, and that excavation continued in response to a falling water-table controlled by the relative fall of sea-level. Alternatively, they may have been cut under periglacial conditions when the ground was frozen and precipitation could not be absorbed. The latter origin has been postulated for features similar to those described above, that dissect raised beaches in Sweden (von Post, 1956) and outwash terraces in southern Germany (Büdel, 1944).

Lateglacial raised beach surfaces may also be interrupted by kettle holes. Pitted beaches are common in parts of Antarctica (Nichols, 1961), but the only clear example in the Earn-Tay area is fragment 33.

ACKNOWLEDGEMENTS

Thanks are due to the former Department of Scientific and Industrial Research for the award of a Research Studentship 1962-5, to the University of Exeter for grants towards fieldwork costs since then, and to Dr. J. B. Sissons and Dr. D. E. Smith for discussion, encouragement and field assistance over the past fourteen years.

The "Oban-Ford Moraine": A Reappraisal

J. M. GRAY (Queen Mary College, London)
and
D. G. SUTHERLAND (University of Edinburgh)

ABSTRACT

The detailed morphometric study carried out by the authors on the constituent parts of the "Oban-Ford Moraine", western Scotland is described. It is concluded that the ice-limits previously described as belonging to this stage are predominantly composed of fluvioglacial landforms rather than moraines and that the associated outwash spreads are related to different, non-synchronous sea-levels. The evidence indicates that the ice-limits do not mark a readvance or significant, synchronous, recessional stage as previously supposed but more probably represent only fairly brief, diachronous halts in the general recession from a period of more extensive ice cover prior to the Loch Lomond Readvance at a time when relative sea-level was falling rapidly. The Loch Lomond Readvance may have reached the south-west end of Loch Awe and if so was more extensive in the south-west Highlands than recently supposed.

INTRODUCTION

Recent research on Scottish glacial stages has been focused on the Loch Lomond Readvance (*e.g.* Sugden, 1970; Clapperton and Sugden, 1972, this volume; Sissons and Grant, 1972; Sissons, 1972a, 1974a, 1974b, 1975b, this volume; Sissons *et al.*, 1973; Thompson, 1972; Gray and Brooks, 1972; Gray, 1975c; Sugden and Clapperton, 1975. The comparatively minor attention that has been paid to earlier glacial stages has been mainly concerned with questioning the validity of the Aberdeen-Lammermuir and Perth readvances (*e.g.* Sissons, 1972a, 1974a; Paterson, 1974; Armstrong *et al.*, 1975). Perhaps the only pre-Loch Lomond Readvance glacial stage remaining virtually unchallenged in the literature is the "Oban-Ford Moraine".

The name "Oban-Ford Moraine" was introduced by Synge (1966) and Synge and Stephens (1966) who believed that in western Scotland Charlesworth's (1955) Stage M is made up of two stages. The younger one, found north of Oban and equivalent to the Loch Lomond Readvance, was related to a sea-level below the Main Postglacial Shoreline (*i.e.* below about 13 m O.D. at Oban). The earlier one, the "Oban-Ford Moraine", was considered to be a readvance or significant, synchronous, recessional stage and is now believed to be approximately equivalent to the Perth stage (F. M. Synge, pers. comm.). It was said to be associated with deltas at about 40-43 m at and south of Oban, though in some valleys a lower outwash delta at about 33.5 m was said to occur (Synge, 1966). This paper seeks to re-examine the evidence for the "Oban-Ford Moraine", beginning in the Ford area and working northwards to Oban (Fig. 1).

METHODS

The deposits in the valleys between Lochgilphead and Oban were first mapped from aerial photographs at about 1:10,000 scale and the maps were subsequently checked and modified during various phases of fieldwork. Section evidence was also noted and in places augering and boring were carried out.

All outwash, kame and river terrace fragments were accurately levelled from O.D. Newlyn bench marks using a Hilger and Watts autoset level. Altitudes were obtained at 50-100 m intervals in lines parallel to the back edges but sufficiently far out to avoid any mass movement deposits or backswamp depressions. Heights were also obtained on kame

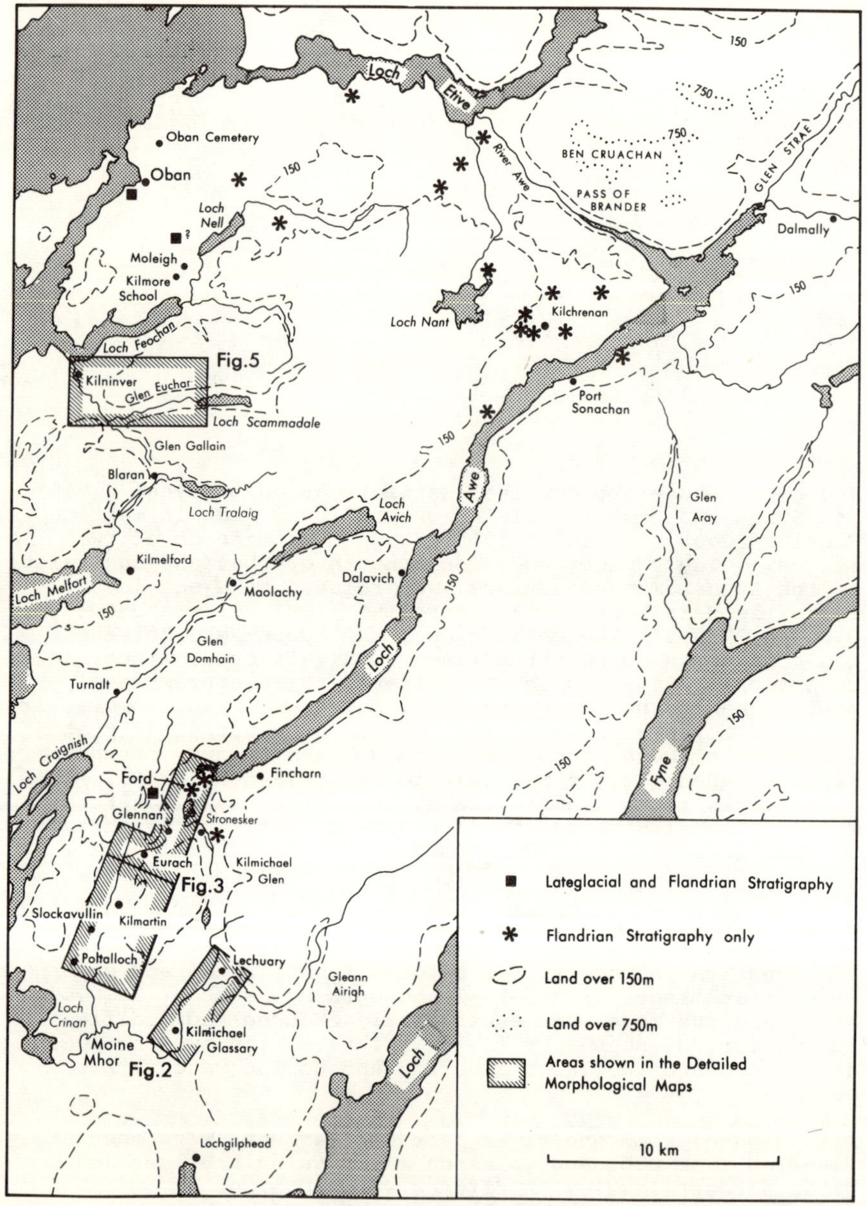

Fig. 1. Location map and positions of kettle holes stratigraphically investigated.

summits, channels, raised beaches and shingle ridges as well as on the present flood plain. In all about 1200 points were levelled, just over 800 being in the Ford/Kilmartin/Kilmichael area. This contrasts with the relatively small number obtained by Synge (1966) by Abney levelling from spot heights and bench marks.

In valleys with extensive terrace systems the altitudes obtained were plotted on height/distance diagrams. The possible errors in such cartographic representations of terrace heights have been outlined by Kirby (1969). The lines of projection clearly simplify the depositional conditions and to be strictly correct the points should be projected onto lines representing the course of the river at each terrace level. This is rarely known and usually straight line sections are chosen as generalisations of the river's course at all terrace levels. The braiding characteristics of rivers depositing outwash trains reduces the probability of this type of error being important. Other errors arise when lines of levelled heights do not correspond exactly with the alignment of the projection plane. This will result in an over-representation of the gradient of the feature by a factor of $(1 - \sec \theta)$ where θ is the difference in bearing between the levelled line and the projection plane. However, where θ is small (say less than 30°) the projection error does not markedly alter the gradient on the diagram (Kirby, 1969).

Correlation of fragments on height/distance diagrams is usually by eye on the basis of continuity along the valley and similar relative elevation above present river level. Correlation across valley is achieved when fragment heights are seen to be superimposed on the diagram.

LOCH AWE - LOCH CRINAN

Kilmichael Valley

Synge (1966) believed that coarse outwash gravels debouching from Gleann Airidh were deposited in a delta at 141 ft (43 m) at Lechuary. Farther down valley he noted that above Kilmichael "thin drift has been washed off the bedrock below 131 ft (39.9 m)".

It appears, however, that Synge placed the sea-level associated with the Lechuary terraces too high, for on the east side of the valley (Fig. 2) the terraces slope south-westward from about 46 m O.D. at the mouth of Gleann Airigh (T6) down to about 37 m O.D. 800 m south-west of Lechuary (T7). Farther south there is a marked upper limit of thick deposits, especially on the east side of the valley, and levelling of several terrace fragments (S1-6) shows that this limit is virtually horizontal at 36 - 38 m O.D. This is

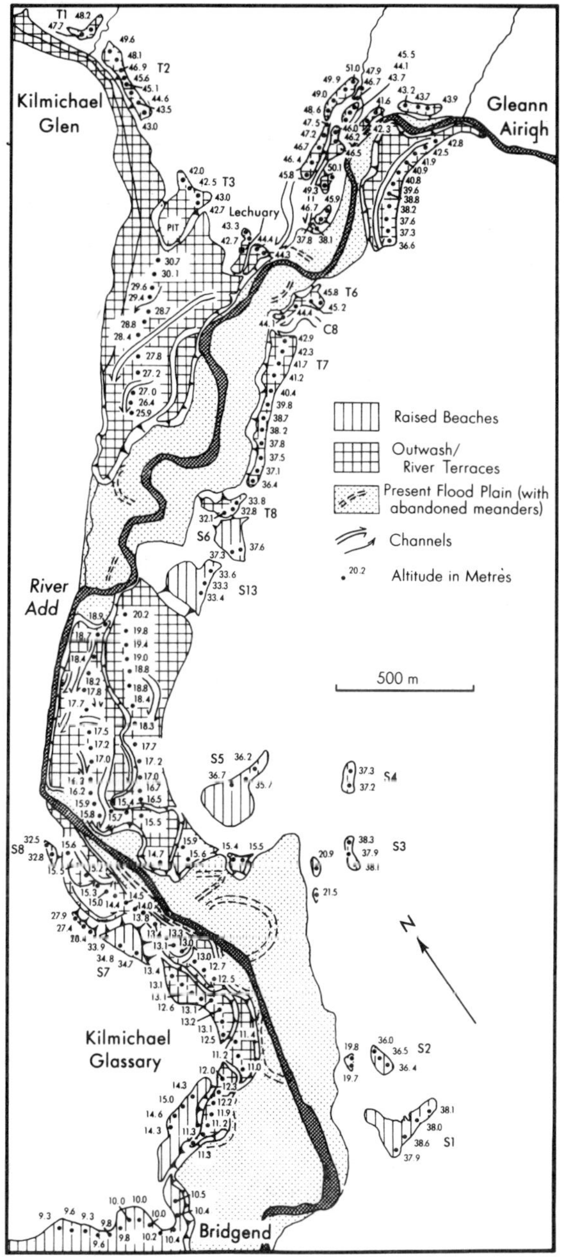

Fig. 2. Morphological map of the Kilmichael valley.

believed to have been the level of the sea at the time the Lechuary outwash terraces were deposited. Fragments S7, S8, S13, T8 and the channel C8 relate to the initial stages of the subsequent fall in sea-level as may a good section towards the lower end of T7 showing several metres of silt with clay, sand, and sand and fine gravel lenses, overlain by 2 - 3 m of sand and gravel. This sequence is compatible with the extension of proglacial outwash over fine marine deposits during a regression of the sea.

From 46 m O.D. at the entrance to Gleann Airigh (T6) the outwash terraces also slope down towards the north for about 500 m to a large fragment (T3) in which there is a disused gravel pit. Good sections in this pit show gently folded beds of poorly sorted sand and gravel with layers of fine silt and sand in places. North of this fragment the terraces (T1-2) rise up Kilmichael Glen implying a second source of material, but the fact that the terraces are not continuous up Kilmichael Glen makes it impossible to correlate them with an ice-limit.

Ford/Kilmartin Valley

The terraces in this valley were described by Hill *et al.* (1905), Charlesworth (1955) and Synge (1966). Synge described the material around Ford as "kame-and-kettle topography" but later in his paper referred to it as an "end-moraine" related to terrace fragments that descend from 192 ft (58.5 m) at Eurach to terminate in a delta at 133 ft (40.5 m) at Kilmartin. Synge also noted a "gravel bench" sloping from 95 ft (29.0 m) at Baluachraig to 80 ft (24.4 m) at Ballymeanoch where shallow channels on it pass below a "beach ridge" rising to 85 ft (25.9 m). Also mentioned were lower terraces at 60 - 70 ft (19 - 22 m) and about 46 ft (14 m), both dissected by shallow river channels. The sequence was interpreted as the result of a fairly simple pattern of falling sea-level with deposition at various levels and subsequent dissection as sea-level fell.

The morphological features of the Ford/Kilmartin valley as mapped by the present authors are shown in Fig. 3 which is generalised because of the small size of some terrace fragments. Figure 4 is a height/distance diagram of the terrace fragment altitudes. Because of changes in valley direction the diagram was constructed in four parts, the position of the projection planes being shown in Fig. 3. Terraces near the intersection of 2 planes were projected into whichever one appeared most appropriate to the orientation and position of each terrace. The morphological map, the height/distance diagram and additional levelled altitudes not shown on Figs. 3 and 4 permit the following interpretations. The marked alignment of terrace fragments T18-28 makes it probable that they all correlate and they are

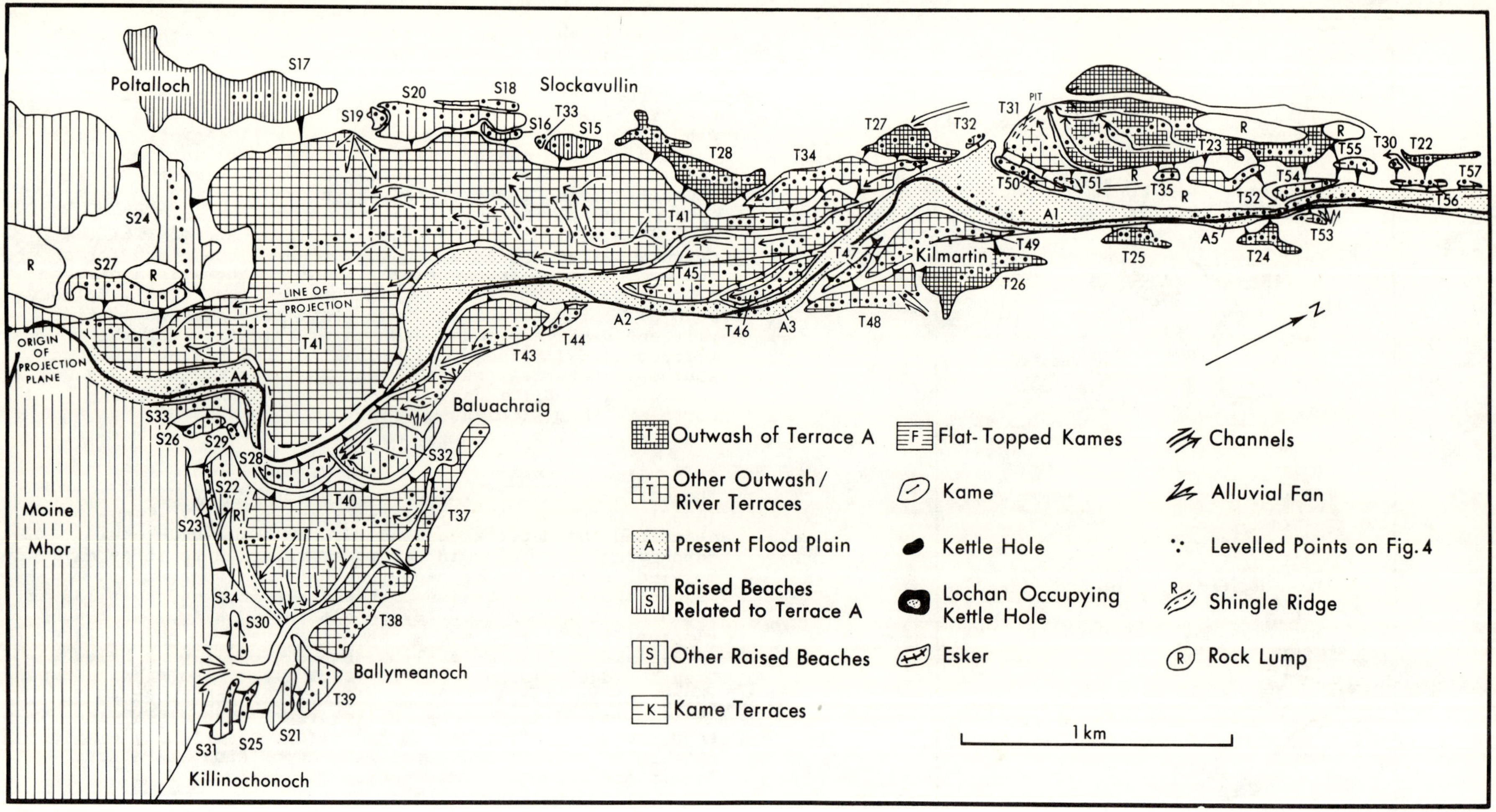

Fig. 3. Morphological map of

here collectively termed Terrace A. As Synge stated this feature begins at Eurach, the absence of correlative terraces farther up valley making it probable that the ice-front lay not far north of Eurach during the deposition of the original valley train. From an altitude of 55 m O.D. at Eurach (T18) Terrace A slopes down valley to 40 m O.D. at Kilmartin (T26) and 36 m O.D. at Slockavullin (T28). Although Terrace A has declined regularly for 4 km, south of Slockavullin the highest terrace fragments are approximately level for 1.7 km (S15-17). This strongly suggests that Terrace A was deposited to a sea-level of 35 - 36 m O.D. As the sea regressed the dissection of the valley train and the laterally constricting rock outcrops along the river course have resulted in relatively few terrace fragments (T30-36) being preserved between Terrace A and the valley floor terraces.

The extensive terrace fragments (T37-40) that make up Synge's

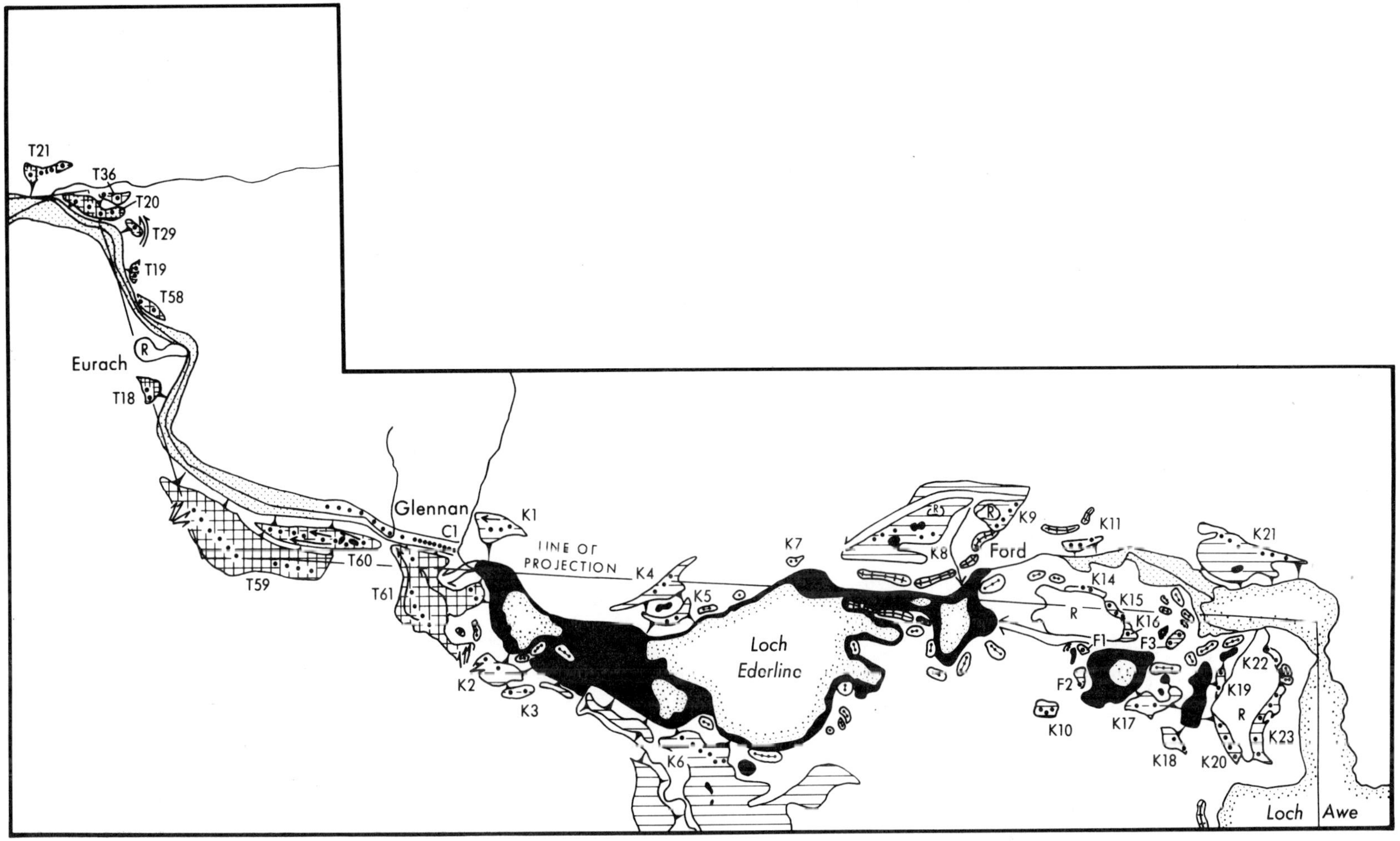

the Ford/Kilmartin valley.

"gravel bench" (and whose gradients are exaggerated on Fig. 4 due to the fact that they were levelled on lines oblique to the projection plane) indicate that as sea-level fell from the 35 - 36 m O.D. level, the Kilmartin valley was still a drainage course for meltwaters. The largest fragment (T40), which is crossed by several channels, appears to relate to a sea-level at 24 - 25 m O.D. and, as stated by Synge (1966), is overlain by a shingle ridge (R1) at its southern edge. Between Killinochonoch and Poltalloch there are a number of small terraces representing raised beaches some of which are dissected. The most dominant shoreline levels are 21 - 22 m O.D. (S21-24), 17½ - 19 m O.D. (S25-27) and 15 - 16 m O.D. (S28-31).

The terraces on the valley floor constitute a convergent system originating in the kame-and-kettle topography between Loch Awe and Glennan (Fig. 3). South of the watershed at Glennan the outwash terrace fragments (T59-61) are up to 10 m

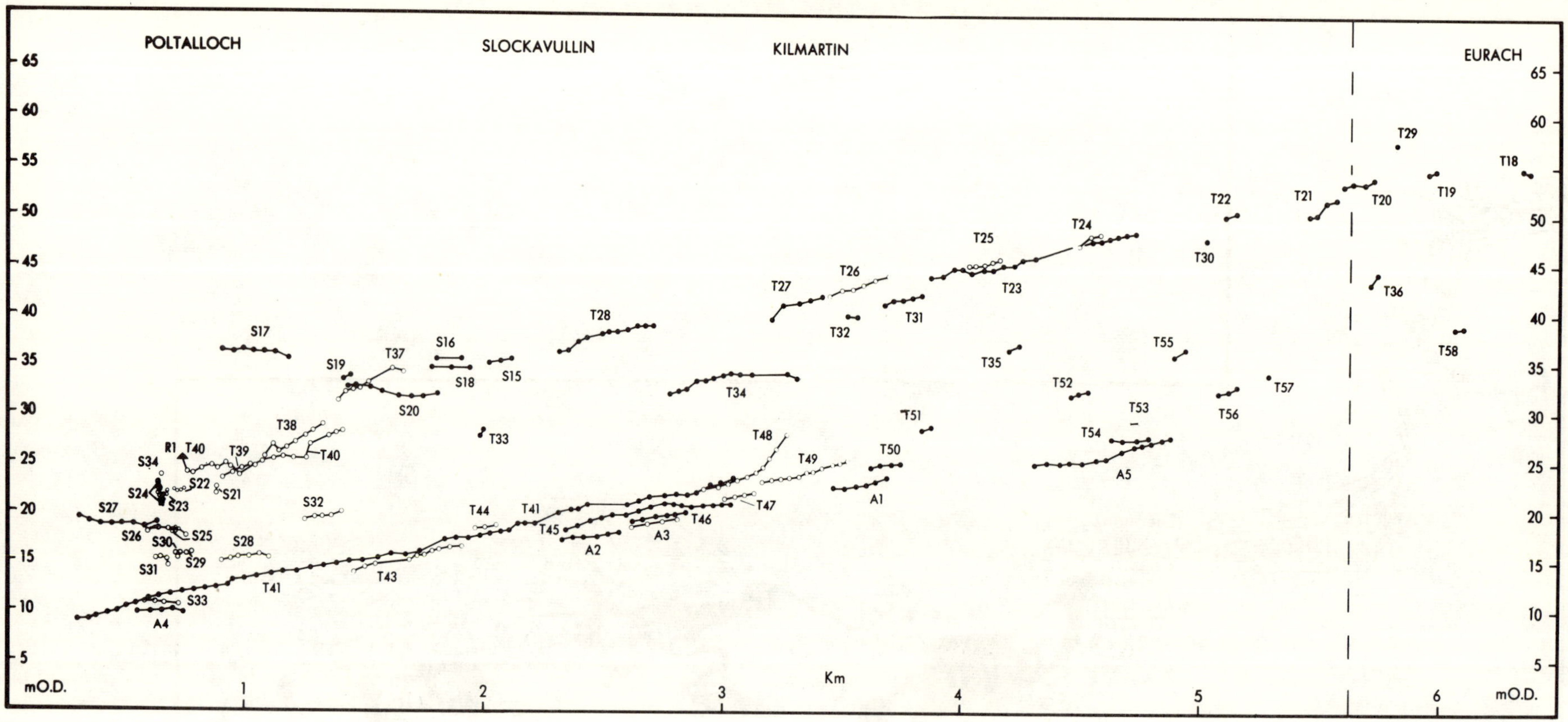

Fig. 4. Height/distance diagram of terrace fragment altitudes in the Ford/Kilmartin

above the present flood plain but down valley they converge with the latter and have been traced down to a level of 9 - 10 m O.D. near Poltalloch where they disappear below the Flandrian marine deposits of the Moine Mhor. Hence the valley floor terraces were related to a sea-level at or below 9 - 10 m O.D. Thus there is evidence that between the time the ice front stood just up valley from Eurach and the time it was situated less than 1 km away immediately north of the Glennan watershed, sea-level had fallen from 35 - 36 m O.D. to below 9 - 10 m O.D. However, the ice position at Glennan was not necessarily a slightly later stage in the retreat of the Eurach ice front.

Extent of the Loch Lomond Readvance in Loch Awe. At this point it becomes relevant to consider the extent of the Loch Lomond Readvance in Loch Awe. This readvance has been delimited by mapping and radiocarbon dating both to the north of the study area at Loch Creran (Peacock, 1971b) and to the south-east at Loch Lomond (Sissons, 1967a). It is reasonable to suppose that this readvance was also a major event in the area between these sites but its extent is much more difficult to delimit. In the Loch Awe area there are three main hypotheses when the loch's drainage pattern is considered:-

1. The ice, advancing down Glen Strae and the Dalmally valley, stopped short of the Pass of Brander so that the loch overflowed by this route as it still does today (Fig. 1).
2. The ice blocked the Pass of Brander and extended some distance down Loch Awe.
3. The ice reached the watershed at Glennan.

These hypotheses will now be considered in turn.

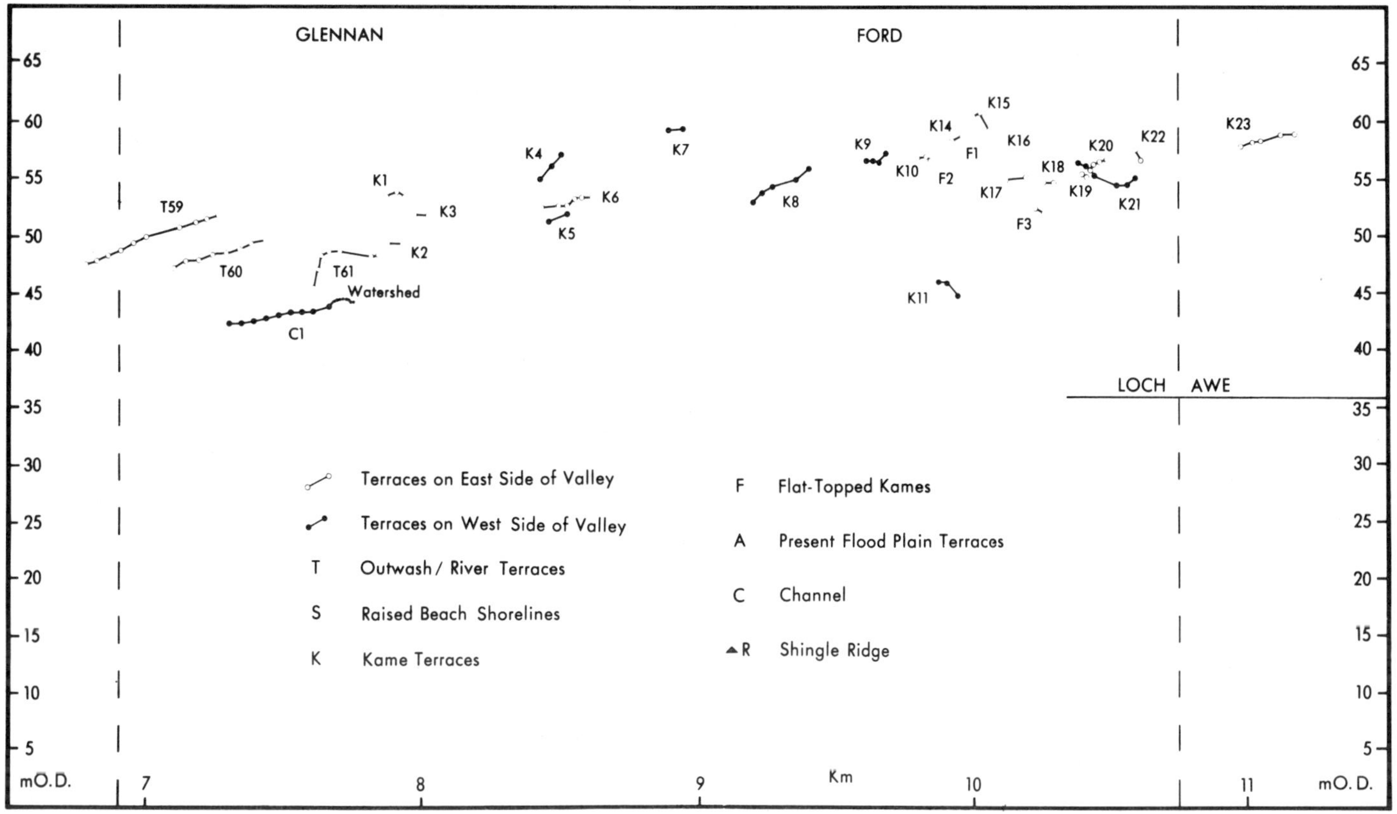

valley. Intersections of the projection planes are indicated by vertical dashed lines.

1. Against this hypothesis is the fact that hummocky moraine, which has formed the basis of delimitation of the Loch Lomond Readvance in the central Grampians (Thompson, 1972; Sissons, 1974b), is very well developed around Dalmally at the north-east end of Loch Awe. Morphological mapping south-westwards along the sides of the loch has revealed no evidence of an ice-limit. Less well developed hummocky moraines occur on both ends of the Pass of Brander, in many of the Ben Cruachan corries and at the head of Glen Aray (Fig. 1). Contemporaneity of all these moraines would indicate that the Loch Lomond Readvance ice passed some distance down Loch Awe.
2. If the ice did pass down the loch it is difficult to conceive of water from Loch Awe finding its way to the sea by a subglacial route through the Pass of Brander and along Loch Etive which was also ice filled at the time (Gray, 1975c). It seems much more likely that the water level would have risen to overflow the Glennan watershed (at *c*. 45 m O.D.) down to the sea at Loch Crinan. This seems improbable however since, other than K11, there is no evidence of notches in drift or shorelines either in the Ford area or farther up the loch at the level of the Glennan watershed.* Neither has augering at the south end of the loch revealed any lake clays mantling the slopes below the watershed level. Furthermore the watershed at Glennan is in drift and shows no evidence of having been a major overflow channel subsequent to its deglaciation.

* The terraces that are present around Loch Awe, *e.g.* at Dalavich and Fincharn (Fig. 1), are more probably kame terraces or, in the case of the terraces a few metres above the present loch, due to Flandrian changes in the loch level.

3. Having put aside the first two hypotheses the possibility that the Loch Lomond Readvance ice reached the south end of Loch Awe and that its stagnation is represented by the fluvioglacial deposits there must be considered. This would present no drainage problems in Loch Awe for at the maximum of the readvance the meltwater would issue directly into the Kilmartin valley thus forming the outwash terrace system on the valley floor. The grading of these terraces to a sea-level below the highest Flandrian marine deposits accords with evidence from many parts of Scotland regarding sea-level at the time of the Loch Lomond Readvance (McCann, 1966a; Sissons, 1967a; Gray, 1975c). As the ice stagnated, overflow would probably continue over the Glennan watershed at first but by the time the Ford kames and eskers became free of ice the overflow in the Pass of Brander had evidently come into operation. There is, however, another piece of evidence favouring this hypothesis. Although a Lateglacial stratigraphy (Sissons *et al.*, 1973) was found in a former lake site in a valley 2 km west of Ford (Fig. 1), so far the authors have been unable to find a similar stratigraphy anywhere along the shores of Loch Awe. At present twelve promising former lake sites between 4 and 12 m deep have been investigated around the loch.* The three at Ford itself lend support to hypothesis 3 while the nine around Kilchrenan and Port Sonachan (Fig. 1) favour either hypothesis 2 or 3. Eleven of the sites contain either a thin clay layer at the bottom (assumed to represent the Loch Lomond Stadial/Flandrian transition) giving way to thick Flandrian peat deposits, or contain peat deposits only. The twelfth, at Stronesker in the tributary valley east of Glennan, shows 3.5 m of basal grey clay (with no evidence of a lithostratigraphic subdivision) overlain by 8.5 m of peat. In this case it is likely that the thick clay represents accumulation in a lake dammed by the ice and kame terraces in the main valley.

Thus until a Lateglacial stratigraphy is discovered along the shores of Loch Awe it must be hypothesised that the Loch Lomond Readvance extended to Ford thus accounting for the absence of Lateglacial deposits in the lake sites around the loch. It is hoped that palynologists and other specialists will turn their attention to this neglected area of Scotland where morphological evidence (such as hummocky moraines) that might help in delimiting the extent of the Loch Lomond Readvance is much more poorly developed than elsewhere in Scotland.

* Sites investigated south of Loch Etive are also included in Fig. 1.

LOCH AVICH - LOCH CRAIGNISH

Between Loch Avich and Loch Craignish Synge (1966) described a "kame moraine" occurring just north of the watershed at Maolachy and inferred that this was related to a delta at 132 ft (40.3 m) behind the farm of Turnalt, 6 - 7 km to the south-west (Fig. 1). Although there is a small esker, a few kames and some short kame terraces at various altitudes around Maolachy, it is not possible to determine the contemporaneous sea-level since Glen Domhain, a long, deep valley, gorge-like in places, and retaining only a few poorly developed terrace fragments, occupies most of the area between Maolachy and Turnalt. The terraces around Turnalt are very well developed but they slope from 47 m O.D. at the exit from Glen Domhain to 34 m O.D. 200 m west of Turnalt.** Several large abandoned channels originate on the terrace fragment near Turnalt and become more and more deeply incised into it down valley implying a falling sea-level while meltwaters were still passing down Glen Domhain and across the terrace. The ice-front position at the time the Turnalt terraces were formed is not known.

LOCH TRALAIG - LOCH MELFORT

At the seaward end of Loch Tralaig Synge (1966) described a "series of morainic ridges" crossing the valley between the loch and Blaran, 1 km west of the loch outlet (Fig. 1). However, examination of this area indicates that the ridges are in fact resistant bands of epidiorite trending NNE - SSW and are marked as such on Geological Survey 1 inch Sheet No. 37 (drift edition). Although there are a few small mounds in which no bedrock could be found it is concluded that there is no evidence of a significant ice-limit at this locality.

LOCH SCAMMADALE - LOCH FEOCHAN

In Glen Euchar, which runs from Loch Scammadale to the sea at Loch Feochan (Fig. 1), well developed outwash terraces have been described by Kynaston *et al.* (1908), Bailey *et al.* (1924), Charlesworth (1955) and Synge (1966). Synge described the highest series of terraces as "emerging from the moraine that impounds Loch Scammadale" and sloping down

** Altitudes around Turnalt were levelled from an O.D. Liverpool bench mark and subsequently corrected to O.D. Newlyn by adding 0.66 m, this being the mean difference between the O.D. Liverpool and O.D. Newlyn values of four bench marks at Kilmelford. The Liverpool altitude of the Turnalt bench mark is precisely the altitude of the terrace as given by Synge (1966).

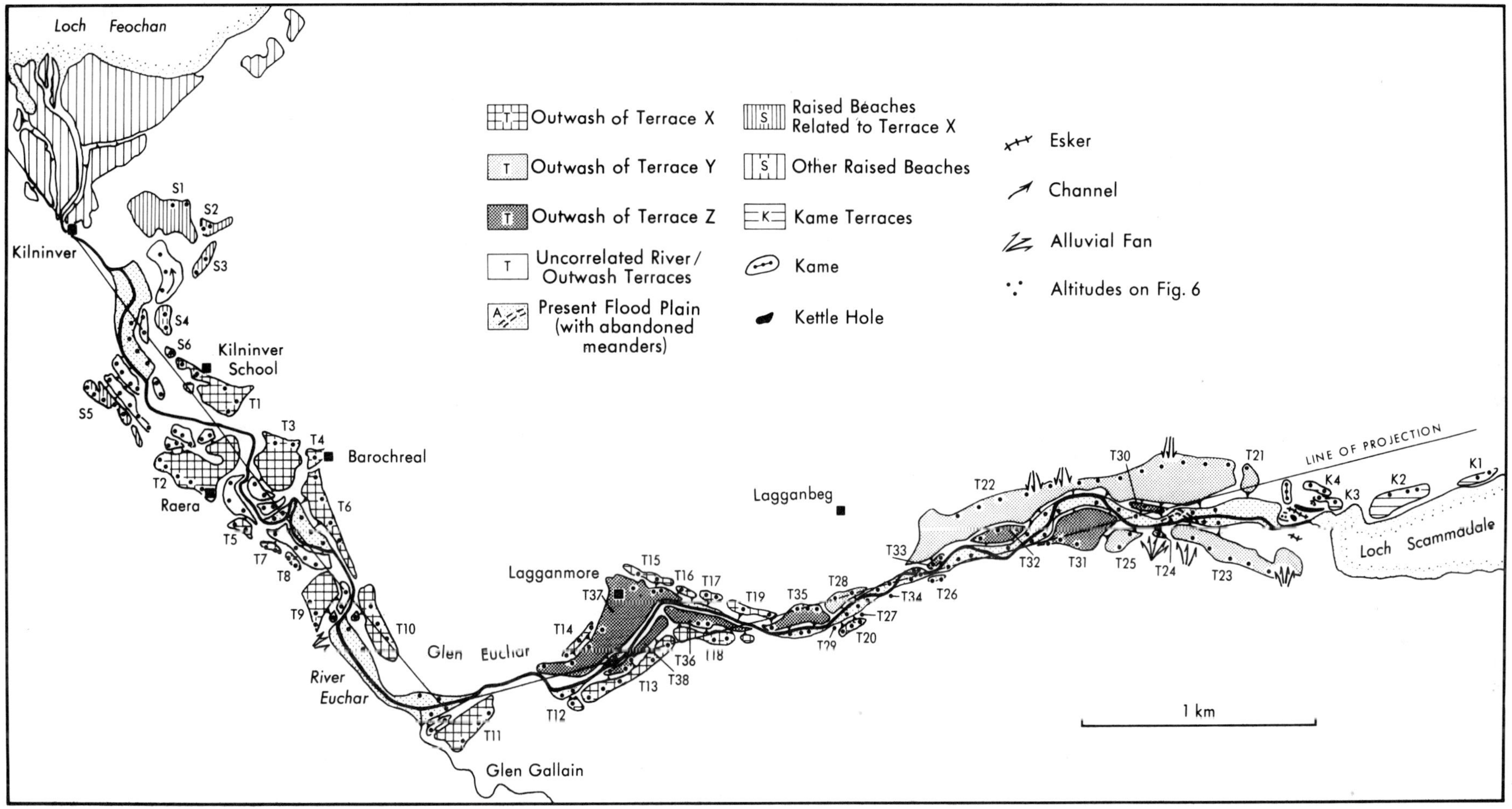

Fig. 5. Morphological map of Glen Euchar.

valley from 234 ft (71.4 m) at the loch to near the coast at Kilninver where it was said to merge with a delta at 132 ft (40.3 m) thus agreeing with the previous authors who had linked it with the "100 ft beach".

The morphological features of Glen Euchar as mapped by the present authors are shown in Fig. 5 which is generalised because of the small size of some terrace fragments. About 250 heights were obtained on the Glen Euchar terraces and the relevant measurements are shown in a height/distance diagram in Fig. 6. Because of the sharp bend in the glen about half way between Kilninver and Loch Scammadale the diagram was constructed in two sections.

The highest terrace fragments up valley from Kilninver mentioned by previous workers are well developed and stand out clearly in Fig. 6. At Kilninver the fragments (S1-6 and the lower part of T1) are almost horizontal at about 41 m O.D. for a distance of 600 - 700 m. This horizontality is interpreted as being the result of marine deposition and the

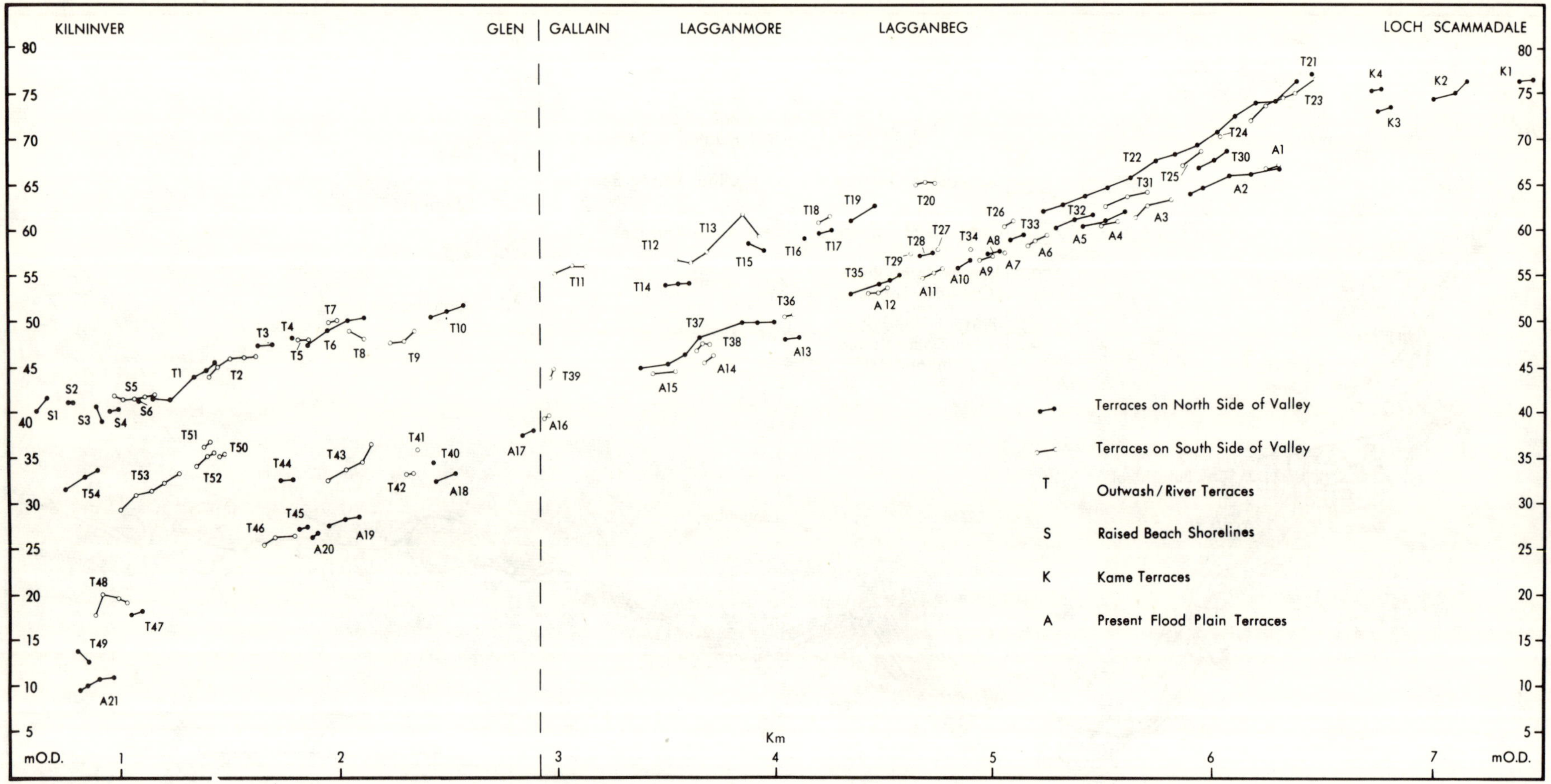

Fig. 6. Height/distance diagram of terrace fragment altitudes in Glen Euchar. Intersection of the projection planes is shown by a vertical dashed line.

conclusion arrived at by previous authors that the highest terrace fragments grade to a high sea-level is upheld.

The change from subaerial to deltaic deposition took place near Kilninver School for upstream from here the highest terrace fragments begin to rise steadily and on the basis of Fig. 6 the fragments can be correlated as far as Lagganbeg beyond which high terrace fragments are absent. Fragments T1-20 have been collectively termed Terrace X, though T9 and T14 and possibly a few others appear a little low and may be slightly younger. The best developed fragments of Terrace X are T3 at Barochreal and T2 at Raera (Fig. 5). These are large, flat features extending 250 m out into the valley and between which the river has cut down vertically and now occupies a narrow sinuous gorge about 25 m below their surfaces.

As there is no evidence that the fragments of Terrace X continue east of Lagganbeg Farm the valley constriction south of the farm is thought to be the position of the ice at the time of formation of Terrace X when sea-level stood at 41 m O.D. Thus, contrary to the interpretation of previous workers, the high terraces at Kilninver are not synchronous with those found around the outlet end of Loch Scammadale.

The terrace fragments that begin amongst the group of small kames, kettles and eskers impounding Loch Scammadale slope down valley as far as Lagganbeg and the fragments (T21-29) have been collectively termed Terrace Y. Although Terrace Y cannot be traced farther down valley it appears that this terrace is graded to a sea-level considerably lower than the 41 m O.D. of Terrace X.

Traceable slightly farther down valley to Lagganmore and lying between Terrace Y and the present flood plain (A1-21) are a few, narrow, probably synchronous terrace fragments (T30-38) here collectively termed Terrace Z. The kame terraces K1-3 probably relate to this stage which like Terrace Y, is thought to mark a later stage in deglaciation when the ice front became stabilised at the lower end of Loch Scammadale.

Down valley from the rock gorge east of the confluence with Glen Gallain there are several terrace fragments (T39-54) between Terrace X and the present flood plain, though correlation of these fragments is extremely difficult. Terrace X having been formed, the river then apparently cut down leaving large remnants of the high terrace and becoming confined to a narrow valley. Thus almost all the remnants below Terrace X in the Euchar valley below Glen Gallain are very narrow. Furthermore they are all unpaired across the valley probably due to the swing of the river from one side of the valley to the other during downcutting.

Figure 6 indicates clearly the convergent upstream nature of the Glen Euchar terrace long profiles. With a stable base-level, terrace height/distance diagrams should show divergent upstream profiles due to downcutting by the headwaters. The difference in Glen Euchar is probably due to lowering of the base-level while the discharge level was fairly stable at the lower end of Loch Scammadale. The impressive downcutting of the River Euchar into a narrow gorge and the formation of narrow unpaired, discontinuous terrace fragments is seen as further evidence of the rapid fall of sea-level.

HEAD OF LOCH FEOCHAN

A terrace at Moleigh Farm between the head of Loch Feochan and Loch Nell (Fig. 1) was described by Synge (1966) as an ice-contact deposit of sand and gravel sloping from 141 to 136 ft (43.0 to 41.7 m) in 200 m. Contrary to Synge and Stephens (1966) statement, however, the feature does not impound Loch Nell. The deposits at the seaward end of Loch Nell are believed to be related to the Loch Lomond Readvance (Gray 1972a, 1975c).

The present study has shown that the Moleigh terrace slopes from about 44 m O.D. at the top of an ice-contact slope to about 41 m O.D. where it runs into a 150 m long channel to open out at Kilmore School. Sea-level at this time must have been at or below 41 m O.D. but, as there is no evidence of a delta at a lower level, it is thought that the small terrace on which Kilmore School is built possibly indicates a sea-level at about 41 m O.D. at the time of deposition of the Moleigh outwash terrace.

OBAN CEMETERY

Synge (1966) described the ground around Oban Cemetery (Fig. 1) as "an area of kame and kettle moraine" beyond which a spread of outwash gravel slopes southwards from 163 ft (49.7 m) to 136 ft (41.4 m).

The area of sand and gravel at and north of Oban Cemetery has a gently undulating surface expression pitted in places by kettle holes and crossed by a few short eskers. Outwash terraces at two levels begin just southwest of the cemetery, and at about 42 m O.D. run into a small, narrow meltwater channel in rock. At about 39 m O.D. the channel opens out onto a flat-bottomed valley where 23 boreholes were put down recently in connection with the construction of Park School, Oban. These show that the valley is filled by "dense" sand and gravel, coarse in places and exceeding 3 m in thickness, that is considered to be a continuation of the outwash deposits. At 33 m O.D. the terrace disappears into another channel which after a short distance becomes obscured by the Corran housing scheme of Oban and it is difficult to determine the probable sea-level at the time. The main point, however, is that the fluvioglacial deposits at Oban Cemetery are related to a sea-level well below the 41.4 m level implied by Synge (1966), for although the outwash slopes down to that level, as indicated above it continues southwestwards as a series of channels and gravel-filled valleys to well below 41.4 m. Examination of aerial photographs taken before the housing scheme was built, discussion with local people and inspection of the area as it is today, indicate that the scheme was built on a flat, depositional area at about 29 m O.D. This may well have been the level of the sea at the time of formation of the Oban Cemetery deposits, for beyond this flat, the present river plunges steeply down to present sea-level.

CONCLUSIONS

There are two main conclusions to emerge from this study

regarding the "Oban-Ford Moraine". First, the constituent parts of the "Oban-Ford Moraine" cannot be described as moraines. At Loch Tralaig the ridges consist of bedrock and elsewhere fluvioglacial material predominates in the form of kame-and-kettle topography.

Secondly, the large number of altitudes obtained in this study has allowed detailed analysis of the relationships between ice-limits, outwash terraces and sea-levels. It is difficult to relate some of the ice-limits to sea-levels but several of those that can are related to levels below the 40 - 43 m O.D. suggested by Synge (1966) and Synge and Stephens (1966). As well as outwash plains grading to 40 - 43 m O.D. Synge (1966) also mentioned raised beaches at about the same level but these appear to have been subjectively chosen to fit with the postulated 40 - 43 m level. A detailed study, outwith the scope of the present one, would be required before correlations could be made between outwash deltas and regional Lateglacial shorelines. Detailed discussion of the various raised beaches is therefore not undertaken here but it may be remarked that Synge and Stephens' work in this field has been discussed elsewhere (Sissons, 1967b; Gray, 1972a, 1974b). It seems probable that the outwash related to 35-36 m O.D. in the Kilmartin valley was synchronous with the outwash descending to about 36 - 38 m O.D. in the neighbouring Kilmichael valley. Differential isostatic uplift also makes it possible that both outwash periods correlate with the 41 m O.D. sea-level in Glen Euchar and/or the 41 m O.D. sea-level at Moleigh, but much more work is needed on the pattern of uplift of Lateglacial shorelines in Argyll before such correlations can be substantiated. The main point is that due to differential isostatic uplift coincidence of altitude of a number of outwash "deltas" over a wide area is not by itself a valid correlative technique.

The evidence discussed in this paper indicates that the ice-limits do not mark a readvance or significant, synchronous, recessional stage but more probably represent only fairly brief, diachronous halts in the general recession from a period of more extensive ice cover prior to the Loch Lomond Readvance. The quickly falling relative sea-level at this time is seen as being due to the rapid isostatic rebound consequent on deglaciation. The exception to the above discussion is the kame-and-kettle topography at Ford which, it has been argued, formed just inside the limit of the Loch Lomond Readvance. If so then this readvance was more extensive in the south-west Highlands than recently supposed (*e.g.* Sissons, 1974c, Fig. 1). It is clear that although Synge (1966) and Synge and Stephens (1966) were correct in demonstrating that, contrary to Charlesworth (1955), the ice-limits described in this paper are generally related to an earlier period than those of the Loch Lomond Readvance north of Oban, they represent a much less significant event than previously supposed.

ACKNOWLEDGEMENTS

We are grateful to D. Cameron, G. Innes and I. Smith for levelling assistance, to the Natural Environment Research Council and the Carnegie Trust for the Universities of Scotland for financial assistance and to Dr. J. B. Sissons for commenting on drafts of this paper.

The Loch Lomond Readvance in the Northern Mainland of Scotland

J. B. SISSONS (University of Edinburgh)

ABSTRACT

The limits of glaciers that are inferred to represent the Loch Lomond Readvance in northern Scotland are shown on a series of maps along with the evidence on which these limits are based. The great majority of the glaciers produced one or several end moraines. Other evidence includes hummocky moraines, fluted moraines, boulder limits, drift limits and the lower limit of well-developed periglacial features. The surfaces of the former glaciers are contoured and equilibrium firn lines calculated and briefly discussed. The glaciers remained active during a period of limited (probably oscillatory) retreat from their maximal extent, following which there appears to have been extensive development of dead ice.

INTRODUCTION

The northern mainland of Scotland contains abundant evidence of a build-up of glaciers following ice-sheet decay. Much of the evidence is so clear that it is remarkable that little of it has been described and that no attempt has been previously made to map it systematically over the area as a whole. This must be mainly attributed to the remoteness of the area for, had the same evidence been present in, say, the southern half of England, it would be difficult to imagine it not having been described and discussed repeatedly for a century or more.

The area possesses excellent end moraines, fluted moraines and hummocky moraines and it also abounds in well-developed periglacial features. The full description of these various forms is impracticable in the space available and would in any case be tedious. The principal evidence from which the glaciers have been reconstructed is therefore presented in a series of maps (Figs. 2-10). The main purpose of the following text is to amplify the mapped data, indicating the limitations of the maps where necessary and drawing attention to some of the outstanding features of the area. The glaciers will normally be referred to by number, some glaciers having been split into two or more hydrological units (Fig. 1).

LITERATURE

The glaciation of the whole or part of northern Scotland has been considered by various workers including Peach *et al.* (1913), Read *et al.* (1926), Thompson (1950), Charlesworth (1955), Kirk and Godwin (1963), Godard (1965), Kirk *et al.* (1966) and Jackes (1973). Former glacier limits have been mapped by some of these authors but not all these limits, especially the numerous ones drawn by Charlesworth, appear to be supported by field evidence. The most valuable contribution is that of Peach *et al.*: they identified several of the former glaciers referred to below, although their belief in an orderly sequence of events comprising ice-sheet, confluent glacier, valley glacier, and corrie glacier stages prevented their recognising the contemporaneity of large valley glaciers and small corrie glaciers.

END MORAINES

The limits of the great majority of the former glaciers identified in northern Scotland are partly marked by end moraines, the termini of many being delimited for a

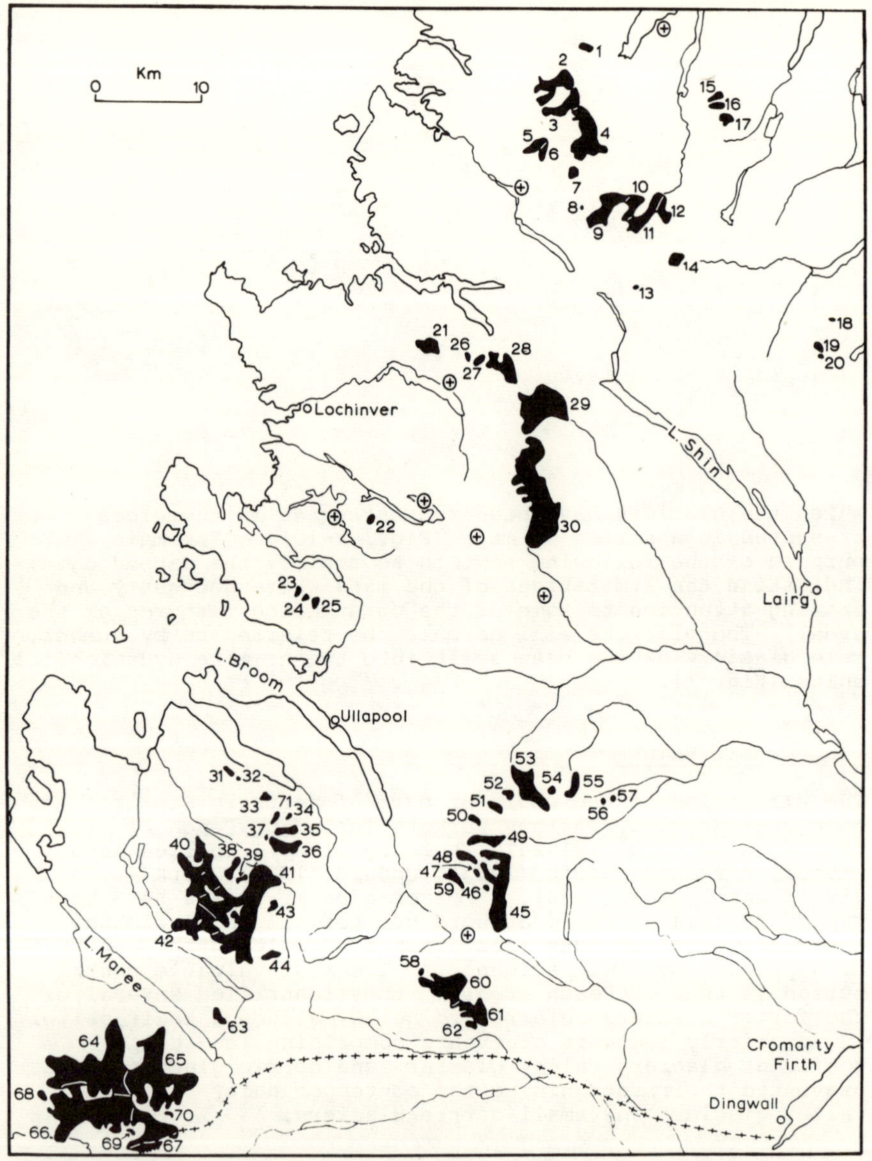

Fig. 1. Former glaciers and Lateglacial pollen sites in the northern mainland of Scotland. The broken line indicates the southern limit of the area investigated.

considerable distance by such features. The most massive moraine, 20-25 m high, 200 m broad and 800 m long, forms a conspicuous rampart that holds back Loch a'Gharbhrain and marks the greatest extent of glacier 45 (Fig. 7). A similar massive barrier, 800 m long and 150 m wide, records the lateral margin of this glacier where it crossed a side valley: this moraine consists of a series of ridges and depressions, the outermost ridge, with a distal slope up to 12 m high, being the most conspicuous element. Another massive feature, damming up Loch Ailsh in Glen Oykel, is continued along the slopes of the glen by two small lateral moraines, these features collectively indicating the former glacier limits over a distance of 5 km (Fig. 5). Similar but less impressive broad end moraines are associated with glacier 68 (Fig. 9) and with the small ice tongue immediately to the north.

The moraine associated with glacier 39 (Fig. 6) has a distal slope up to 40 m high standing at an angle of 30-35 degrees, the highest such feature in northern Scotland. The inner slope of this moraine is no more than 8 m high, however, partly because the glacier that formed the moraine was descending a steep slope and partly because of debris deposited by the glacier within the morainic arc. These factors were responsible for some other asymmetrical end moraines, such as those formed by glaciers 69 and 70 (Fig. 9), 22 (Fig. 10) and 35 (Fig. 6). The last has a steep frontal slope up to 30 m high but its inner slope, although approaching 10 m at one point, is typically only 1-3 m high: a thick accumulation of glacial debris within the morainic arc contrasts with the extensive areas of bare bedrock outside. Many lateral moraines are asymmetrical in cross profile owing to their having been built on steep valley sides, the most pronounced example being associated with the western margin of glacier 53 (Fig. 7) where it barred the mouth of the valley containing glacier 52. The proximal side of the moraine is some 30 m high while its distal slope rises only 5 m: local damming of the stream draining glacier 52 is demonstrated by a terrace 3 m above the present stream.

The former limits of many small glaciers are accurately defined by text-book examples of arcuate end moraines varying in length from a few hundred metres to more than 2 km. Mostly they form ridges 1-8 m high that are normally littered with boulders and sometimes (*e.g.* 47 and 48, Fig. 7) appear to be composed entirely of boulders. The most impressive boulder belt is associated with the northern margin of glacier 2 (Fig. 2): here a great quantity of boulders up to 5 m long occupies a zone up to 200 m wide that descends from an altitude of over 500 m to 100 m in a distance of 2½ km. The outer edge of the boulder zone is sharply defined and is locally characterised by linear concentrations of boulders

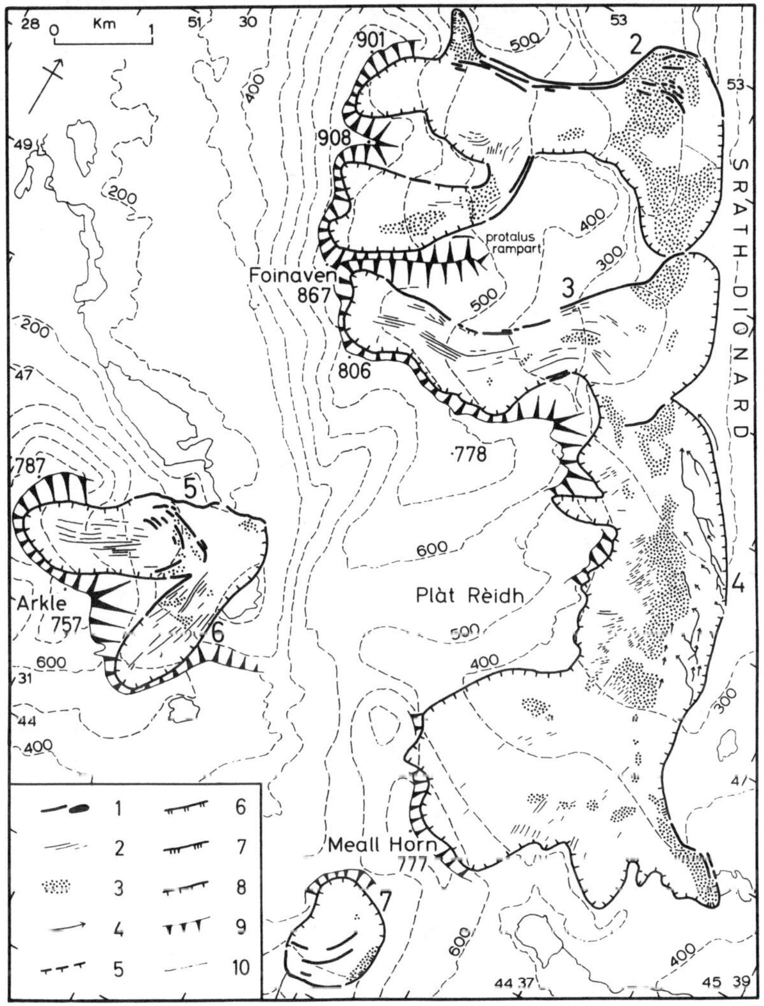

Fig. 2. Former glaciers in the vicinity of Foinaven, Arkle and Meall Horn. Symbols (also used in Figs. 3-10) are as follows. 1, End moraines. 2, Fluted moraines and other linear ridges not firmly identifiable as end moraines. 3, Hummocky moraines. 4, Meltwater channel. 5, Former ice limit inferred from boulder spread. 6, Former ice limit inferred from periglacial features. 7, Former ice limit based on limit of hummocky moraines or on limit of drift sheet. 8, Interpolated ice limit. 9, Very steep slopes of glacier source areas. 10, Contours on land and former glacier surfaces, interval 100 m.

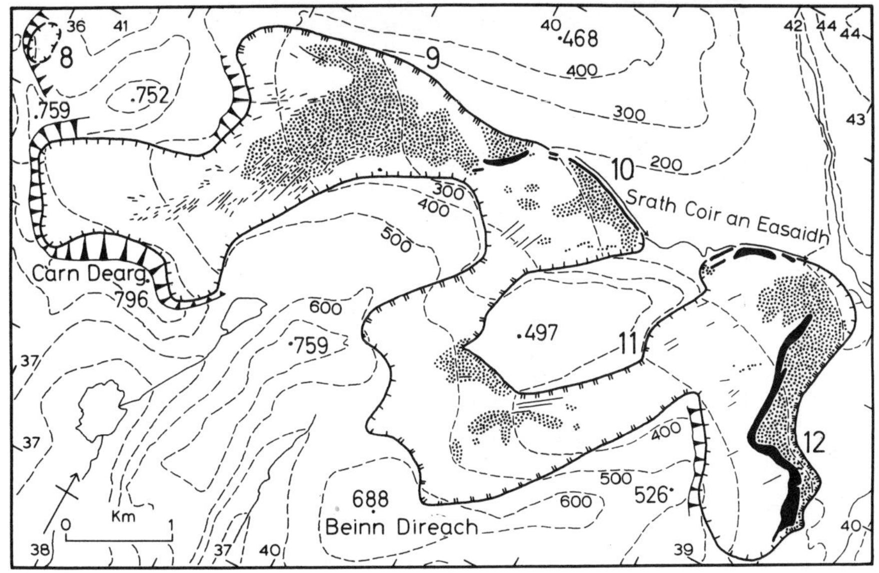

Fig. 3. Former glaciers near Carn Dearg and Beinn Direach.

representing an intermittent lateral moraine, while other concentrations within the boulder belt represent similar features.

A visually unimpressive but equally precise record of a former ice limit is supplied by the glacier that emanated from the northern corrie of Arkle (5, Fig. 2). Here great quantities of quartzite stones (including small boulders) were carried by the ice from the corrie onto the sloping surface of Lewisian Gneiss that leads down from the corrie mouth. The northern edge of the zone of quartzite erratics is partly marked by a tiny moraine that occasionally reaches 3 m high but is typically about 0.5 m; elsewhere the quartzite debris (sometimes present as a single layer of touching stones lying on bare Lewisian Gneiss) ends along a line that can be defined with an accuracy of 1-2 m. The whole of this limit is shown in Fig. 2 as an end moraine since it is impracticable at publication scale to indicate the changes from lateral moraine to boulder limit (and the precise distinction between the two is arbitrary).

The margins of a number of other former glaciers are well defined on one or both sides by lateral moraines. For example, an end moraine along the shore of Loch na Seilg turns sharply away from the shore at both ends to run steeply uphill to the corrie in which glacier 15 (Fig. 10) accumulated. The former terminus of the large glacier that

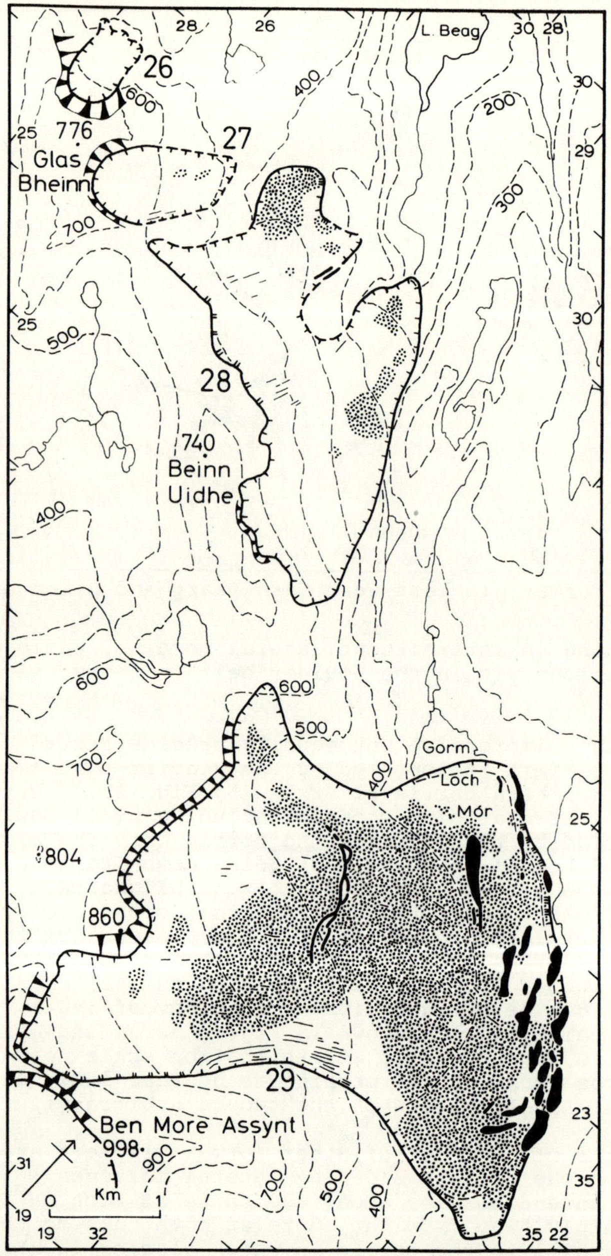

Fig. 4. Former glaciers north of Ben More Assynt.

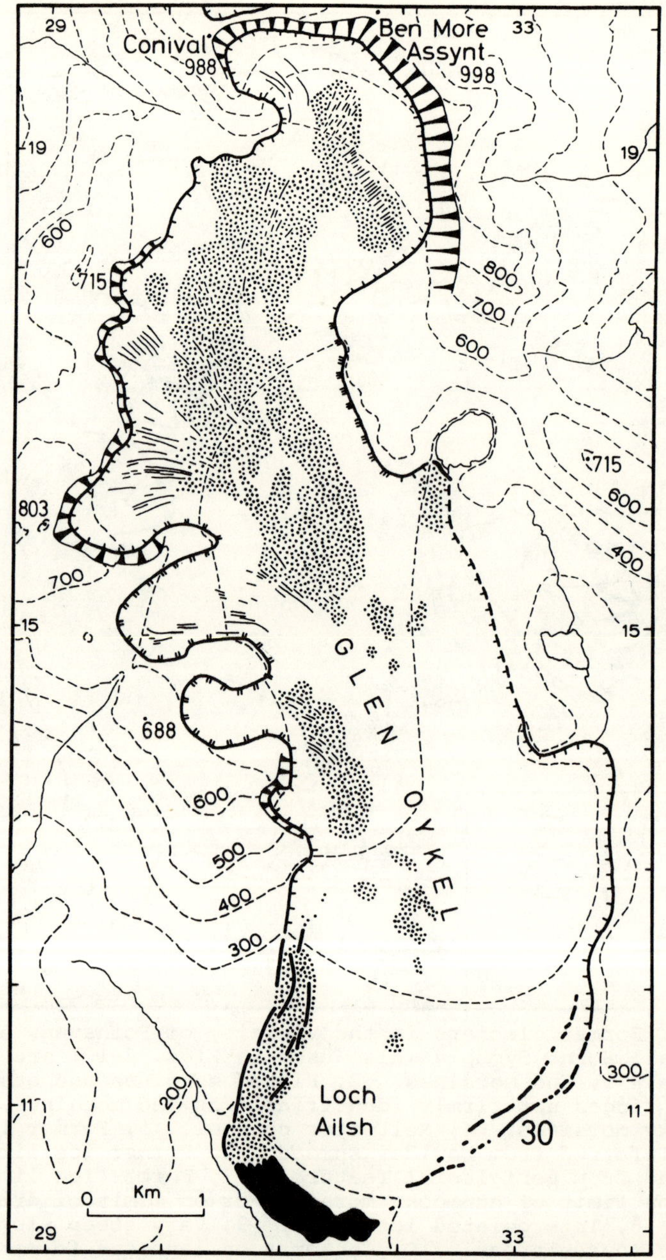

Fig. 5. The former glacier in Glen Oykel.

Fig. 6. Former glaciers south of Little Loch Broom.

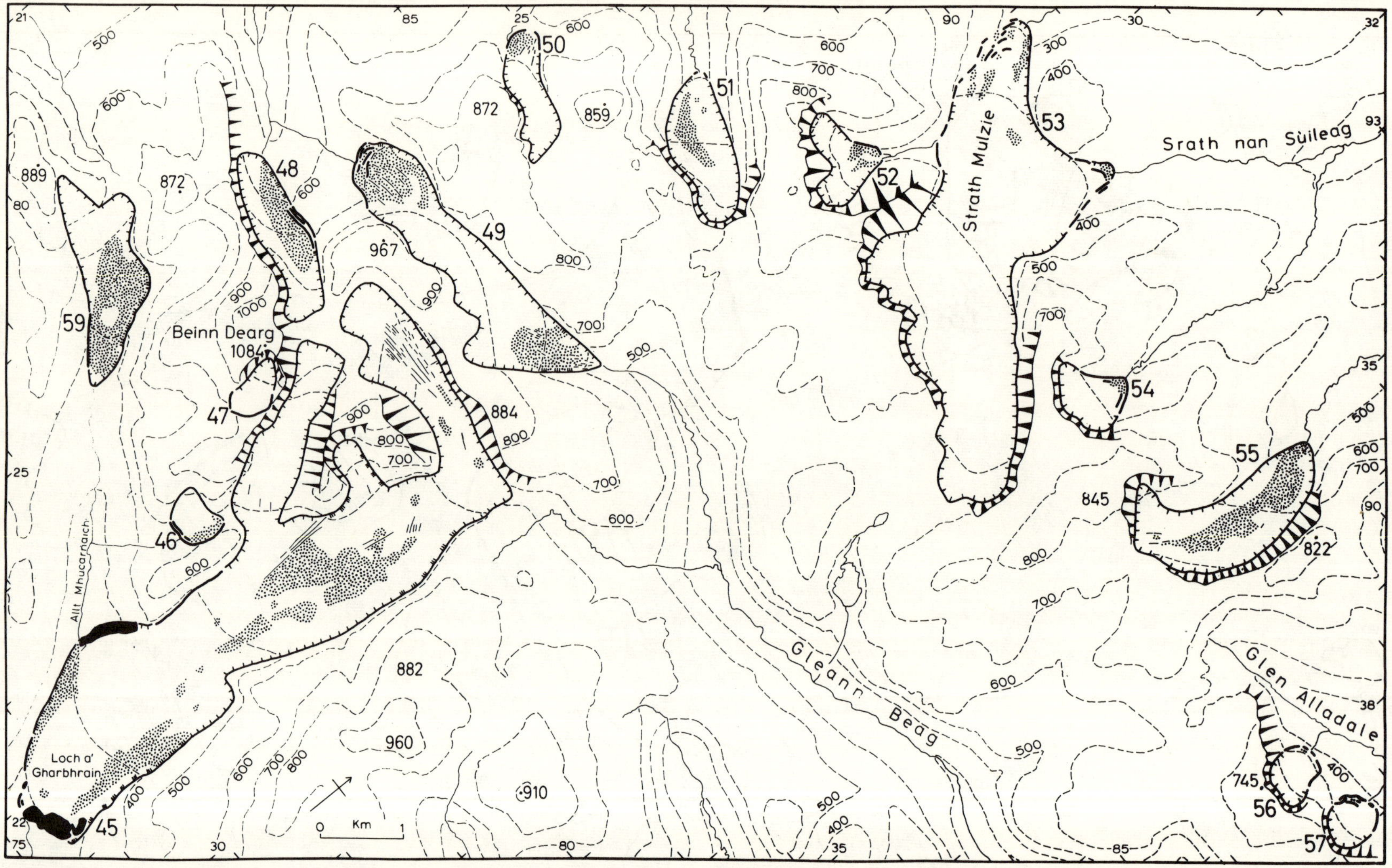

Fig. 7. Former glaciers in the Beinn Dearg area.

occupied Gleann na Muice (41, Fig. 6) and blocked Strath na Sealga to cause considerable fluvial aggradation (Peach *et al.*, 1913) is marked by a long curving end moraine whose apparent continuation climbs gently up the eastern side of the glen as a clear lateral. A small terminal moraine on the floor of Glen Torridon at about 50 m altitude is continued after a short gap by a narrow lateral boulder moraine that can be followed obliquely up the side of the glen (67, Fig. 9). Farther west glacier 66 extended just beyond the present shoreline. Its western limit is continuously defined by a moraine that climbs from sea-level to a corrie entrance at nearly 500 m altitude. The eastern edge of the same ice mass is recorded by a lateral moraine that rises gradually along the slope of the Liathach ridge to end at

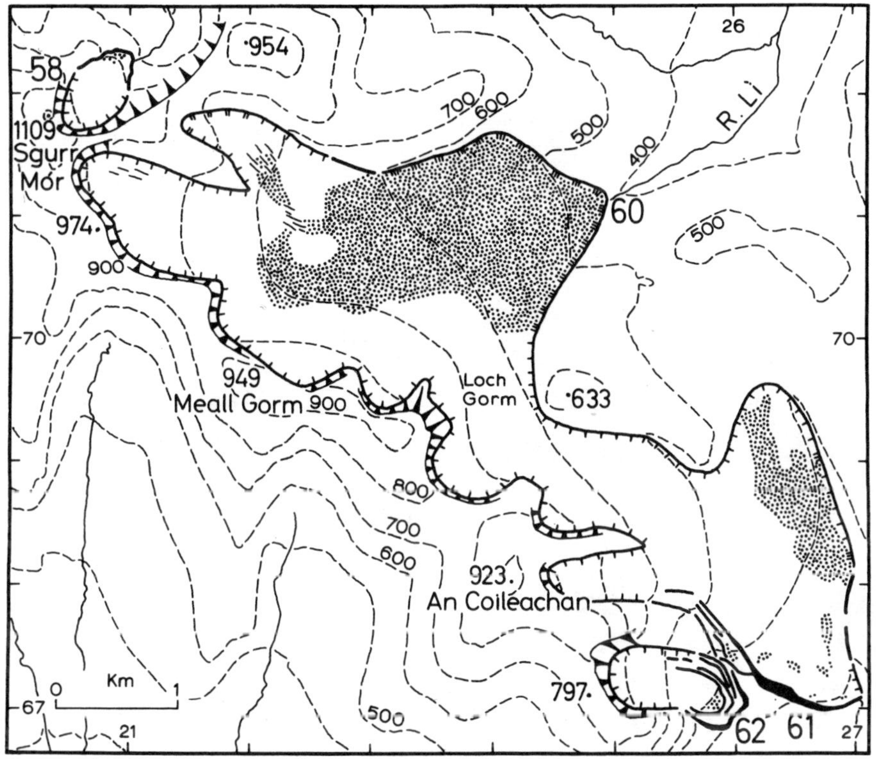

Fig. 8. Former glaciers in the vicinity of Sgurr Mòr and An Coileachan.

about the same altitude 3 km back from the glacier snout.

Multiple end moraines were produced by many of the former glaciers of northern Scotland. They are more common than indicated in Figs. 2-10 since the scale of the maps prevents the representation of some minor features. For example, former margins of glacier 58 are defined by two boulder moraines normally 1-2 m high and a third low boulder moraine is present in places: since the moraines are only 20 m apart only one can be shown in Fig. 8. The maximum number of end moraines produced by any one glacier is uncertain. For example, within the clear outermost end moraine of glacier 5 curving belts of quartzite boulders clearly record several positions of the former ice margin (Fig. 2) but there are hints of other ones. The greatest number of *clear* moraines found in association with a former glacier is five. Thus glacier 62 (Fig. 8) produced an arcuate asymmetrical end moraine with a frontal slope up to about 14 m high, this being followed by four small arcuate boulder moraines of which the last is the best developed. Glacier 38 is shown in Fig. 6 as having three long arcuate end moraines with an indication of a fourth over a short distance; the innermost moraine, however, is a double feature. Five end moraines also appear to be represented in the boulder area associated with the terminal zone of glacier 33 (Fig. 6). Very clear end moraines with abundant large boulders record three small glaciers that developed on the steep northern slope of Ben More Coigach and spread out slightly at the base of the mountain (23-25, Fig. 10). The easternmost glacier has left four distinct end moraines, the outermost (4-6 m high) being the largest, although the adjacent glaciers are respectively recorded by only one and two clear end moraines. In contrast to these clear features the main evidence for glacier 7 consists of four low linear mounds whose identification is difficult owing to extensive peat. Yet the gently curving pattern of the mounds (Fig. 2) and their relation to a likely glacier source area strongly suggest that they are end moraines.

The end moraines associated with glacier 14 (Fig. 10) are unusual in that the innermost is by far the biggest, attaining a height of 10-12 m and having a much greater mass than the two small moraines situated outside it. The terminal zone of glacier 12 (Fig. 3) is rather similar: here the ice limit is marked by the limit of hummocky moraines, 100-400 m within which is a large end moraine over 2 km long and 5-8 m high that has a steep proximal slope. With these two principal exceptions, the normal pattern in northern Scotland is for the outermost moraine, corresponding with the limit of the readvance, to be the most prominent. Where several end moraines are present the innermost is often the next best developed feature. The broad massive moraines (*e.g.* glaciers 30 and 45) probably equate with the several moraines of some glaciers and their normally well-defined, often steep inner and outer slopes suggest stable glacier margin positions corresponding with the outermost and innermost moraines elsewhere. The relationship of the broad end moraine in Glen Oykel to the two lesser end moraines that continue it (Fig. 5) accords with this suggestion.

Figures 2-10 show the end moraines as associated with the farthest limits of the former glaciers or located within a

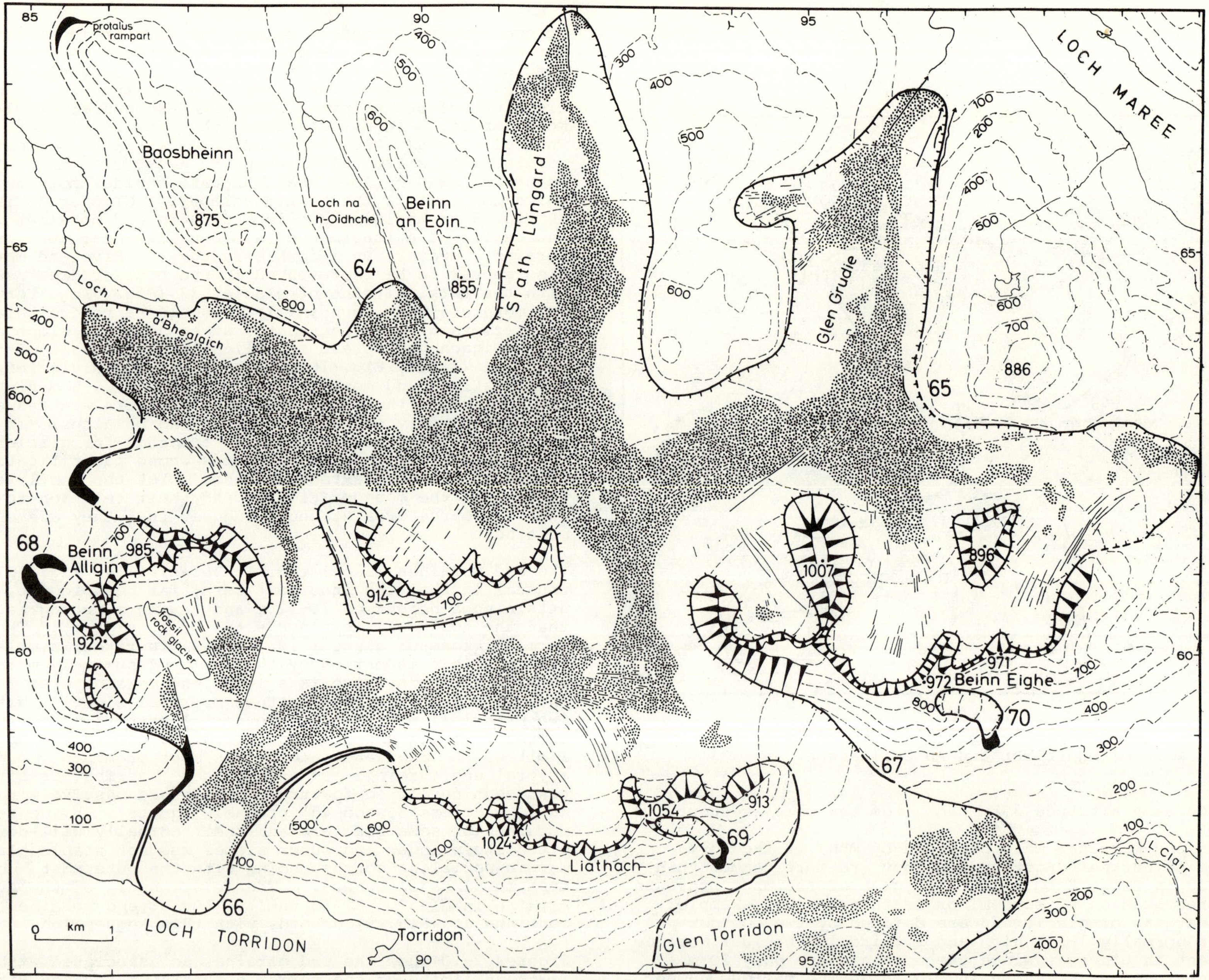

Fig. 9. Former glaciers between lochs Torridon and Maree.

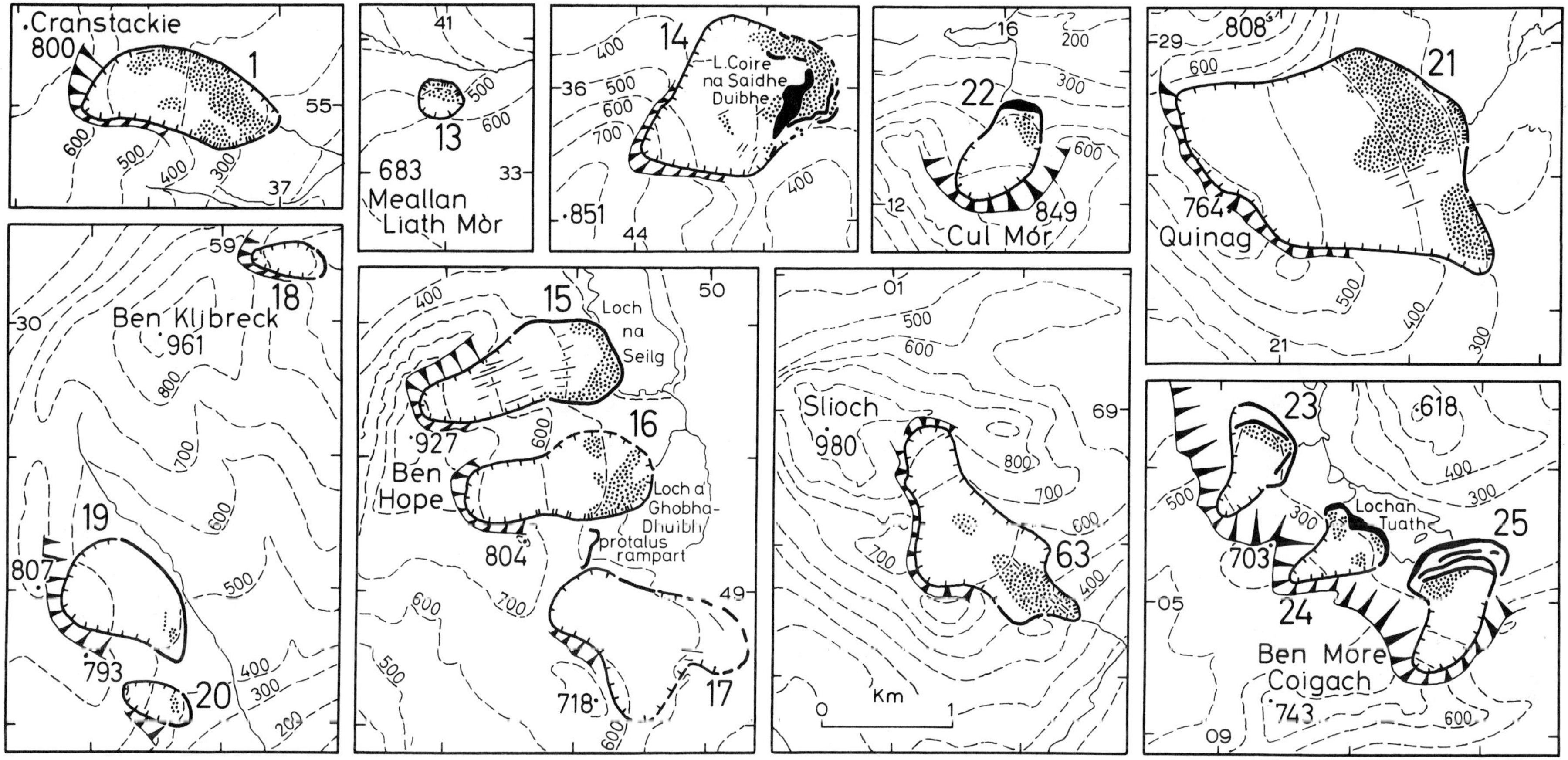

Fig. 10. Small former glaciers at various locations in northern Scotland.

short distance of these limits. Back from these positions clear end moraines are absent with the possible exception of a feature associated with glacier 29 (Fig. 4). Bulky end-moraine ridges surrounding and rising out of a string of lochs mark the greatest extent of this glacier as well as slightly less extensive positions. Some 2 km back from the maximal limit is a pronounced steep-sided ridge bearing numerous boulders up to 5 m long. At three places the ridge becomes a double feature enclosing a steep-sided depression. The plan of the ridge (Fig. 4) might suggest it is an esker, but this interpretation is considered improbable in view of the general trend of the feature and because its large size appears inconsistent with the limited evidence of fluvio-glacial activity in the ground covered by glacier 29. The only reasonable alternative explanation is that the ridge is a moraine, which view accords with its trend and the abundance of large boulders. The deep depressions clearly indicate the sites of dead-ice blocks, suggesting that parts of the ridge were originally ice-cored. The feature is thus unlike other moraine ridges referred to above.

HUMMOCKY MORAINES

Hummocky moraines comprising boulder-strewn mounds of fresh appearance separated by peaty hollows are common in parts of northern Scotland. Typically the individual mounds are a few metres high and they sometimes occur in great numbers. A simple distribution of hummocky moraines, indicating erosion in glacier source areas and deposition towards glacier snouts, occurs in association with many small former glaciers, for the hummocks cover the ground within the end-moraine arcs and die out up slope. With the exception of glacier 12 (mentioned above), the mounds end abruptly at the associated end moraines. This suggests that where end moraines are absent or of limited extent the down-valley termination of hummocky moraines may be used to reconstruct former glacier limits.

For example, in the area covered by Fig. 7 end moraines show that small glaciers existed in several valleys and corries facing between north-west and north-east. In valley 55, however, which is bounded on its eastern and southern sides by a very steep, partly vertical slope often 300 m high, there is no end moraine. The valley floor is choked with mounds, including a few fluted moraines aligned down valley, thus suggesting the former presence of a glacier whose farthest extent corresponded with the down-valley termination of the hummocky belt. Another example is glacier 49, the western limit of which is clearly marked by an end moraine composed of boulders surrounding a hummocky terrain of similar composition. The former presence of this ice tongue suggests that an associated ice tongue flowed eastwards, for the valley that was occupied by the western part of glacier 49 leads through to the Glen Beag drainage basin. Hummocky and fluted moraines with a sharp down-valley limit support this suggestion.

An area where interpretation of some of the morphological evidence is difficult is included in Fig. 8. In the north-west of this area the former presence of corrie glacier 58 is proved by end moraines. At the southern end of the area the series of end moraines produced by glacier 62, and the long arcuate moraine of 61 continued up slope by several laterals towards a corrie source, provide further conclusive evidence. Yet in the series of corries that lies between glaciers 58 and 62 there is no comparable evidence. In the low ground to which the corries drain, especially in the valley of the River Lì, there are extensive areas of large, fresh-looking, boulder-strewn morainic hummocks. However, there are also other groups of morainic mounds farther down the Lì valley, of varying degrees of clarity, some of them situated far beyond any glacial limits that could be reasonably correlated with the limits in Fig. 8 that are accurately established by end moraines. Within the area of uncertainty the only line based on morphological evidence that appears to represent a possible limit of a former glacier in the Lì valley is the down-valley limit of the major area of large morainic hummocks. This line is therefore shown in Fig. 8, although it is regarded as tentative.

The glacier that formed in the Slioch corrie (63, Fig. 10) built a clear moraine along part of its lateral margin but its terminal zone, which was located on a very steep slope, lacks definite end moraines. Instead there is here a precisely-defined area of drift (with abundant boulders) that forms low mounds and ridges. The limit of a glacier that occupied the Cranstackie corrie (1, Fig. 10) is marked by an end moraine for only 150 m, but the extent of the ice tongue beyond the corrie is proved by spreads and heaps of large boulders with a clearly-defined outer limit. In the eastern corrie of Quinag (21, Fig. 10) there is again only a short end moraine but hummocky moraines with numerous boulders are interpreted as related to a former ice limit.

Pronounced hummocky moraines associated with glacier 29 (Fig. 4), made more impressive by many boulders up to 5 m long, include a few mounds with a central depression suggesting a former ice core. By far the most extensive area of hummocky moraines covers valley floors and lower slopes north of Ben Eighe and Liathach (Fig. 9). Boulders, most of them blocks of Torridon Sandstone, but with Cambrian Quartzite locally dominant, occur in enormous numbers: a rough estimate indicates that surface boulders alone number millions. At the head of Srath Lungard the morainic relief is exceptionally strong, with massive mounds up to 15 m high. In Glen Grudie and Srath Lungard, and by lochs na h-Oidche and a'Bhealaich, mounds are absent beyond a line which, though not definable as accurately as with many small glaciers, can usually be placed in a zone about 100-200 m broad. Hummocky moraines have also been utilised in delimiting the ice tongues that existed north-east of Ben Eighe and by Loch Clair: both these limits are approximate and might be more accurately defined by detailed studies of erratics. It is worthy of note that the four north-facing and two east-facing ice tongues of the major ice mass shown in Fig. 9 failed to produce end moraines, whereas the south and west-facing tongues of the same ice mass built excellent end moraines. This contrast suggests differences in glacier regimes related to aspect.

FLUTED MORAINES

Fluted moraines and other linear ridges not firmly identifiable as end moraines are grouped together in Figs. 2-10. A few of the ridges are probably minor lateral moraines and a few more are medial moraines, the best individual example of the latter being the 2-km-long ridge east of Ben Alligin (Fig. 9). The ridges are often difficult to map in the field but stand out on vertical aerial photographs or when viewed from adjacent high ground. Many ridges extend for only 50-100 m but a few are over 500 m long. Individual ridges are straight or gently curved in plan. The ridges normally occur in groups, which record the last direction of ice movement in their vicinity. Often they extend to higher altitudes than the hummocky moraines (*e.g.* Figs. 2, 3, 5, 9) and demonstrate ice-flow directions out of the corries. The frequent occurrence of fluted moraines in places where glaciers flowed down steep slopes suggests an association with rapidly-moving ice.

Fluted moraines are common in the area covered by Fig. 9, the clearest ones being formed of quartzite debris (*e.g.* north-east of Ben Eighe). One of the most pronounced groups of fluted moraines in northern Scotland is in the source area of glacier 45 (Fig. 7), where they form a ridge-and-furrow landscape with an amplitude of up to 5 m. A few short ridges of similar height, composed of Torridon Sandstone boulders, were formed on steeply-sloping ground by glacier 39 (Fig. 6). At the southern foot of Ben More Assynt many of the fluted moraines produced by glacier 30 (Fig. 5) locally climb up hill, although down-hill situations prevail generally.

The fluted moraines of Srath Dionard (Fig. 2) are particularly significant in that they lead obliquely outwards from the precipitous rock wall that here forms the western side of the strath. This demonstrates glacier accumulation at an exceptionally low altitude, for the strath floor is here at only slightly above 100 m. This doubtless relates to the extensive plateau situated to the west and south-west, from which snow would have been blown to accumulate beneath the rock wall. Another interesting feature of this area is that, on the eastern side of the strath floor, pronounced dead-ice topography with a relief amplitude of up to 25 m has fluted moraines superimposed on it, the trend of the fluting being unaffected by the presence of the dead-ice terrain. This clearly implies that active ice over-rode dead-ice topography in which the dead ice and dead-ice moraine on the one hand, and active ice on the other, was a smooth surface. It may be added that this relationship between fluting and dead-ice moraine occurs in deposits attributed to the Loch Lomond Readvance in the Loch Skene corrie in the central Southern Uplands (Sissons, unpublished).

All the fluted moraines identified in northern Scotland lie inside the inferred readvance limit with the possible exception of one small area situated in the east-facing corrie of Ben Hee (NC/4334). Here several long, low, parallel drift ridges lead outwards from the base of a huge landslide that has wrecked the corrie backwall. However, the ridges are unlike normal fluted moraines in their lack of relation to local relief. Their origin is therefore uncertain, but some connection with the landslide is suspected.

BOULDER LIMITS

Many of the glacier limits inferred from end moraines and hummocky moraines correspond with an abrupt change from ground with abundant boulders (sometimes entirely surfaced with them) to ground with scattered boulders. In a few places where boulders are not generally piled on top of one another (and hence have not been mapped as moraines) a sudden reduction in their frequency appears to mark a former ice limit and is mapped as a boulder limit in Figs. 2-10. A special case of such limits applies to glaciers 26-28 (Fig. 4). Here the plateau of Glas Bheinn and Beinn Uidhe, and the upper part of the steep slope that limits it north-eastwards, are composed of Cambrian Quartzite, whereas the ground farther north-east is formed of Lewisian Gneiss. The limit of quartzite boulders carried onto the gneiss by glacier ice is clearest in the north-facing corrie of Beinn Uidhe (26) where it allows the former ice limit to be identified to an accuracy of 1-2 m. Farther south abundant quartzite boulders resting on the gneiss, in places forming distinct white bands trending north-eastwards, have been used to determine parts of the limits of glaciers 27 and 28. Near the eastern margin of glacier 28, however, a belt of quartzite bedrock descends from the plateau to the valley floor, so that hereabouts the simple pattern is likely to be increasingly complicated by quartzite boulders carried north-westwards under preceding ice-sheet conditions. Detailed boulder studies would probably enable the limits of glaciers 27 and 28 to be defined more accurately.

DRIFT LIMITS

A particularly good example of a drift limit defines the northern margin of glacier 36 (Fig. 6) for 2 km, the debris laid down by the glacier ending along a clear line beyond which is bare bedrock. This drift limit corresponds with a low lateral moraine and with the limit of hummocky moraines. Drift limits associated with many other glaciers in northern Scotland show similar relationships. Such drift limits are not shown in Figs. 2-10 since other symbols are given precedence.

In some areas sheets of drift lacking definite constructional forms are associated with the former glaciers. Peat frequently conceals the limits of such sheets but in a few instances a limit can be mapped and is shown in Figs. 2-10. Probably the best examples are related to the northern margin of glacier 9 (Fig. 3) and to the eastern margins of glaciers 30 (Fig. 5) and 45 (Fig. 7). In all three of these areas hummocky moraines on low ground merge on the lower valley slopes into a till sheet whose linear upper limit is believed to mark the former lateral limit of the relevant glacier.

PERIGLACIAL FEATURES

Periglacial features are extremely widespread on high ground in northern Scotland and in places, especially in the far north on quartzite bedrock, extend down to low altitudes. They include very extensive areas of frost-shattered bedrock, as well as solifluction lobes and terraces, stone polygons and protalus ramparts. Minor periglacial features are currently active and White and Mottershead (1972) have demonstrated the movement of a solifluction lobe in the latter half of the Flandrian. However, these authors, in common with other workers, consider the great majority of the periglacial features to have been formed before the Flandrian under severe climatic conditions. That the bulk of the periglacial features are fossil is indicated by the almost complete absence of major features from ground inferred from evidence summarised above to have been covered by the readvance glaciers. The fresh appearance of the hummocky moraines, the straight or gently-curving courses of the delicate fluted moraines, and extensive glaciated bed-rock surfaces unaffected by major frost action, all indicate the lack of major periglacial activity since the glaciers disappeared. Hence where the lower limit of well-developed periglacial features on hill slopes can be identified it may be used to help define the upper limit of former glaciers.

The relation between periglacial and glacial evidence is not perfect, however, and requires discussion. A massive accumulation of very coarse debris interpreted as a fossil rock glacier lies within the ground covered by glacier 66 (Fig. 9). It has been suggested that this feature resulted from reactivation of dead ice following a major landslide (Sissons, 1975a): the feature is thus attributed to exceptional circumstances. More significant is a fossil frost wedge just inside the limit of glacier 41 (Fig. 6). The wedge is in horizontally-bedded gravels and tapers down from the present ground surface, where it is about 75 cm wide, for 3 m. The wedge is filled with sand and a few vertical pebbles, the sand probably having been derived from a bed of sand that has since been eroded from the site of the wedge but still overlies the gravels immediately adjacent to it. The wedge implies that severe climatic conditions associated with the readvance continued after the ice reached its greatest extent. Since the series of end moraines associated with the terminal zone of many of the former glaciers of northern Scotland suggest that a severe climate prevailed for a time after the maximum, it is inferred that the frost wedge developed during this time. It may seem that a corollary of this is that periglacial phenomena cannot be used to infer the upper limits of the former glaciers on the higher ground. However, since the end moraine zones at the glacier snouts are of limited width, especially in comparison with the dimensions of the larger glaciers (where periglacial features are most useful in inferring former ice limits), contemporaneous downwasting of the higher parts of the glaciers is likely to have been slight.

An example of well-developed solifluction lobes of quartzite debris inside a former glacier limit occurs at altitudes of about 300-400 m along the steep south-facing slope of the valley occupied by glacier 3 (Fig. 2). Since White and Mottershead have demonstrated Flandrian solifluction of similar material on nearby Arkle, the features in the ground covered by glacier 3 may be of similar age. Down-slope movement of debris in the latter area has been limited, for small lateral and fluted moraines can be identified even though they are distorted. In this context the evidence provided by a small lateral moraine near the head of the same valley at an altitude of 500-600 m is particularly striking. Here the ground inside and outside the moraine almost lacks vegetation and is entirely covered by flattish quartzite stones that are so loose that walking across the ground is accompanied by a constant rattling of the stones. The loose debris also forms the lateral moraine, which is 30-50 cm high on its up-slope side, 3-5 m wide, and extends for 800 m falling regularly in altitude down valley. The clarity of this delicate feature composed of debris very susceptible to solifluction must relate in part to the moderate angle of the slope on which it rests, but it demonstrates that in some localities even at high altitudes periglacial processes have been ineffective during the Flandrian.

Former glacier limits based on the extent of periglacial features are distinguished by a separate symbol in Figs. 2-10. On the ground such limits appear as a line or, more often, as a narrow zone in which there is a rapid change, often from intensely frost-shattered ground to ground in which glaciated rock surfaces of fresh appearance are common. One of the areas in which such contrasts are most clearly seen is in association with glaciers 10 and 11 (Fig. 4). A rather different relationship is exemplified by the western side of glacier 41 (Fig. 6). Here the east-facing slope of

Beinn Dearg Mór is littered with numerous large blocks of Torridon Sandstone derived from the rock faces on this slope, especially on its higher parts. Below a line that inclines gently northwards along the hill side boulders are far less numerous: the line,which is occasionally marked by a band of Torridon Sandstone boulders, clearly represents the former ice margin for northwards it is continued by the well-defined end moraine delimiting glacier 41. It thus appears that a periglacial climate enabled the boulders to move down slope to the glacier margin, where they were carried away; their down-slope termination implies that when the glacier decayed conditions became unfavourable for significant movement of the boulders and have since remained so.

RECONSTRUCTION OF THE GLACIERS

Although the evidence summarised above and illustrated in Figs. 2-10 provides a firm basis on which to commence reconstruction of all but a few of the former glaciers, most of the ice limits shown on the maps are classified as interpolated. While some of these interpolations are tentative, the constraints imposed on the great majority by various factors mean that they are unlikely to be seriously in error. The source areas of almost all the glaciers are obvious; since these are normally corries or other sites bounded by very steep slopes, even considerable errors in estimating former ice-surface altitudes would have little effect on mapped glacier limits in these localities. The simple requirement that a glacier surface should slope up continuously towards its source area also restricts speculation. In reconstructing the glaciers the probability that their surfaces were more steeply inclined where their beds were steeper, particularly where glaciers were thin, was taken into account.

The glacier surfaces were contoured at 50 m intervals (only the 100 m contours being shown in Figs. 2-10). The lowest contour on a glacier was related to the form of the snout, successive contours on the lower part of the glacier being drawn with diminishing up-valley concavity. On the higher parts of the glaciers the contours were normally drawn with up-valley convexity, the amount usually increasing towards the head of a glacier where contour form was related to the bounding slopes and to probable areas of major snow accumulation. The latter factor is important, for the distribution of rock walls in some corries (and occasionally elsewhere) shows that glacier accumulation was not symmetrically distributed in these corries. For example, the large corries in which glaciers 35 and 36 (Fig. 6) accumulated have precipitous rock walls facing north and east but their south-facing sides, although very steep, are largely uniform slopes veneered with debris. Glacier accumulation beneath the east- and north-facing slopes is thus indicated and glacier surface contours were drawn accordingly.

Systems of fluted moraines were also taken into account in drawing the contours, for the latter will usually be at right-angles to the flow lines recorded by such moraines. A disadvantage of this approach is that, where fluted moraines continued to be formed as glaciers decayed, their directions may differ from those of ice flow when glaciers were at their maximal extent. Yet the patterns of some fluted moraine systems, such as those on the western side of Glen Oykel (Fig. 5) and on the northern side of Liathach (Fig. 9), strongly suggest that these systems were formed when the ice was very extensive: had they been produced under restricted ice cover one would have expected them to trend less obliquely down slope. This in turn suggests the possibility of extensive ice stagnation in northern Scotland following the formation of the end moraines that characterise the terminal zone of many glaciers. The absence of end moraines back from this terminal zone (excepting the anomalous feature associated with glacier 29) accords with this interpretation. So also do the many areas of hummocky moraines that include numerous closed depressions pointing to the former presence of dead ice.

The contouring of the glacier surfaces was found to be a valuable exercise, for it necessitated detailed consideration of the form of each glacier and its relation to general relief and to morphological details. During the contouring it was often found necessary to make minor adjustments to the interpolated ice margins.

In summary, it is believed that the great majority of the former glacier limits have been determined with considerable accuracy. Spreads of ice-transported boulders occasionally define limits to a probable accuracy of 1-2 m, most end moraines probably permit an accuracy of 10 m, while many other plotted limits (including many interpolated ones where ground slopes are steep) are believed to be accurate to within 50-100 m. There are, however, certain areas of very considerable uncertainty in which detailed studies are required. It may also be noted that a few of the extremely small areas shown as glaciers in Figs. 2-10 may in fact have been permanent snow beds. This is very probable for numbers 8, 13, 20 and 37, for ground form at these sites would have allowed snow to accumulate to only a limited depth.

Some former glaciers may not have been identified. Numerous likely sites in which initial aerial photograph study revealed no clear morphological evidence of former local glaciers were subsequently examined in the field and confirmed as lacking such evidence. (Areas investigated included the Ben Wyvis upland mass.) However, it was not practicable to visit every such site in northern Scotland. Furthermore, it is possible that in some instances critical evidence is concealed by peat while in others glaciers may have failed to leave clear morphological evidence.

DATE OF THE READVANCE

All the former glaciers shown in Figs. 2-10 are interpreted as having been associated with a single readvance, for the restriction of end moraines to the terminal zones of the glaciers (with the possible exception of glacier 29) permits no other reasonable conclusion. In other words, if it were argued that the end moraine belts of some glaciers relate to a particular readvance and that those of other glaciers belong to a quite different readvance, it would be extremely difficult to explain why no glacier possesses two end-moraine belts representing the two readvances.

Nine Lateglacial pollen sites, all of them on relatively low ground, are known in the area and are marked in Fig. 1 (Kirk and Godwin, 1963; Pennington *et al.*, 1972; H. J. B. Birks, pers. comm.; W. Tutin, pers. comm.). In terms of a down-wasting ice-sheet these sites point to total deglaciation of the area before the Loch Lomond Readvance. The radiocarbon dates of 12,810 ± 155 B.P. from the Lateglacial Interstadial deposits at Loch Droma (Kirk and Godwin, 1963) and 12,956 ± 240 B.P. from the lowest 10 cm of core containing measureable carbon at Cam Loch (W. Pennington, this volume) strongly suggest that the ice-sheet had disappeared by 13,000 B.P. The Loch Lomond Readvance thus represents a subsequent return to glacial conditions (which means it should strictly be called the Loch Lomond Advance).

DISTRIBUTION OF THE GLACIERS

The Loch Lomond Readvance glaciers of northern Scotland show large variations in size and altitude and include various glacier types. The most massive ice accumulation (64-67) was situated in the extreme south-west between lochs Torridon and Maree, where some of the highest ground lies close to the west coast. Considerable valley glaciers, each supplied from several sources, are represented by 30, 41 and 45, each 9-10 km long. Glaciers 40-42 were partly nourished by a small plateau ice-cap, although this was not of simple form since steeply-sloping high ground that rises above the plateau was also a glacier source area. Glaciers 2 and 3 descended steeply to Srath Dionard to spread out as piedmont glaciers. The end moraine of glacier 10 shows that it had a similar terminus, which merged on its western side with glacier 9. Independent corrie glaciers included

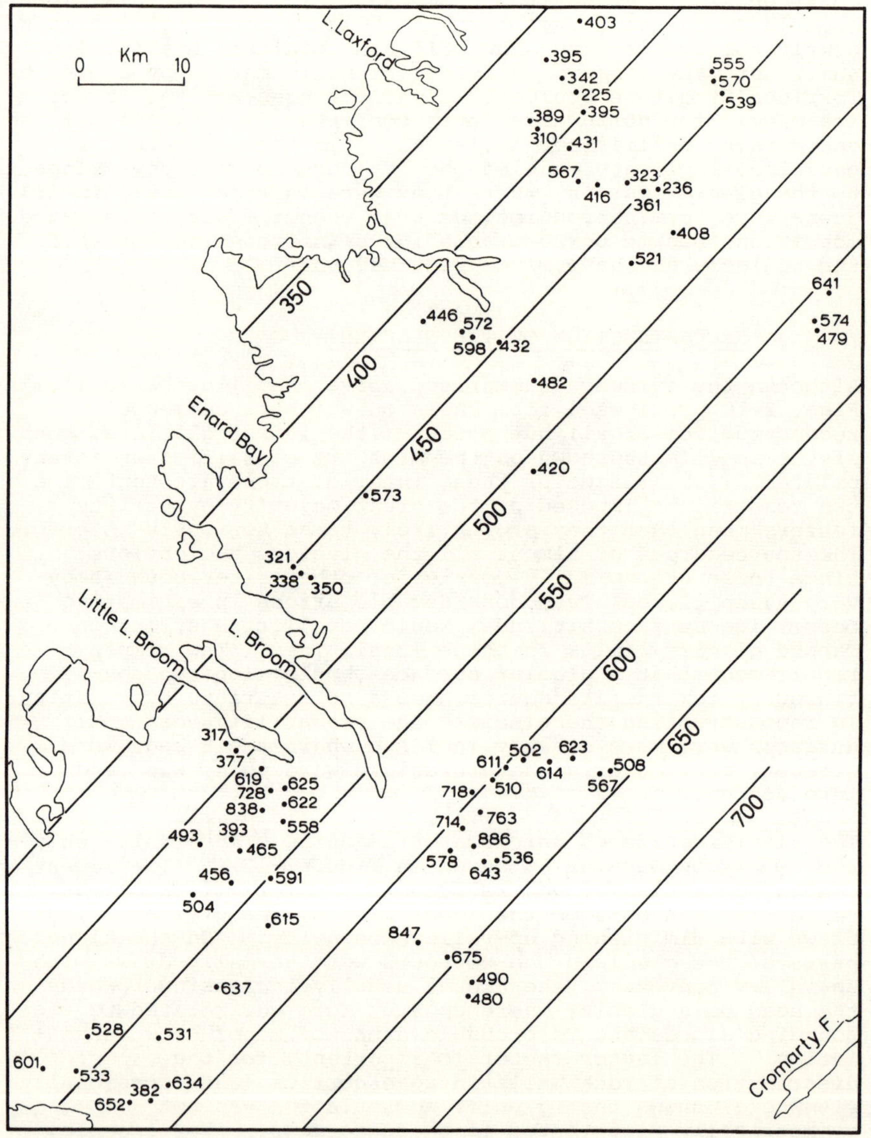

Fig. 11. Calculated firn line altitudes for the former glaciers of northern Scotland when at their greatest extent and a linear trend surface through these values.

small ones less than a kilometre long that were restricted to their corries (*e.g.* 22, 26, 54) and larger ones that extended beyond their corries to flow steeply down slope (*e.g.* 15, 16, 38, 39). Yet other glaciers accumulated beneath rock walls that are straight or slightly irregular in plan (*e.g.* 12, 31, 48, part of 4).

Firn lines for the glaciers when at their greatest extent have been calculated by the method given in Sissons (1974b). They are plotted in Fig. 11 and generalised by a linear trend surface.* The inland rise in the former regional firn line suggests that snowfall diminished inland. Yet there are marked local deviations from this general trend. One factor involved in these deviations appears to have been the amount of snow blown onto the glaciers. The very low altitude of part of glacier 4 (Fig. 2), to the west and south-west of which is an extensive plateau, has already been mentioned. Glacier 12, which has a calculated firn-line altitude of only 236 m, developed beneath a rock slope that terminates a considerable plateau lying to the south-west. Glacier 14, which appears large in relation to the poorly-formed corrie from which it flowed, was also favoured by a plateau to the south-west.

There is a marked contrast in glacier dimensions and altitudes on opposite sides of some of the individual mountain areas. Thus glacier 30, despite southerly aspect, was of large dimensions and contrasts with glaciers 26 and 27. The largest glacier in the Beinn Dearg area (45) flowed due south: much smaller glaciers existed at higher altitudes to the north and north-west of Beinn Dearg. In the mountain mass that nourished glaciers 58 and 60 to 62, glaciers 61 and 62, despite unfavourable aspect, were larger and descended to a much lower altitude than 58: west of 58 several north-facing sites lack morphological evidence for local glaciers. Such relationships suggest that, super-imposed on a general inland decrease in precipitation, there was a marked south to north (or SE-NW) decrease in at least some of the individual mountain masses. Such a pattern of precipitation distribution seems most likely to have been related to fronts advancing from the south-west.

Further discussion of the glacier firn lines and the factors that influenced them would be unprofitable without detailed analysis. Such analysis is being carried out using methods already applied to the former glaciers of the south-east Grampians (Sissons and Sutherland, 1976).

*For this calculation glacier 4 was divided into two parts.

ACKNOWLEDGEMENTS

The writer wishes to thank Mr. D. G. Sutherland for providing the trend surface shown in Fig. 11.

Stratigraphical and Faunal Evidence for Lateglacial and Early Flandrian Environments in South-West Scotland

W. W. BISHOP (Queen Mary College, London)
and
G. R. COOPE (University of Birmingham)

ABSTRACT

The environmental history of Lateglacial and early Flandrian times in south-west Scotland is interpreted from detailed local stratigraphic investigation at ten localities between the Solway Firth and the Firth of Clyde. Individual sequences are calibrated by radiocarbon dates. Assemblages of fossil coleoptera from five of the localities are analysed to provide additional information of palaeoenvironments and to reconstruct former climatic conditions.

A consistent pattern of climatic changes emerges that differs from the traditional view in both the timing and intensity of the thermal maximum during the Lateglacial. For the period between 11,000 and 10,000 years B.P. the insects and the nature of the deposits indicate an extremely cold climate nearing that of modern polar deserts. A thoroughly thermophilous insect assemblage shows that the climate of south-west Scotland was at least as warm as that in the same area at the present day by about 9,500 B.P.

Relative movements of land and sea are reconstructed from the stratigraphic sequences at different localities. Differences in the timing of Lateglacial sea-level changes between the Solway and Clyde estuaries are ascribed to differential ice loading and rates of isostatic recovery.

GENERAL INTRODUCTION

Details of the geometry of Lateglacial and early Flandrian sediments and of their geomorphological setting are recorded for selected localities in south-west Scotland. Sites from which samples were obtained for radiocarbon dating and for investigation of fossil insect content are specified. More than thirty dates from the area were published in *Radiocarbon* over the period between 1960 and 1969, together with preliminary data on pollen content of associated samples. Analyses of the insect faunas have been completed only recently. The two parts of the paper, dealing respectively with the stratigraphy and insect faunas, are presented separately but palaeoenvironmental implications are drawn from the combined evidence of both studies.

PART I STRATIGRAPHICAL EVIDENCE (WWB)

INTRODUCTION

A selected series of localities in south-west Scotland was investigated to establish the geomorphological and stratigraphical relationships between glacial, fluvioglacial, fluvial, lacustrine and estuarine sediments. The last category included compressed peats, carse clays and raised bogs as part of the "raised beach" complex. Both types of peat frequently contained large pieces of wood and occasionally the compressed peat yielded stumps of trees still in position of growth representing former submerged forests now being re-exposed from beneath the carse deposits. Areas for study were chosen by revisiting, where possible, localities recorded in early literature (*e.g.* Bryce, 1865; Horne *et al.*, 1896; Jamieson, 1865; Crosskey, 1874; Robertson, 1881; Richey *et al.*, 1930), and by using the local knowledge and help of several individuals and groups.

The method of recording was to map both natural and temporary exposures by use of a plane table and to construct accurate cross-sections drawn to true scale by use of a grid of metre

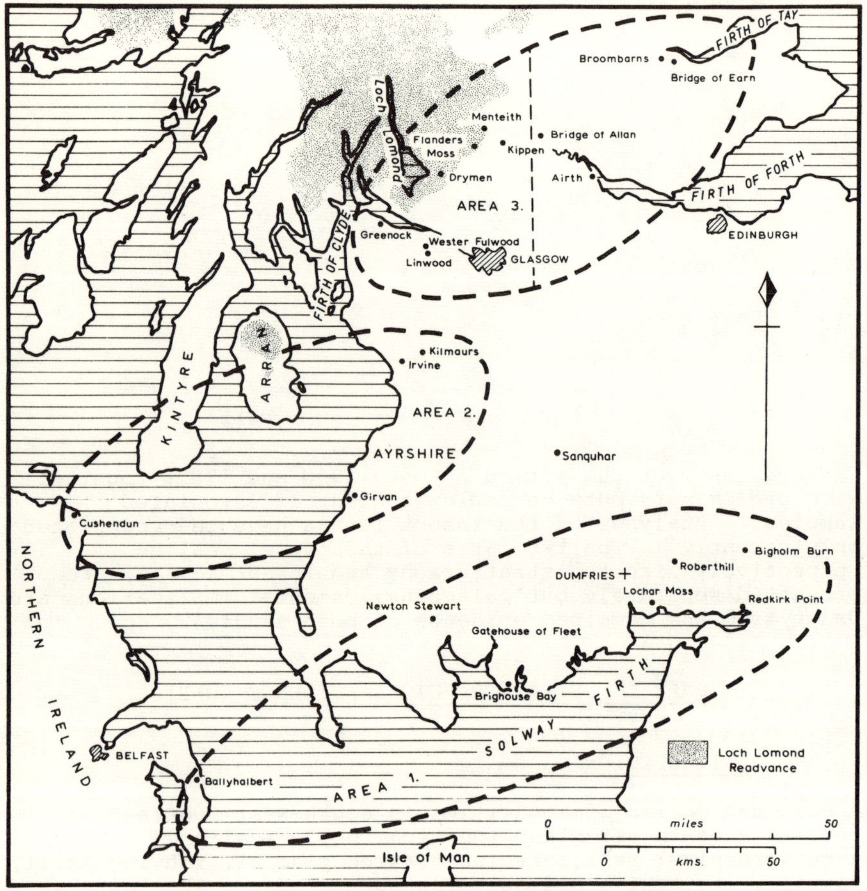

Fig. 1. Map of southern Scotland showing localities mentioned in the text and three principal areas of study.

squares constructed on exposures as a guide in transferring stratigraphical data onto graph paper. The plane table proved indispensible for intertidal localities which changed rapidly as exposures were obscured and others cleared by successive winter storms. The resulting maps and sections record combined observations taken over a number of years. A dumpy level and a self-compensating level were used to establish accurate height control from Ordnance Survey bench marks and spot heights (Newlyn datum).

The aim was to produce an accurate strato-geometry for the lithological units and then to attempt a reconstruction of the stratigraphic history for each locality. Extensive use has been made of radiocarbon dates established on samples collected during the work. Pollen data are available for several localities and the evidence from fossil insects is described in the second part of this paper.

For convenience the descriptions of localities are grouped into the following three areas, *viz.* 1. the Solway Firth, 2. the Ayrshire coast, and 3. the Firth of Clyde (Fig. 1).

SOLWAY FIRTH

Roberthill (NY/110797)

This locality was investigated as a rescue operation in 1960 immediately prior to the exposures on the east bank of the River Annan, Dumfriesshire, being obscured by large blocks of sandstone to prevent further erosion. A description has been published (Bishop, 1963) of the 3 m thick sequence outcropping in two breached anticlines with approximate wavelength of 110 m, amplitude of 5 m and axes striking NE-SW.

The lowest lithological unit was a clean, medium sandy gravel of which only the upper 35 cm were visible in the core of the more southerly of the two anticlines. The deposit is interpreted as glacial outwash and consists mainly of material derived from local Silurian and New Red Sandstone outcrops.

The overlying deposits have a thin pebble gravel at the base but pass up through well bedded sands and silts into a light grey, gritty clay at the top of the 2 m thick sequence. Traces and stringers of plant material occur throughout but a bedded detrital peat band 20 cm thick (Roberthill Beetle Sample), divided into an upper and lower section by a 3 cm thick sandy stringer, occurs in the middle of the sequence. A piece of wood from the lower section of this peat yielded a radiocarbon date of 12,940 ± 250 B.P.; (Q-643; Godwin and Willis, 1964). Four samples for pollen analysis (G5, G7, G8 and G10) were collected in 1960. Two were from the lower and upper parts of the peat bed respectively (G7 and G8), G5 was from a seam rich in plant material near the base of the sequence and G10 from a plant-bearing clay near the top of the sequence. The samples were studied by Moar (1963, pp. 133-34) who tentatively assigned the deposits to Jessen-Godwin pollen zone I on the basis of low values of tree pollen and an assemblage of pollen grains of herbaceous and aquatic plants. The presence of mildly thermophilous plants such as crowberry (*Empetrum nigrum*) and shoreweed (*Littorella lacustris*) suggested oceanic affinities.

The stratigraphic and pollen evidence, when considered in conjunction with the local morphological setting, suggest the existence in Lateglacial times of a lake at about 48 m O.D., some 5 to 6 km from north to south and 3 to 4 km from east to west, lying south-east of the surviving Castle Loch and other lochans near the village of Lochmaben (NY/083823). The lake appears to have been the result of slight glacial overdeepening behind a resistant rock step. With later downcutting of the rock lip at the lake outlet (near NY/110760) the present meandering course of the River Annan was superimposed across the lake sediments. River flood-plain silts up to 70 cm thick now truncate the anticlinal crests.

The deformation which produced the folding may have resulted from the growth of hydrolaccoliths of ground ice in the underlying outwash deposits rather than, as suggested by Bishop (1963), from disturbance following readvance of glacier ice into the lake basin from the north-west near the present location of Lochmaben.

Another deposit of peaty silts and sands overlying coarse gravels which occurred to the west and 40 m upstream of the last outcrop of the lacustrine sequence also contained a piece of tree stump. Heartwood from the stump yielded two radiocarbon dates of 3,847 ± 60 B.P. and 3,911 ± 59 B.P. (Birm-5 and Birm-11; Shotton *et al.*, 1968). The deposit appears to be a recent infill of a former meander loop of the River Annan incised into the Lateglacial lacustrine sequence.

Thus there are two organic deposits at Roberthill of different age and implying very different environmental conditions. The younger ages do not raise any doubts concerning the older date as suggested by Jardine and Peacock (*in* Mitchell *et al.*, 1973, p.56).

Redkirk Point (NY/301652)

At the head of the Solway Firth, on the north bank 2 km south-west of Gretna village, three compacted and buckled peats are exposed from time to time on the foreshore from 200 to 400 m west of Redkirk Point. The degree of superficial cover by beach cobbles and sand varies with the direction and strength of wave and current action in relation to the state of the tide and the flow in the channel of the River Esk during winter storms. The plane table plan (Fig. 2) is a composite survey covering the years between 1960 and 1967. Some of the structures have been removed by erosion or considerably modified since the mapping was completed.

The succession and strato-geometry are shown in Fig. 3. Thickness and height above O.D. of particular beds are best obtained from this diagram as the deposits show lateral change of facies and the peats and other strata thin towards the east against a ridge of underlying New Red Sandstone. To the east and north-east a lens of red, sandy and pebbly glacial till thickens to an observed maximum of 1.5 m. Erosion in a shallow valley has removed all trace of till in the western part of the section. Along the axis of the valley later sediments rest directly upon sandstone. The sequence in the vicinity of the channel axis (Fig. 3) is as follows:

8.	Disturbed ground and soil profile	0.15 m
7.	Sandy silts and alluvium	0.40 m
6.	Grey clays and silts - undisturbed typical carse clays. Many carbonaceous traces of plant rhizomes and a thin peat (Peat 3).	3.00 m
5.	Grey silts and fine sands with several thin discontinuous peat lenses - rhizomes in east. Bedding disturbed in the vicinity of valley in sandstone.	3.00 m
4.	Peat 2. Highly compacted and locally disturbed woody peat. Tree stumps in position of growth in the east (Q-637).	0.15 m
3.	Grey silts and fine sands with bedding highly disturbed.	1.60 m
2.	Peat 1. (Birm-40, Birm-41, Q-815) (Redkirk Beetle Samples B and C). Highly compacted and locally distorted peat. Local peat lens (Peat 1A) immediately below Peat 1.	0.25 m
1.	Carbonaceous silts and fine sands. (Redkirk Beetle Sample A).	0.30 m
	Red sandy and pebbly till.	1.50 m
	New Red Sandstone	

The details of the outcrop plan are plotted in Fig. 2 on which can be seen structural patterns produced by planation of the gently sloping foreshore across anticlines and synclines produced by buckling of the peat strata, with axial planes striking broadly along the direction of the valley. The folds may be interpreted as the result of compaction of the peats beneath the load supplied by the carse deposits

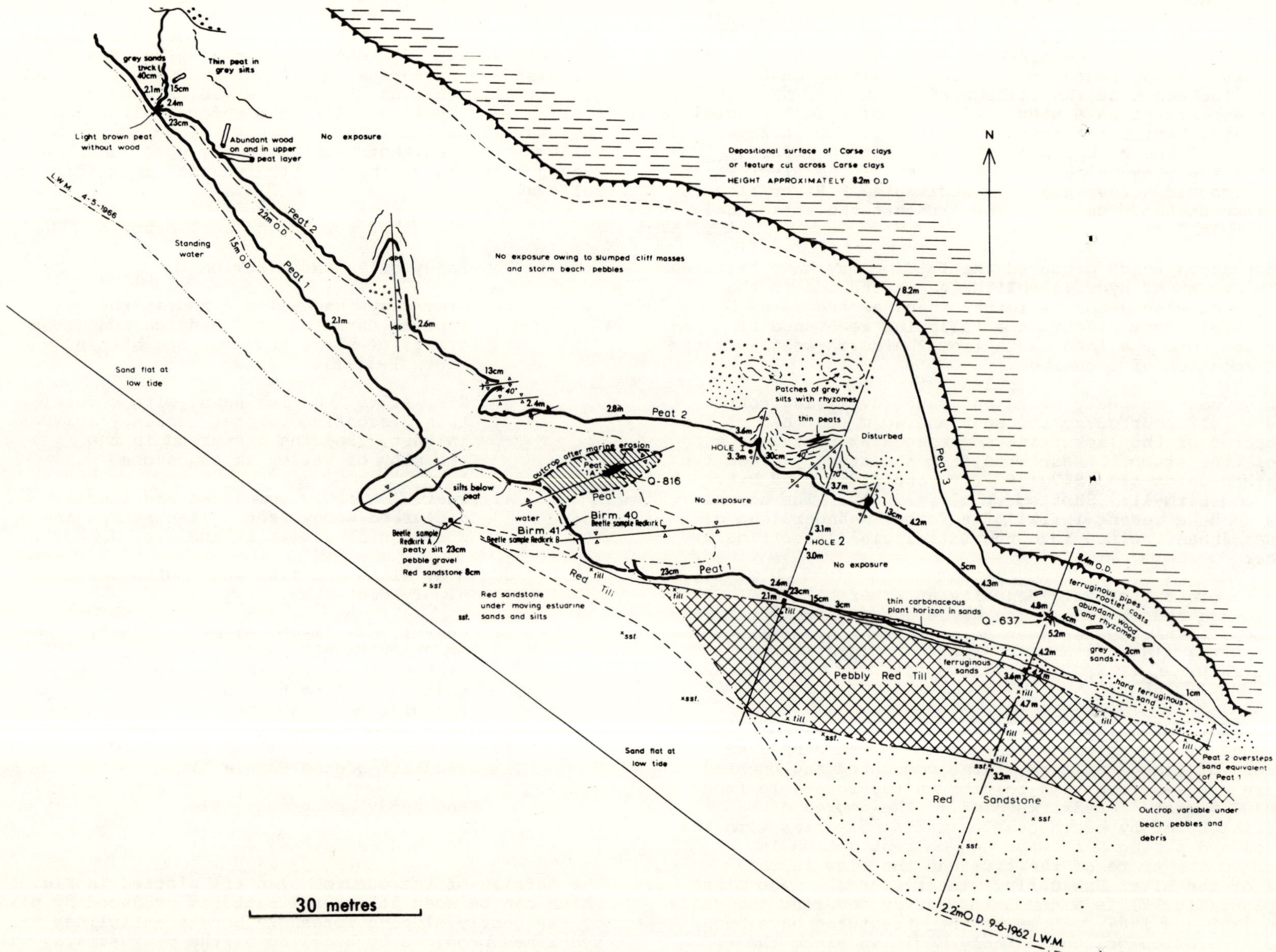

Fig. 2. Redkirk Point, Solway Firth, Dumfriesshire. Plane table map of Lateglacial deposits exposed on foreshore at low tide. Altitudes and thicknesses in metres and centimetres.

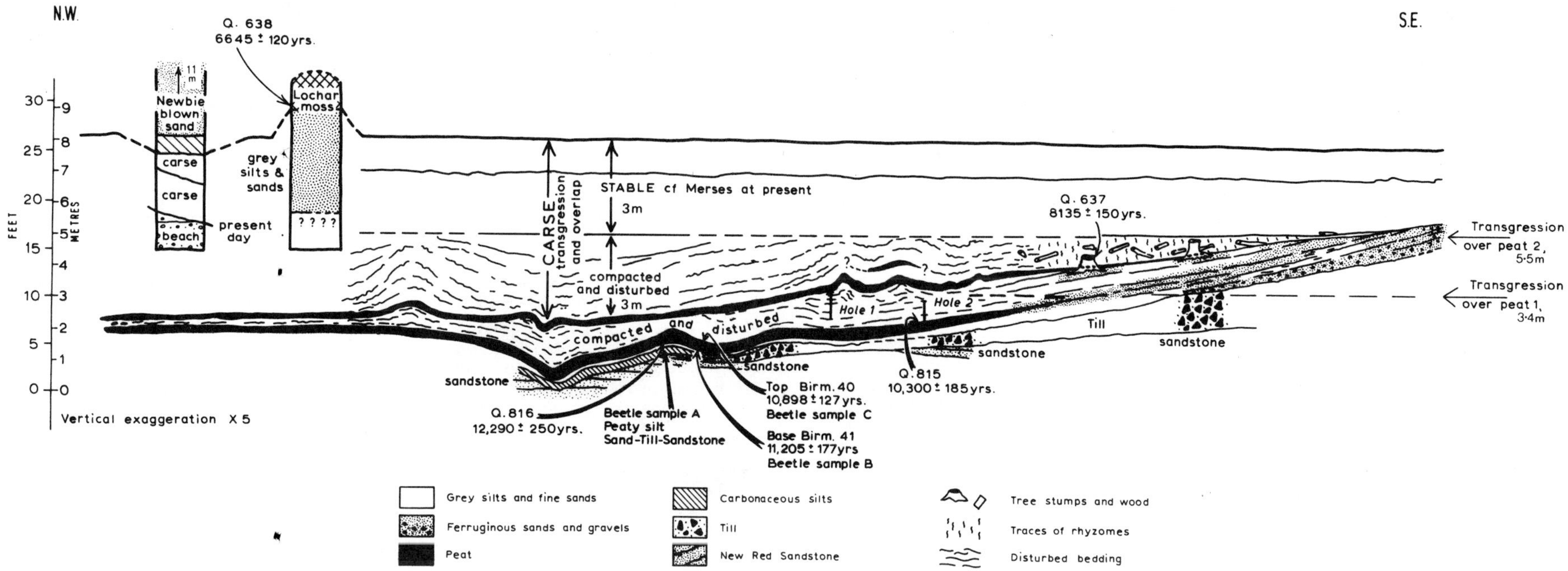

Fig. 3. Cross section of Lateglacial deposits exposed at Redkirk Point showing location of radiocarbon dated and fossil insect samples.

within the broad valley form cut in sandstone and till. The compacted peats thicken consistently into the valley from locations to the east at which they are first discernible (Fig. 3). If a minimum compaction ratio of 5:1 is assumed (Bloom, 1964) then Peats 1 and 2 may be inferred to have had original thicknesses in the order of 125 cm and 75 cm respectively. The intervening silts and clays show thickening and thinning as the result of flow between the tough but pliant peats.

The maximum altitudes at which the feather edges of peat lenses occur, when combined with the probable original thickness of the waterlogged and uncompacted peats suggest that transgression of water bodies in which the clays and silts were deposited occurred at altitudes of about 3.4 m and 5.5 m O.D. respectively (Fig. 3). It seems probable that the local rises of base level occurred in sympathy with rising sea-level in the Solway Firth impeding the fresh water inflow from the numerous rivers tributary to the Firth. In the case of Peat 2 the effect of the rising water level was to kill trees growing on the margin of the peat-filled hollow as represented by the stumps and logs illustrated in Fig. 3. It should be noted that the altitude quoted by Jardine (1964, Fig. 1) for the dated tree stump (Q-637) is about 2 m too low.

Peat Bed 1A occurs as a local lens just below the base of Peat 1 and is brought up on the crest of an anticlinal fold (Fig. 2) but is not intersected by the line of section in Fig. 3. A further undeformed thin peat band (Peat 3) occurs within the carse clays (Fig. 2) but was not sampled for dating.

The series of samples of wood and peat taken for radiocarbon dating (Table 1) provide a time-scale for the sequence of events shown in Figs. 2 and 3. The sequence of events represented at Redkirk Point seems to indicate a phase of low sea-level following glaciation during which a broad valley was eroded into the till sheet and underlying sandstone. Sedimentation commenced more than 12,000 years ago with deposition of carbonaceous sands of fluvial origin which lie well below present high tide level.

TABLE 1 Radiocarbon dates from Redkirk Point

Sample No.	Stratigraphic position	Age (C14 yrs B.P.)
Q-637	Peat 2. Wood from outer rings of tree stump in growth position, 5 m O.D. (Godwin and Willis, 1962).	8,135 ± 150
Q-815	Peat 1. Wood (*Populus*) from eroded top of peat. (Godwin *et al.*, 1965).	10,300 ± 185
Birm-40	Peat 1. Peat sample from top 3 cm of peat band. (Shotton *et al.*, 1968).	10,898 ± 127
Birm-41	Peat 1. Peat sample from bottom 6 cm of peat band. (Shotton *et al.*, 1968).	11,205 ± 177
Q-816	Peat 1A. Peat sample from inlier on crest of anticline. (Godwin *et al.*, 1965).	12,290 ± 250

A local peat lens (1A) containing limited arboreal pollen (*Betula, Pinus, Salix*) abundant herbaceous pollen typical of the Lateglacial including *Selaginella, Empetrum, Thalictrum* and *cf. Saussurea alpina* (Godwin and Switsur, 1966), is overlain by Peat 1.* This bed probably attained a thickness of almost 2 m and the radiocarbon dates suggest that peat growth commenced before 11,000 B.P. and may have continued as late as about 10,300 B.P. before a rising sea-level to the west caused local ponding. This commenced at about 3.5 m O.D. and was followed by transgression, deposition of overlying silts and sands and the commencement of compaction. Both peats 1 and 1A show fretted and water-worn margins at their junctions with the silts, occasionally with long strands and finer filaments of vegetable matter interfingering with the water-laid deposits.

Local stabilization or a fall of water level permitted vegetation to establish itself again and Peat 2 was formed with large trees along the margin of the area of accumulation. The trees were finally killed by a transgression at about 5.5 m O.D. 8,135 ± 150 years ago. This is in keeping with a date of 8,120 ± 135 B.P. (Q-214; Godwin and Willis, 1960) from wood in a similar situation 2 m above O.D. at Ballyhalbert, Northern Ireland. Continued rise of water level resulted in further transgression and deposition of about 6 m of carse clays. The major period of compaction seems to have occurred during the first half of the period of carse clay deposition and is represented in deformation of the peats and silts (Fig. 3).

The thin Peat 3 was not dated and no further evidence of age occurs locally at Redkirk Point. However, 25 km farther west the writer, in company with his extra mural class, dug a hole in the raised bog of Lochar Moss at Cockpool Plantation (NY/055680). Peat from the bottom of the hole, resting upon the surface of the carse clays at 9.25 m O.D. gave an age of 6,645 ± 120 B.P. (Q-638; Godwin and Willis, 1962). This provides a local terminal date for the withdrawal of the water at the close of the carse clay deposition which is supported by a further date (6,470 ± 280 B.P., Birm-324; Shotton and Williams, 1973) for a locality 6 km east of Cockpool Plantation (NY/118657).

Bigholm Burn (NY/316812)

An important section is exposed by stream erosion on the east bank of Bigholm Burn, Dumfriesshire about 6 km south-west of Langholm. Pollen investigations were carried out by Moar (1964) and radiocarbon dates for four samples collected by him were published by Godwin and Willis (1964). Dates for 3 further samples are given in Godwin *et al.* (1965). Figure 4 shows the detailed stratigraphy of the Bigholm Burn section.

The oldest deposit, a grey pebbly and clayey till largely derived from local Silurian rocks, is seen only at the south-eastern end of the section to a thickness of 1.5 m. The till is overlain by 1 m of ferruginous gravel and a further 2 m of crudely-bedded, grey, pebble gravel with a dip of 10 to 15° towards the north-west. The section is capped by 15 cm of pale grey silts overlain by up to 2 m of peat. The gravels are sub-angular and are interpreted as solifluction gravels. To the west a section of 15 m lacks exposure, except for the upper peat, because of bank slip. There is then an outcrop 15 m in length which is the exposure sampled by Moar. A thin, blue-grey clay (seen to 20 cm) appears above water level and is overlain by a slightly compressed, undulating and deformed bed of sedge peat up to 32 cm thick (Bigholm Burn Beetle Sample A). This is overlain by grey, sub-angular, solifluction gravel dipping from 15 to 25° towards the north-west and with ferruginous staining immediately above the peat. The tight folds seen in Fig. 4 are the result of deformation following loading of the saturated peat and silt by the solifluction gravels. At one point a tight upfold has been broken at its crest and

*Godwin and Switsur (1966) record that Peat 1A is above Peat 1 but this observation was made before it was known that 1A outcrops on the axis of an eroded anticline (Fig. 2).

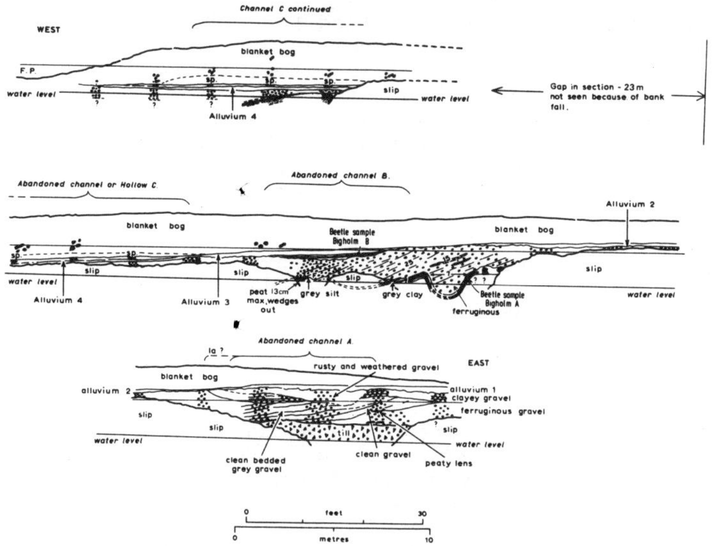

Fig. 4. Cross section of Lateglacial deposits exposed on the eastern bank of Bigholm Burn, Dumfriesshire.

discrete eroded blocks of the peat and clay are incorporated with the gravels.

Farther to the north-west the top of the gravel dips away gently westwards. The uneven gravel surface is overlain in pockets by lenses of dark grey silt (Bigholm Burn Beetle Sample B). This is overlain unconformably by a brown organic mud which is pale when dry and has a blocky structure. This in turn passes up into woody peat of a blanket bog (Fig. 4).

The whole sequence is calibrated by seven radiocarbon dates (Godwin and Willis, 1964; Godwin *et al.*, 1965) shown in order of superposition in Table 2.

The evidence suggests that during Lateglacial times pools developed on an irregular surface of pebbly till. Blue-grey clay with traces of lamination accumulated in shallow water by erosion of the fines from till slopes. About 11,500 years ago the water surface was dominated by sedge peat with *Carex* rootlets, *Chara* and *Ranunculus* (Bigholm Burn Beetle Sample A).

Later, coarse angular solifluction gravel was deposited, either as a primary result of accelerated slope movement, or following secondary washing of material fed into streams by increased freeze-thaw on surrounding slopes. The gravel load slightly compressed the peat and caused it to buckle locally into small folds (Fig. 4). At least one tight upfold ruptured and a diapir of underlying clay with peat blocks was incorporated into the base of the gravels. A sample from one such eroded peat mass yielded an age of 10,820 ± 170 B.P. (Q-695). This date may record the end phase of peat formation but may be too young due to percolation of humic water from the overlying blanket bog through the open lattice of the solifluction gravel.

TABLE 2 Radiocarbon dates from Bigholm Burn

Sample No.	Age (C14 years B.P.)	Stratigraphic position
Q-702	5,475 ± 120	In blanket bog above Q-701
Q-701	7,640 ± 160	Base of woody peat (just above Q-700)
Q-700	7,735 ± 155	Base of woody peat
Q-699	8,650 ± 165	Transition between sedge and woody peat
Q-697	9,590 ± 170 9,470 ± 170	Organic mud resting on silts and solifluction gravel
Q-695	10,820 ± 170	Eroded block of lower peat (Q-694) caught up in solifluction gravel
Q-694	11,820 ± 180 11,580 ± 180	Compacted and deformed sedge peat below solifluction gravel (Bigholm Burn Beetle Sample A)

Towards the close of the cold phase (Loch Lomond Stadial), marked by the solifluction gravel, small ponds, in which thin organic silts accumulated (Bigholm Burn Beetle Sample B), were established on the irregular surface of the gravel spreads. These silts appear to be intimately associated with the solifluction gravels and faunal evidence (see Part II), suggests continuation of cold conditions.

The silts are overlain by discontinuous lenses of muddy sediments that resemble alluvium. They decline in altitude from south-east to north-west and are interpreted as relating to lateral movement of the stream course across the gravel surface in a general westward direction. This had the dual effect of eroding all but remnant patches of the organic silts and producing a westward migration of the locus of deposition of alluvial sediment. (Alluvium 1 to 4 of Fig. 4).

The deposits rest unconformably upon the gravel sequence and may represent a considerable period of time. It is possible that the dating sample Q-697, collected by Moar (1964), which yielded an age of about 9,500 years for organic mud resting on silts and solifluction gravel, was collected from Alluvium 3 (Fig. 4) or that contamination from above may have resulted in an age which appears at least 400 years too young.

Eventually peat growth was established upon the alluvial surface. Moar noted a transition from *Phragmites* peat (Q-699; 8,650 ± 165 B.P.) and other dates (Q-700, Q-701) were obtained for samples of woody (*Betula*) peat which are overlain by blanket bog. There is a possibility of contamination of some or all three of these samples owing to rootlet penetration or enrichment by humic acid percolation from above.

Sanquhar (NS/783107)

No stratigraphic observations were made at this locality by W.W.B. but some details are given by G.R.C. in Part II.

Brighouse Bay (NX/634452)

Peat up to 30 cm thick can be found over an area more than 15 m square by digging at low tide through a thin (10 - 15 cm) cover of beach sand on the foreshore of Brighouse Bay, Kirkcudbrightshire. The locality is approximately 125 m south-west of the slipway at Rockvale. The deposit only emerges above sea-level for a few hours at low tide and is about 4 m below mean high water mark at about O.D. A date of 9,640 ± 180 B.P. (Q-398; Godwin and Willis, 1961) for woody peat with a pollen spectrum dominated by pine, birch and hazel, provides evidence that sea-level in the Kirkcudbright area of the Solway Firth was below O.D. at that time. This is in keeping with an age of 9,390 ± 130 B.P. for organic detritus at South Carse, south of Dumfries (NX/988594), at -1 m O.D. (Birm-323; Shotton and Williams, 1973).

AYRSHIRE COAST

Girvan and Irvine (NX/191987, NX/204994 and NS/337373)

Two localities near Girvan, Ayrshire, with thin peaty beds overlain by Flandrian raised beach deposits have been described by Jardine (1975). Near the water tank at Girvan railway station (NX/191987) two dates of 9,020 ± 150 B.P. (Q-640; Godwin and Willis, 1962) and 8,400 ± 200 B.P. (Birm-187; Shotton and Williams, 1971) bracket the organic strata. The sample that yielded the older of the two dates was collected by the writer from a 15 cm bed of grey silt and sand, containing abundant plant material, overlying ochreous sands and overlain by 15 cm of grey clay with organic debris. The material dated comprised wood fragments, seeds and hazel nuts (*Corylus avellana*) from a thin horizon in the bottom 5 cm of the peaty sand at 1.6 m O.D. The younger date for organic detritus (Birm-187) is recorded in Jardine (1975) as from the upper 5 cm of the same organic bed as Q-640.

A sample from a borehole at Enoch Farm (NX/204994) provided a date of 9,362 ± 150 B.P. (Q-641; Godwin and Willis, 1962) for organic detritus from a 30 cm peaty stratum 5.15 m O.D. Both Girvan localities provide evidence similar to that from the Solway Firth providing a limiting date for the early Flandrian rise of sea-level.

At Dundonald Burn near Shewalton Bridge, Irvine, Ayrshire (NS/337373) a woody peat 24 cm thick rests upon grey pebbly till and is overlain by bedded clean sands which have eroded the top of the compressed peat. The peat surface is at an altitude of 6.3 m O.D. and a piece of wood yielded ages of 9,620 ± 150 and 9,530 ± 130 B.P. for two assays (Q-642; Godwin and Willis, 1962). Raised beach gravels overlie the deposit and have a flat top at 13.3 m O.D. The raised beach deposits are locally overlain by wind blown sand which forms stabilised dune features to the east of the large open bay which was occupied by sea in early Flandrian times.

Kilmaurs (NS/405395)

In November 1864 Bryce re-visited the Old Woodhill Quarry (sometimes incorrectly referred to as Greenhill Quarry) near Kilmaurs, Ayrshire. Finding the face levelled and trees growing he engaged four workmen to open a trench (Bryce, 1865). No further observations were made at this locality until almost 100 years later in the period April to June 1961 when preliminary investigations followed by a "rescue dig" were carried out. The trees mentioned by Bryce were felled during the period 1939-1944 and the quarry was used subsequently by Kilmarnock Corporation as a refuse tip. The lower part of the quarry was completely covered by early 1961 and in succeeding years the whole upper face has been obscured by debris and rendered inaccessible.

The first fossil mammals were found at Woodhill in December 1816 and the first published references to the site were in the *Ayr Advertiser* during January 1817. The following description was given of the finding of a mammoth tusk. "The tusk was found between a stratum of extremely dingy, dark coloured clay, that lay incumbent upon it, and a stratum of gravel with rolled stones immediately below it ...".

Other references to early work are given in Bryce (1865) and in Richey *et al.* (1930, p.325), recording later laboratory investigations on specimens collected between about 1816 and 1831. The key reference remains Bryce's (1865) paper, in which he records details of the section exposed in his Woodhill trench. It is unfortunate that this was dug almost 50 years after the first discoveries by which time the quarry had been rehabilitated for some years.

Geikie (1863, p.71) appreciated the importance of the locality and wrote "It is singular that in this little Ayrshire Valley, within the compass of a few yards, there should occur a greater number of mammalian remains than have been obtained from the drift of all the rest of Scotland". The most important collection of these fossils is in the Geology Department of the Hunterian Museum, University of Glasgow (Gregory and Currie, 1928).

The reinvestigation of Woodhill Quarry in 1961 commenced in April when three shallow pits were dug into undisturbed till slopes visible above the rising tide of rubbish. As one pit yielded shell fragments and a few foraminifera from a washed auger sample at a depth of 2 m it was decided to put down a deep trench in that vicinity using a mechanical excavator.

In 1961 a 1.3 m wide trench aligned approximately north - south was cut by drag-line to an average depth of 4 m for a length of 11 m with a maximum depth of 5 m. An auger hole at the deepest point proved similar deposits for a further 1 m to a total depth of 6 m.

A horizontal datum-line was established, a grid set up on the face and the detailed section shown in Fig. 5 was recorded from the eastern face. Owing to the westward slope of the surface and the presence of disturbed ground as a result of the old quarrying, a much less satisfactory section was exposed on the western face of the trench which was largely in unconsolidated refuse.

Unfortunately the solid sandstone that was originally quarried was not reached by the excavator or in the auger hole. It seems probable that the drift thickens northwards off a solid sandstone hill. Water problems and the unsafe nature of the trench walls made it necessary to abandon work and fill in the trench after recording and sampling had been completed.

Survey with a level and staff established accurate altitudes above O.D. for the section and the highest point on the old quarry face below which Bryce's 1865 section was taken (see below and Fig. 5). The 1865 and 1961 sequences and tentative correlations between them are shown in Table 3, imperial units being retained to facilitate comparison with the early literature.

In the 1961 excavation fragments of mollusca and foraminifera were first seen at a level of 142 ft O.D. (bed 2) and shell fragments occurred in the clay down into bed 1 to 132 ft O.D. It seems probable that many of the foraminifera and mollusca recorded in the old literature were derived from sandy bands in the till or from "shelly boulder-clay".

One particularly carbonaceous silty sand lens in bed 1, at a depth of 135 - 136 ft O.D., yielded abundant coal fragments and derived coal-measure spores but also pollen grains and fragments of rhizomes. However, in 1961 no remains of mammals, seeds or beetles were found. The mammalian fauna from beneath the till at Kilmaurs, probably from bed 3 of Bryce, included several tusks of mammoth (*Elephas primigenius*) and antlers of reindeer (*Rangifer tarandus*) (Gregory and Currie, 1928). A radiocarbon date of 13,700 (+1300, -1700) B.P. for one mammoth tusk (Hunterian Museum specimen v.5120) appeared to suggest a fauna of Late Devensian age (Sissons, 1967c). However, the thickness and nature of the overlying till is more compatible with the date of >40,000 years (Birm-99; Shotton and Williams, 1970) for a large fragment of reindeer antler (Hunterian Museum specimen v.5187). Records in the Hunterian Museum state that a pair of reindeer antlers and a piece of mammoth tusk, the sources of both dates, were presented by Thomas Fulton in 1829 after they were found 34 ft below the surface at Kilmaurs. This enables some comparison to be made with the section recorded by Bryce.

The origin of these specimens was investigated further by Young and Craig (1869) who washed a sample of sandy clay "the size of a duck egg", sent to the Museum with the specimens. They obtained 300 seed cases and seeds including *Potamogeton perfoliatus*?, *Ranunculus aquatilis* and *Hippuris*. Investigation of crevices in the tusk yielded more examples of the same species of seed and the wing case and thorax of the beetle *Notiophilus aquaticus*, Linn.

There are many records in the literature of marine bivalves from Woodhill Quarry and other localities in pits and mine shafts in the vicinity. In most cases it is not clear whether they were derived from till or other deposits. The majority represent taxa that might have been dredged by ice from the glacial clays of the Firth of Clyde. Gregory and Currie (1928) record abundant *Pecten islandica*, *Tellina calcarea* and *Leda oblonga* as representatives of arctic forms extinct in British seas. Details of invertebrate assemblages are not repeated here except for those in a paper by Bennie (1885) who described the contents of two small parcels of clay located in 1878 in the Museum attached to the Library of the Royal Society of Edinburgh. These were found along

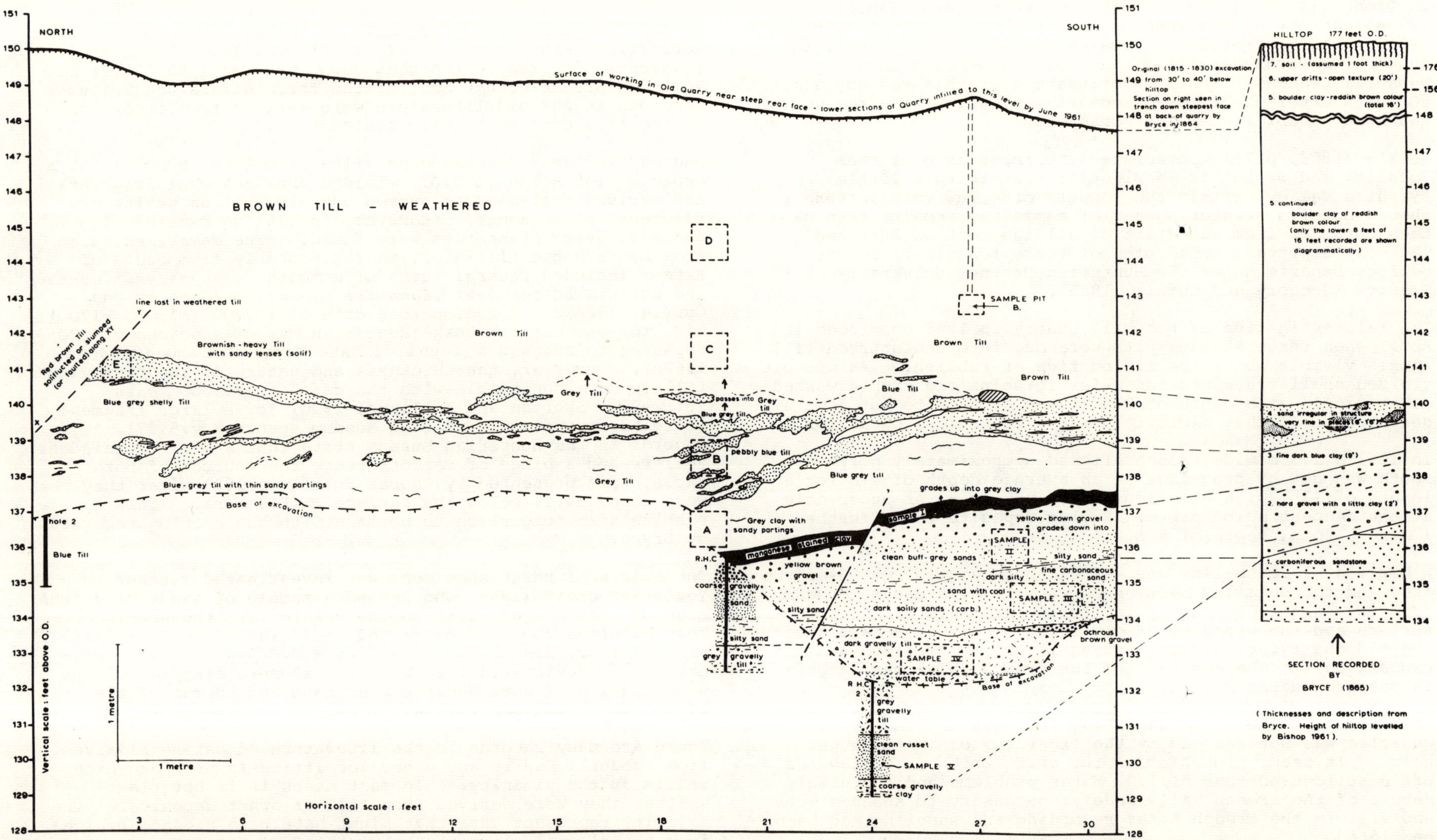

Fig. 5. Excavated section at the Old Woodhill Quarry, Kilmaurs, Ayrshire, showing details of glacial deposits visible in June 1961 and tentative correlations with the succession of Bryce (1865). Altitudes and thicknesses in feet and inches.

TABLE 3 Tentative correlations between the 1865 and 1961 sequences in Old Woodhill Quarry, Kilmaurs

1961 SEQUENCE

	Ground level	148' O.D.
3.	Soil 9" to 1'	147' O.D.
2.	Loose-textured, weathered, brown till gradually passing down into typical brown till with undulating base - in its turn passes down into (1) 6' to 7'	c.140' O.D.
1.	Tough, blue till with many irregular lenses of grey silty sand and gravel (up to 1'6" in thickness). In places the clay of the till appeared poorly bedded and banded.	
	Seen to 8'	c.132' O.D.
	Augered to 11'	c.129' O.D.

1865 SEQUENCE

	Ground level	177' O.D.
7.	Soil (assumed c.1' WWB)	
6.	Upper drift - open texture 20'	156' O.D.
5.	Red-brown "boulder clay" 16'	140' O.D.
4.	Sand and gravel 6" - 1'6" (mollusca and foraminifera reported to be from here)	c.139' O.D.
3.	Fine, dark blue clay - small quartz and other pebbles 9" (mammalia, seeds and beetle reported to be from here)	c.138' O.D.
2.	Hard gravel and little clay 2'	c.136' O.D.
1.	Carboniferous sandstone	

with a mammoth tusk from 17½ ft below the surface (Bald, 1822). Bennie washed the two samples and noted that one spoke of the sea and the other of the land. One sample, a clay of "marine origin much like Paisley or Glasgow", yielded 9 species of foraminifera suggesting origin in cold shallow water, and 10 ostracod species of "arctic character". The other sample was coarser grained, more red in colour and yielded an assemblage of seeds remarkably similar to that recorded by Young and Craig (1869), including *Potamogeton*, *Ranunculus aquatilis*, *Carex*?, *Polygonum*? and *Chara* plus the body of one beetle, fragments of another beetle and other organisms. Craig (1888) added *Myriophyllum speciatum* and *Potentilla* to the list.

The mammal fauna, associated flora and beetle remains seem to be established as older than 40,000 years and from a well vegetated, moist to freshwater environment, as indicated by the presence of pondweed, water crowfoot and mare's-tail. Although this probably represents an early interstadial of the Devensian glaciation it may indicate cooling conditions at the close of the Ipswichian interglacial. The section excavated in 1961, although failing to reveal the source bed of the terrestrial fossils, showed a crudely stratified grey till (interpreted as lodgement till) composed of tough, grey clay largely derived from a marine source area with intercalated fluvioglacial deposits towards the base. This contained comminuted marine bivalves and foraminifera together with masses and lenses of sand with marine mollusca, some of which survived intact in their sand matrix.

The upper part of the till sequence (upper drift of Bryce) resembles ablation till but has certainly been affected by later solifluction. It may be flow till formed during deglaciation, with further modification by intense freeze-thaw during the Loch Lomond Stadial. At that time large glaciers extended to near present sea-level only 40 km to the north (Fig. 1) in the type area of Loch Lomond and corrie glaciers developed on the Isle of Arran to the west (Sissons, 1976b).

The evidence in Cowden Glen, (NS/459567) between Kilmaurs and Glasgow (Gregory and Currie, 1928), shows how readily modification of existing deposits and formation of pseudo-till can result from slip and flow on steep, moist, till slopes even during the Flandrian. There is no evidence in the Kilmaurs area of any late readvance of ice in the period 12,300 to 11,950 B.P., as invoked by Jardine (1968) to account for red till overlying grey till in the Glasgow area. Such an explanation also seems unlikely in the light of radiocarbon dating evidence (Sissons, 1974c, 1976b).

FIRTH OF CLYDE

Paisley - Linwood area, Renfrewshire

Foundations for building works and electricity pylons, boreholes in connection with new factories, bridge works and alignment of motorways in the Paisley and Linwood area from 1959 to 1962 exposed numerous temporary sections in Lateglacial and Flandrian sediments and yielded new data on the form of the sub-drift topography (Fig. 6). An almost enclosed embayment with valley forms suggesting modified predecessors of the Lochar, Gryfe, Black Cart and White Cart waters existed south-west of the lower Clyde at the close of glaciation. The base level of erosion in the region where the combined valleys trend north-eastwards towards the Clyde was below -30 m O.D. The lower valley system, which appears to have served the combined proto-Black and White Cart Waters, exhibits a close alignment with a zone of weakness along the shatter belt of the Paisley Ruck.

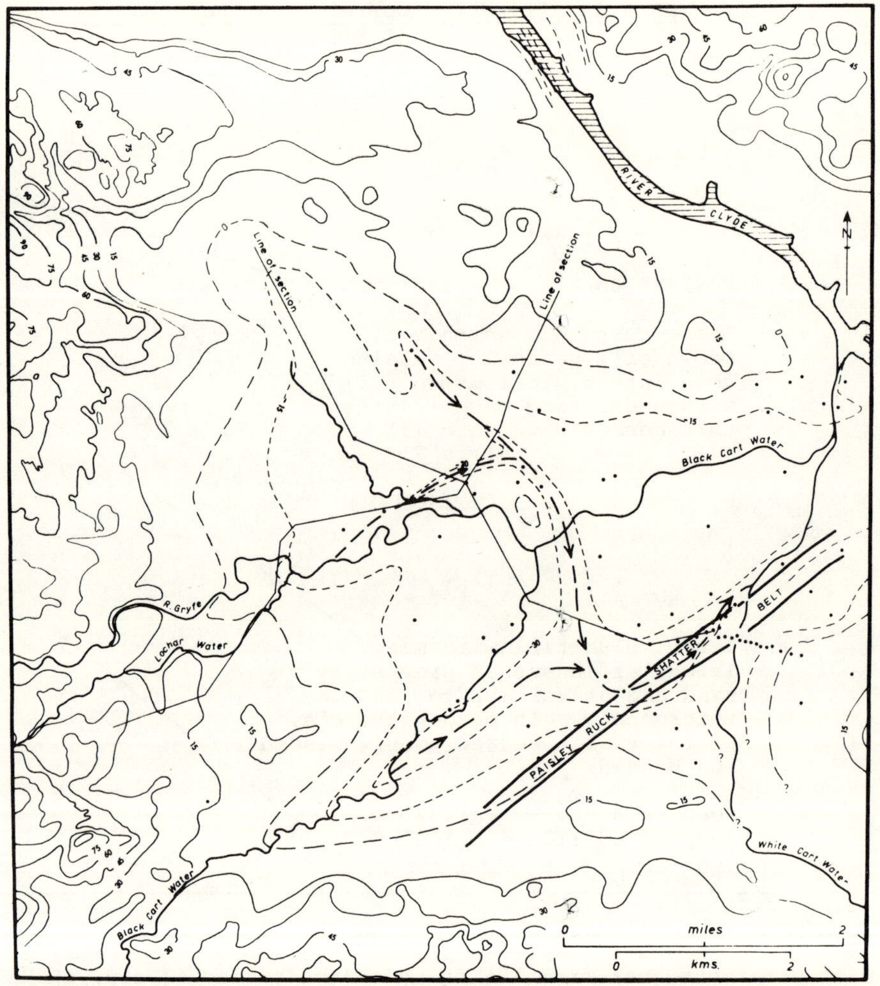

Fig. 6. Paisley area, Renfrewshire: rockhead contours (in metres). Dots = location of boreholes.

Within the embayment an irregular surface of till and hummocky sand and gravel is overlain by dark blue-grey clays of marine origin with a well-preserved and often abundant fauna of cold water mollusca. These deposits were penetrated in many of the boreholes and pits numbered along the lines of section in Fig. 7. A maximum thickness of 21 m was recorded near the centre of the basin in hole 45 which ran into till at -20 m O.D. In the marginal areas of the Paisley-Linwood basin shelly marine clays are virtually at the surface with only a thin cover of soil or solifluction. Clays were recorded up to 6 m O.D. backing out against surrounding till slopes, or hillocks within the basin. In the central area (holes 35A to 46, Fig. 7) the clays are usually overlain by silts, probably of fresh or brackish water origin, which lack marine fossils. These capping

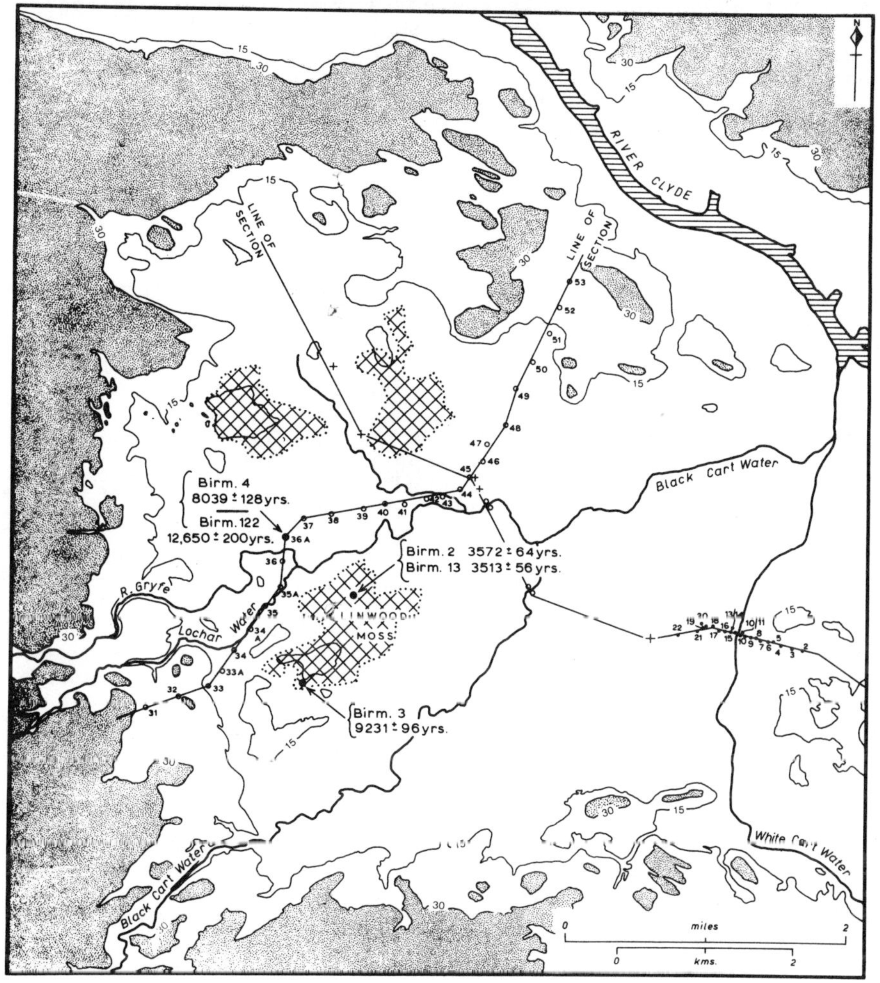

Fig. 7. Paisley area, Renfrewshire: present day topography (contours in metres), distribution of peat mosses and location of boreholes.

silts averaged 6 m in thickness. Following a fall of water level, raised peat bogs were developed upon their ill-drained surface at altitudes ranging from 12 m to 6 m and exceptionally 4.5 m above O.D.

A date of 12,650 ± 200 B.P. was established on the inner portions of large valves of *Arctica islandica* from marine clays 3 m above O.D. in hole 36A at Wester Fulwood (NS/432669; Birm-122a; Shotton *et al.*, 1967). Other similar dates are quoted by Price (1975, Fig. 3) and Peacock (1971a) from the Paisley-Glasgow area.**

In order to establish a limiting date for the emergence above water level of the surface of the silts, various localities in the Linwood Moss area were investigated to find a point where the silt/peat junction and the overlying peat moss had not been disturbed by stripping for fuel or agricultural improvements. Such a locality was found on Clippens Farm (NS/433654, Fig. 7) and a pit 2 m square was dug exposing the sequence shown in Table 4.

TABLE 4 The sequence at NS/433654, Linwood Moss

	Thickness (m)	Depth (m)	Altitude (m O.D.)
4. Dark brown, well humified cotton grass peat - surface disturbed and irregular		Surface	15.10
	0.70	0.70	14.40
3. Dark brown humified peat with little wood	1.80	2.50	12.60
2. Very woody (birch) peat in a matrix of well humified dark brown peat (Wood for radiocarbon dating: Birm-3, 9,231 ± 96 B.P.)	0.38	2.88	12.22
1. Grey brown peaty silts with fine sand partings. Pebbles and one boulder 15 cm x 10 cm. Seen to	0.22	3.10	12.00

The date of 9,231 ± 96 B.P. (Birm-3; Shotton *et al.*, 1967) is close to the beginning of peat colonisation of the silt surface. A further sample (Birm-4) gave an age of 8039 ± 128 B.P. for wood (hazel and birch) from 45 cm of unsorted sand and gravel with abundant organic debris overlying Birm-122a in pit 36A at Wester Fulwood (NS/432669). The

**But see Mangerud and Gullikson (1975) who show that radiocarbon dates on marine shells may be too old by several hundred years.

deposit appears to be an early terrace of the River Gryfe at 3.35 m O.D. resting upon blue-grey clay with stones and marine mollusca indicating cold conditions. It gives evidence of local exposure of the upper surface of the marine clays to river erosion by this date.

A further pit was dug through Linwood Moss at a point where the peat/silt junction was at a lower level (7.5 m O.D.) than the Clippens pit (Fig. 7, NS/439664). The details given in Table 5 show that the history of these raised mosses and of the rise and fall of water level in the Paisley 'embayment' is complex.

TABLE 5 The sequence at NS/439664, Linwood Moss

	Thickness (m)	Depth (m)	Altitude (m O.D.)
		surface	10.00
10. Loose dry bracken	0.07	0.07	9.93
9. Dark brown peat with roots	0.10	0.17	9.83
8. Red brown peat with few roots	0.07	0.24	9.76
7. Dark cotton grass peat with few roots or wood fragments	0.43	0.67	9.33
6. Dark peat with numerous fibrous roots	0.53	1.20	8.80
5. Dark brown fibrous cotton grass peat. Wood more abundant with depth.	0.33	1.53	8.47
4. Humified light red-brown peat. Wood abundant (birch) (radiocarbon samples Birm-2, 3572 ± 64; Birm-13, 3513 ± 56 B.P.)	0.25	1.78	8.22
3. Dark brown peaty silt with only occasional wood	0.10	1.88	8.12
2. Grey silt with sand and plant matter. Occasional wood, rootlets and pebbles.	0.15	2.03	7.97
1. Grey sandy silt with some pebbles and quartz fragments. Wood and plant stringers. Seen to	0.45	2.48	7.52

Dates for an early stage in the development of the peat bog are given by Birm-2 and Birm-13 (Shotton *et al.*, 1968), yielding 3,572 ± 62 B.P. for wood and 3,513 ± 56 B.P. for peat, sampled near the base of the peat section at 8.3 m O.D. This implies erosion of the older silts and clays and modification of margins of earlier peats developed on the exposed surface of Lateglacial deposits. Stabilisation or a slight rise of local sea-level must have taken place between about 8,000 and 4,000 years ago at about 8 m O.D.

Greenock (NS/307752)

A number of papers were written during the second half of the nineteenth century describing temporary exposures in the Paisley area and near Greenock. The latter included sections visible during the digging of the James Watt Dock at Garvel Park (Crosskey, 1874; Robertson, 1881). In 1962 excavations were carried out for a new Graving Dock in Greenock Great Harbour adjacent to the old James Watt Dock (Fig. 8). Preliminary conclusions based on the study of these exposures have been published (Bishop and Dickson, 1970). Figure 9 (2) illustrates the detailed section along the line AB in Fig. 8. The succession is similar to that seen by Crosskey in 1874 (Fig. 9 (1)) and to that described above for the Paisley-Linwood embayment. Till of variable thickness from 0 to over 10 m overlie Carboniferous sandstone, mudstone and shale. The surface of the till is irregular, lying at about -5 to -9 m O.D. It is overlain locally by laminated clays, probably varved, 1 m in thickness, which are succeeded at -5 to -7 m O.D. by 6 to 7 m of dark grey clays and silts yielding cold marine mollusca and algae typical of cold seas between 0 and 20 m depth (Brett and Norton, 1969). Valves of *Astarte sulcata* (Fig. 9) from the clays at -3.5 m O.D. yielded an age of 10,560 ± 180 B.P. (Birm-121; Shotton *et al.*, 1967).

The shelly clays and silts are overlain by black organic sands and silts up to 3 m in thickness occupying broad hollows (Fig. 9 (2)). A monolith collected from the silts in the sampling section between boreholes 46 and 48 (Fig. 8) yielded abundant plant fossils indicative of open ground vegetation dominated by dwarf shrubs, herbs and mosses (Bishop and Dickson, 1970), including *Aulacomnium turgidum*, *Distichium capillaceum*, *Polytrichum alpinum*, *Oxyria digyna*, *Salix herbacea* and *S.* cf. *reticulata*.

The more detailed section CD shows an enlargement without vertical exaggeration of part of the Graving Dock wall including the location of the monolith from which the mosses have been studied by Dickson (1973). This sampled sequence included valves of *Mya truncata* from -1.5 m O.D. (Birm-120; Shotton *et al.*, 1967) dated at 9,890 ± 160 B.P. It seems likely that the carbonaceous silts overlying both the

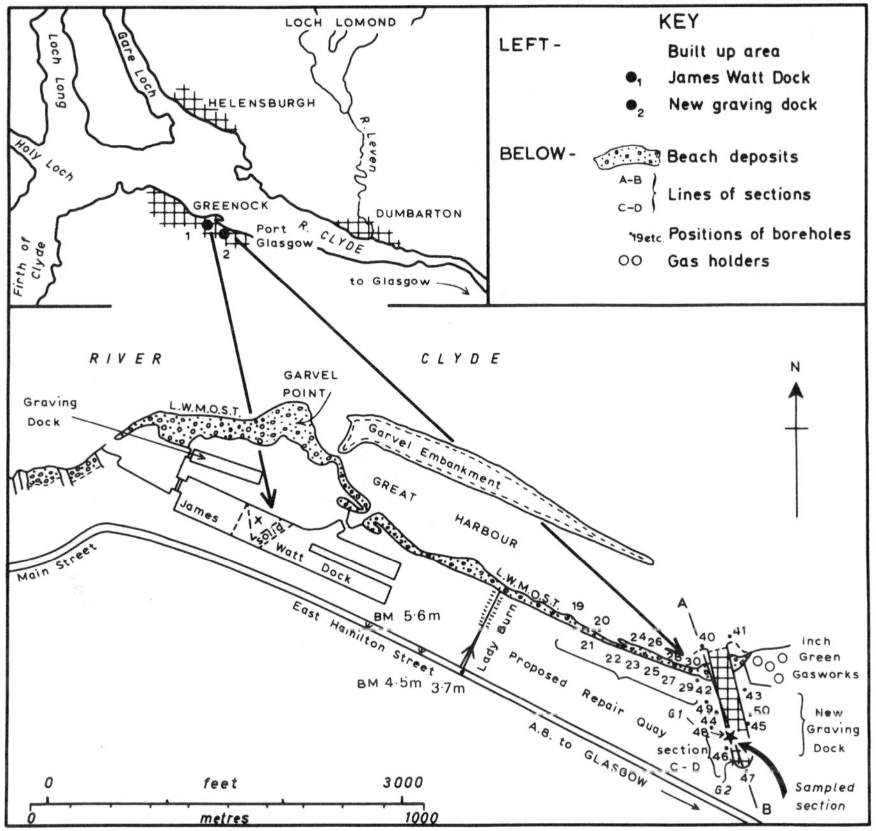

Fig. 8. Locality maps of the Greenock area showing the location of the Great Harbour and New Graving Dock (2) and the James Watt Dock (1). Numbers indicate positions of boreholes and letters lines of section (Fig. 9). Bench mark altitudes in metres. Built-up areas shaded.

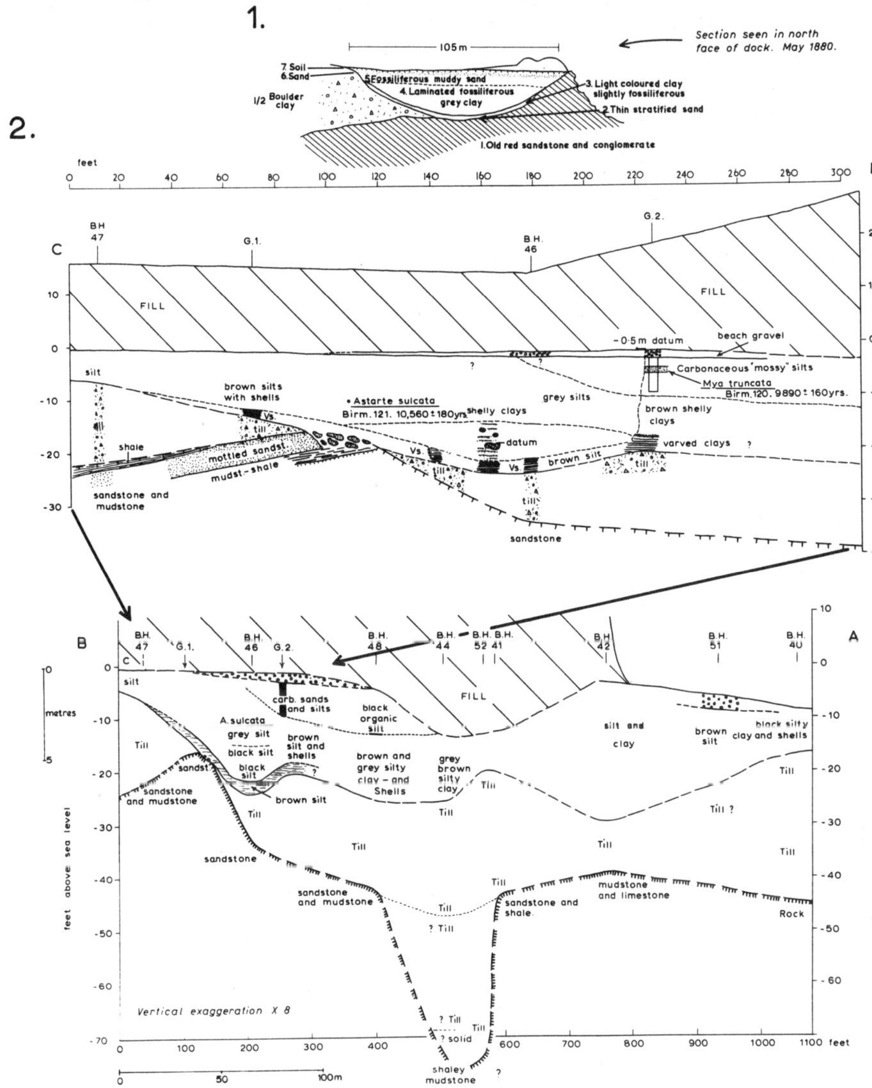

Fig. 9. Excavated section (2) at New Graving Dock, Great Harbour, Greenock, visible in 1962, showing location of dated specimens and other sample sites referred to in text and in Bishop and Dickson (1970); compared with section to the north of the Graving Dock (1) visible in 1880 on the site of the James Watt Dock.

Greenock and Paisley sequences represent similar conditions of formation and broadly contemporaneous periods of deposition. The date from the Clippens Farm pit at Linwood Moss of 9,231 ± 96 B.P. for the establishment of peat on the exposed surface of the silts supports this. The evidence suggests that shallowing and increased input of land-derived plant debris commenced at Greenock at the close of the Lateglacial about 10,000 years ago and continued until about 9,500 years ago (Bishop and Dickson, 1970). At their maximal extent Loch Lomond Readvance glaciers lay only 10 km to the north and east of Garvel Park in the Gare Loch and at Loch Lomond (Sissons, 1974c, 1976b). The placing of the Loch Lomond Readvance into the sequence of deposits described from Paisley-Linwood and Greenock is by reference to limiting dates of 11,700 ± 170 B.P. and 11,800 ± 170 B.P. (Sissons, 1967c) for marine shells caught up in the readvance terminal moraines at Drymen and Menteith respectively. During this cold phase it is easy to envisage only sparse ground cover with numerous arctic mosses clinging to the steep faces of the Carboniferous plateau lavas above Greenock. Erosion of fragments of vegetation combined with downslope movement would account for the abundance of mosses recorded by Robertson (1881) and Dickson (1973, pp.226-28).

PART II FAUNAL EVIDENCE (GRC)

INTRODUCTION

This interpretation of environmental history is based almost entirely on evidence provided by assemblages of fossil coleoptera obtained from a number of sites that have been stratigraphically investigated and, in most cases, radiocarbon dated. The location of these sites is given in Fig. 1. Fossils of coleoptera were by far the most common animal remains in these deposits both in terms of specific diversity and individual abundance. In building up the picture of the environment given here, just over two hundred taxa have been used, the majority of which have been determined specifically. It has thus been possible to establish that, in the course of about 3,000 years, the insect fauna of south-west Scotland changed from one dominated by temperate species to an arctic assemblage and back again to a temperate fauna. The amplitude of the climatic changes that must have been the cause of these alterations of the insect fauna may be estimated from the fact that species, at present found only in areas well to the south or to the north, successively inhabited this part of Scotland. Thus in this relatively brief period, the climate must have varied between conditions at least as warm as those of the present day to conditions of arctic severity.

Fossils of large coleoptera were very conspicuous in the field at almost all the sites investigated, but specimens found in the field are not included in the fossil lists because this would have biased the numbers in favour of those most easily seen and would thus have distorted the environmental interpretation. Bulk samples were therefore taken for processing in the laboratory where all beetle fragments were extracted by sieving and microscopic searching of the residues. As far as was possible samples were taken of comparable size, namely about 5 kg: in one case only a smaller sample, 1 kg, had to suffice (Bigholm Burn B).

In the faunal lists (Tables 6 and 7) the families and genera are arranged taxonomically but within each genus the species are in alphabetical order. The numbers in each column refer to the minimum numbers of individuals present in the sample. This figure is arrived at by counting the maximum number of any particular skeletal element. It is inevitable, therefore, that these figures under-represent the abundance of the commoner species and overemphasize those species that occur as single specimens only. This is because the latter must be given equal status to species represented by a number of specimens none of which happen to be duplicates. Although every effort was made to collect comparative samples by weight, the varying rates of sedimentation and the consequent different degrees of dilution of the fauna in the deposit mean that care must be taken in comparisons of numbers from one column to another. As far as is possible the sites have been arranged so that the oldest deposits are on the left of the table. Similarly for a particular site the lowest levels are also on the left. Since the early Flandrian site at Brighouse Bay has such a different faunal assemblage compared with the Lateglacial sites, its species list is given on a separate table.

Since most of the samples were taken in the summer of 1964, some explanation seems called for to account for the long delay in publication of the results. At that time, the environmental interpretation demanded by these fossil assemblages of beetles was so much at variance with the then accepted picture, particularly with respect to the timing and intensity of the climatic changes, that a great deal of regional investigation of other beetle faunas of the same age had to be undertaken before adequate confidence could be established in the ability of the coleoptera to revolutionise our views of the environments of the Lateglacial period.

FAUNAL SITES

Roberthill

The samples were obtained from a layer of silty detritus mud ("the main peaty horizon" of Bishop, 1963), interbedded within a sequence of well laminated, poorly organic silts that had been arched up into a gentle anticline. This was the same horizon from which a pollen spectrum was obtained by Moar (1963), indicating open vegetation attributable to Jessen-Godwin pollen zone I. This interpretation was supported by a radiocarbon date of 12,940 ± 250 B.P. (Q-643) obtained from a piece of wood from the same stratum. There can thus be no question of this date being too old due to any "hard water error".

TABLE 6 Faunal list and specific abundance of coleoptera from Lateglacial sites in south-west Scotland

	Robert-hill	Redkirk Point			Bigholm Burn		Sanquhar	
		A	B	C	A	B	A	B
Carabidae								
Cicindela sp.		1						
Carabus problematicus Hbst.				1				
Leistus rufescens F.							2	
Nebria gyllenhali Schnh.								1
Pelophila borealis Payk.				1				
Notiophilus aquaticus L.			2	2	1		1	
Blethisa multipunctata L.			4					
**Diacheila arctica* Gyll.			5	8	7			
Elaphrus lapponicus Gyll.			1	2	1			
E. riparius L.	1							
Loricera pilicornis F.	2		2		2			1
Dyschirius globosus Hbst.	1	1		1				
Bembidion doris Panz.			1		1			1
B. callosum Kust.							1	
B. gilvipes Sturm		1						
B. schuppeli Dej.		7						
B. transparens Gebl.			2	1				
**B. velox* L.	2							
Trechus obtusus Er.			1				1	1
T. secalis Payk.			2	3				
Patrobus assimilis Chd.		2	8	5	2		2	6
P. septentrionis Dej.			5	5	9			5
**Chlaenius costulatus* Mtsch.			1	1				
Trichocellus cognatus Gyll.			1		1			
**Amara torrida* Ill.			2	7				1
Pterostichus adstrictus Eschz.			1					
P. diligens Sturm			12	19	2			1
P. nigrita F.			8	6	2			3
P. strenuus Panz.					1			
**Agonum consimile* Gyll.			1	2	9			
A. fuliginosum Panz.		2	2	12	3		1	1
A. gracile Gyll.				1				
Calathus melanocephalus L.			1	2			4	1
**Cymindis angularis* Gyll.		1						
Haliplidae								
Haliplus obliquus F.	7							
Dytiscidae								
**Hydroporus arcticus* Th.					1			2
H. erythrocephalus L.					1			
H. longicornis Sharp			5	10	16			
H. melanarius Sturm			1	4	3			2
H. palustris L.							1	
**H. tartaricus* Lec.			11		?			
H. umbrosus Gyll.			2		8			1
Dytiscidae cont'd								
Hydroporus sp.			4	17	11			6
Deronectes depressus F./*elegans* Pz.	2							
Agabus affinis Payk.				2				
A. arcticus Payk.								1
A. bipustulatus L.							1	1
A. congener Thbg. group						1		
**Ilybius angustior* Gyll.			9	1	4			
I. subaeneus Er.								1
Ilybius sp.			2					
Rantus sp.			1					
**Colymbetes dolabratus* Payk.								1
**C. paykulli* Er.								1
Acilius sulcatus L.							1	
Dytiscus lapponicus Gyll.								1
Gyrinidae								
Gyrinus marinus Gyll.							3	
G. opacus Sahlb.					1			2
Hydrophilidae								
Ochthebius minimus F.	1				6		1	4
Ochthebius sp.	1		1	1				
Hydraena gracilis Germ.		1						
Limnebius sp.					1			2
Helophorus aquaticus L.					7	2		
H. brevipalpis		5						
H. flavipes F.					9			
H. pallipes Gebl.					1			
H. strigifrons Th.					1			
Coelostoma orbiculare F.			2	1				
Cercyon spp.	1		1					9
Hydrobius fuscipes L.			3	1				
Laccobius sp.	1						1	
Enochrus sp.			1					
Chaetarthria seminulum Hbst.							1	1
Silphidae								
Thanatophilus dispar Hbst.			1	1				
Silpha sp.								1
Choleva sp.							1	1
Colon sp.				1				
Liodidae								
Liodes sp.							1	
Agathidium laevigatum Er.			2					
Ptiliidae								
Acrotrichis sp.								1
Staphylinidae								
Eusphalerum minutum F.					2		1	
**Pycnoglypta lurida* Gyll.	2		1	19	5			
Omalium caesum Gr.		1						
O. excavatum Steph.		1						
O. rivulare Payk.				1				

Staphylinidae cont'd							
Olophrum assimile Payk.	2	3	8	9	10	4	3
**O. boreale* Payk.			1				1
O. fuscum Gr.			6	47	26		6
**O. rotundicolle* Sahlb.				1			
Arpedium brachypterum Gr.	4	2	8	74	5	1	2
Acidota crenata F.	1	2	16	5	2	1	
A. cruentata Mnh.			1			1	
**A. quadrata* Zett.			6	1			
Lesteva longelytrata Gze.						1	
L. monticola Kies.			2				
Geodromicus plagiatus F.		2					
Anthophagus caraboides L.			1	1	1		
**Boreaphilus henningianus* Sahlb.	1		4	26	2		2
Trogophloeus sp.		1					
Oxytelus complanatus Er.		1					
O. laquaeatus Mnh.	1						
O. nitidulus Gr.	1						
O. rugosus F.	1	1					
Stenus juno F.			1	2+			
Stenus spp.	4	3	25	240	26	2	3
Euaesthetus laeviusculus Mnh.				2	1		
Lathrobium terminatum Gr.				5	1		2
Lathrobium spp.			1	11	1	1	2
Xantholinus sp.		4					
Philonthus sp.		1	4	8	1		1
Ocypus olens Mull.	1						
Quedius spp.	1	1	5	15	2	2	6
Boletobiinae indet.	1			4	4		
Tachyporus spp.					2		2
Tachinus corticinus Gr.	2	3		2		2	
T. elongatus Gyll.			1		1	1	2
T. rufipes De G.		3					
Tachinus sp.	2	1	1				
Gymnusa brevicollis Malth.			1	5	5		
G. variegata Kies.				2			
Aleocharinae indet.	6	19	3	24	14	2	9
Cantharidae							
Cantharis spp.	2				1		
Elateridae							
Hypnoidus riparius F.		2					
Helodidae							
Gen. *et* sp. indet.	5		1	3	13		3
Dryopidae							
Dryops ernesti Goz.	1					1	
Elminthidae							
Elmis aenea Mull.		3					
Esolus parallelepipedus Mull.		5					
Oulimnius tuberculatus Mull.	1	1					
Limnius volkmark Panz.		2			1		

Georyssidae							
Georyssus crenulatus Rossi	1	1					
Heteroceridae							
**Heterocerus intermedius* Kies.	1						
Byrrhidae							
Simplocaria semistriata F.					1		
Cytilus sericeus Forst.			1	1		1	
Byrrhus sp.							1
Phalacridae							
Phalacrus sp.				1			
Lathridiidae							
Corticarina fuscula Gyll.	1	3		1			
Nitidulidae							
Kateretes bipustulatus Payk.					1		
Coccinellidae							
Scymnus redtenbacheri Muls.		1					
Scymnus sp.					1		
Coccinella septempunctata L.					2	1	
Oedemeridae							
Oedemera sp.	1						
Anthicidae							
Anthicus scoticus Rye	1						
Scarabaeidae							
Aphodius lapponum Gyll.							1
Aphodius spp.	2	1	2	2		3	
Aegialia sabuleti Panz.							2
Cetonia cuprea F.			1				
Chrysomelidae							
Haemonia appendiculata Panz.	16						
Donacia aquatica L.						1	1
D. obscura Gyll.					1	3	2
D. thalassina Germ.							1
D. versicolorea Brahm						2	
Plateumaris discolor Panz.		1			6		
P. sericea L.	1		1	4		7	10
**Adoxus obscurus* L.						2	
Chrysolina marginata L.						1	
Phaedon sp.					1		
Hydrothassa hanoverana F.					2		
Phytodecta pallidus L.							1
P. viminalis L.	1						
Phyllodecta sp.	1						
Galerucella grisescens Joann.	1						
Galeruca tanaceti L.					1		
Phyllotreta flexuosa Ill.					3		
Haltica cf. britteni spp.	1						
Haltica sp.					1		
Crepidodera		2					

Chrysomelidae cont'd							
Mantura rustica L.	1						
Curculionidae							
Apion sp.	1	2	4	4	4	1	2
Otiorrhynchus nodosus Mull.			30	12		3	9
Barynotus squamosus Germ.			1	1			
Sitona crinitus Hbst.	1						
Tropiphorus obtusus Bonsd.			5	5			
Notaris aethiops F.		3	18	20	5	3	1
Phytonomus sp.			1		1		
Cryptorrhynchidius lapathi L.				1		1	
Rhinoncus pericarpius L.			1				
Litodactylus leucogaster Mrsh.			1			4	1
Eubrychius velutus Beck.	6				1		
Phytobius comari Hbst.					3		
Orobites cyaneus L.			1				
Gymnetron beccabungae L.					1		
Miarus campanulae L.	1						
Rhynchaenus salicis L.			1	1	1		

* Species not now present in Britain

Initially the organic horizon was sampled in two parts, an upper and a lower half, but since no significant difference was found in the fossil assemblages, they have been combined together into one column in Table 6. This means that the sample size was effectively doubled, but in spite of this the total fauna is rather restricted in terms of numbers of species. However, the fact that a few species are represented by relatively large numbers of individuals suggests that this paucity in the number of species is not just a function of the dilution of the fossil content of the samples by the high silt content but reflects a truly small fauna and a rather monotonous local environment.

There can be no doubt from the faunal evidence alone that the deposit accumulated in a body of standing or very slowly moving water, an interpretation that is entirely in keeping with both the sedimentological evidence and the geological setting of the site. The relatively large numbers of the subaquatic phytophagous species *Haemonia appendiculata* and *Eubrychius velutus* indicate the abundance of submerged plants, probably *Potamogeton* and *Myriophyllum*. The fact that the pond margin beetle species *Oulimnius tuberculatus*, *Georyssus crenulatus* and *Heterocerus intermedius* are represented by only one individual each suggests that, at the sampling site, the deposits accumulated well out from the water's edge.

On the land adjacent to the water body two habitats seem to have been available. First, the two carnivorous species *Elaphrus riparius* and *Bembidion velox*, hunt over open sandy places where the ground may be quite bare of vegetation. Secondly, patches of leaf litter and damp moss are indicated by most of the small staphylinid beetles. This leaf litter was probably provided by *Salix* bushes, the usual host plant of *Phytodecta viminalis*. This inference is supported by the occurrence of many small *Salix* leaves in the deposit.* The poor representation of other phytophagous coleoptera suggests that the vegetation on land was sparse. However, some hint of its composition is provided by *Plateumaris sericea*, which lives on various Cyperaceae, *Galerucella grisescens* which is recorded from *Lysimachia* and *Hydrocharis morsus-ranae* (Freude *et al.*, 1966), and *Mantura rustica* which lives on species of *Rumex*.

In spite of the evident openness and poverty of the vegetation cover at this time, the fauna does not suggest that the climatic conditions were similar to those of the tundra. The association of *Haliplus obliquus*, *Haemonia appendiculata* and *Eubrychius velutus* as dominant members of the fauna would be quite out of place under such conditions; rather it suggests a climate not much different from that in south-west Scotland today. It is unlikely that conditions were any warmer than this because of the presence of the dominantly northern species *Pycnoglypta lurida* and *Arpedium brachypterium*. The presence in this fauna of a single elytron of the boreal species *Boreaphilus henningianus* is somewhat enigmatic but it is a good flier and could have reached the Roberthill site from the adjacent Southern Uplands. Even this species lives today as far south as southern Finland and the south of Norway in areas where summer temperatures are little different from those of south-west Scotland.

<u>Redkirk Point</u>
At this site the Lateglacial deposits were sampled from three horizons. In Table 6 three samples are designated as follows: Redkirk A is from the lower carbonaceous silt, Redkirk B is from the basal half of the lower peat (Peat 1) and Redkirk C is from the upper half of the same peat. The exact localities of the sampling positions are shown in Figs. 2 and 3. Since the faunas from these samples were significantly different from one another, they will each be considered under a separate heading beginning with the lowest.

<u>Redkirk A.</u> One of the most outstanding features of the assemblage of coleoptera from this sample is the abundance of Elminthid species. These are all beetles characteristic of

*No leaves of *Salix herbacea*, the dwarf willow, were found in the sample, but one leaf undoubtedly of *Betula nana*, the dwarf birch, was discovered.

running water and it is evident that the lower carbonaceous silt represents the deposit of a small stream that had incised through the layer of till into the underlying Triassic bedrock. This interpretation is given added support by the occurrence of *Hydraena gracilis*, a species of fast-flowing rivers or streams. The banks of this stream would have provided habitats for *Bembidion schueppeli* which is a species "almost confined to river banks, on sand mixed with (plant) detritus where the vegetation is sparse" (Lindroth, 1974, p.58). This is also the habitat for the elaterid *Hypnoidus riparius*. Farther away from the stream, dry, sun-exposed banks with very short vegetation must have been available for the carnivorous species *Cicindela* and *Cymindis angularis*. The phytophagous species are conspicuously rare with the noteworthy exception of *Notaris aethiops* which, though usually associated with *Sparganium*, is so widespread in our fossil assemblages that it may tolerate other host plants. *Plateumaris discolor* probably lives on a wide variety of Cyperaceae. Patches of decomposing vegetation must have been present for the numerous staphylinid species.

The climatic inferences that can be drawn from this fauna are very similar to those already made from the Roberthill assemblage, namely a more or less temperate climate little different from that in south-west Scotland today. Another similarity with Roberthill is that this relatively warm climate is associated with a sparse vegetation cover. In this climatic context the present day geographical distribution of certain species may be highlighted. Thus *Bembidion gilvipes* is a species that occurs today throughout England and Ireland but which is absent from Scotland and all but the southern parts of Fennoscandia: *Cymindes angularis* does not occur at all in the British Isles as a living species but again only ranges as far north as southern Fennoscandia: the small genus *Esolus*, though widespread in Scotland today, is absent from all but the southernmost areas of Fennoscandia. To balance these species of relatively southern disposition are several whose geographical distributions are largely northern. Thus, *Bembidion schueppeli* and *Arpedium brachypterum* range as far south as northern England and throughout almost all of Fennoscandia. Taken together these relatively northern and southern groups of species suggest a climate in which the average July temperature was about 15°C and, since the assemblage is western European in character, it must also be inferred that the climate at that time was moderately oceanic. Certainly adequate precipitation must have been available during the summer to maintain a stream acceptable to this ecologically-compact community of beetles.

Redkirk B. The insect assemblage from the basal half of the lower peat came from the same place as the sample that gave a radiocarbon date of 11,205 ± 177 B.P. (Birm-41) (Fig. 3). This rich fauna presents a picture of the local environment that is in sharp contrast to that of the underlying organic silts. It would be inappropriate here to go into the individual environmental requirements of each species of so long a list but the published table should provide an opportunity for those with specialist knowledge to fill in the details of the environmental summary given here. The peat was evidently laid down in a rather acid bog with a light tree cover, possibly of *Salix*, and with much moss and *Carex* surrounding very shallow pools. So consistent is this fauna in its environmental requirements that it suggests that this habitat was widespread and monotonous and there is very little evidence to indicate what the surrounding country must have been like.

Climatically this fauna also presents a consistent picture. There is abundant evidence of a marked deterioration in the climatic conditions compared with those of the underlying organic silts. All the relatively southern species are absent from this fauna and in their place there is a suite of species with decidedly northern ranges. Some of these do not extend as far south as the British Isles. The following species have dominantly northern geographical distributions, *Diacheila arctica* (Fig. 10), *Elaphrus lapponicus*, *Patrobus septentrionis*, *Amara torrida*, *Agonum consimile*, *Hydroporus tartaricus*, *Ilbius anquestior*, *Olophrum boreale*, *Acidota quadrata*, *Boreaphilus henningianus* and *Otiorrhynchus nodosus*. As a group, these species would be characteristic of the *Betula* zone in the Scandinavian mountains, that is at an altitude near to the climatic limit for tree growth. The average July temperature was probably about 12°C. Two species in this assemblage, *Chlaenius costulatus* and *Bembidion transparens* are predominantly eastern in their geographical distribution, in north-east Europe only reaching as far west as Sweden (the latter species has an isolated occurrence in northern Norway). These species hint at a greater degree of climatic continentality than that of the organic silt (Redkirk A). Nevertheless, sufficient precipitation must have been available to support extensive wetlands indicated by this assemblage of coleoptera.

Redkirk C. The insect assemblage from the upper part of the Lateglacial peat was obtained from the same place as the sample that gave a radiocarbon date of 10,898 ± 127 B.P. (Birm-40) (Fig. 3). Qualitatively the fauna is very similar to that of the lower part of the same peat but quantitively there is a consistent increase in the numbers of the northern species and in those of more open habitats. Thus, though the acid bog appears to have persisted, there is clear evidence of further opening out of the tree cover. This is particularly noticeable in the increase in numbers of the omaline staphylinid species *Pycnoglypta lurida*, *Olophrum*

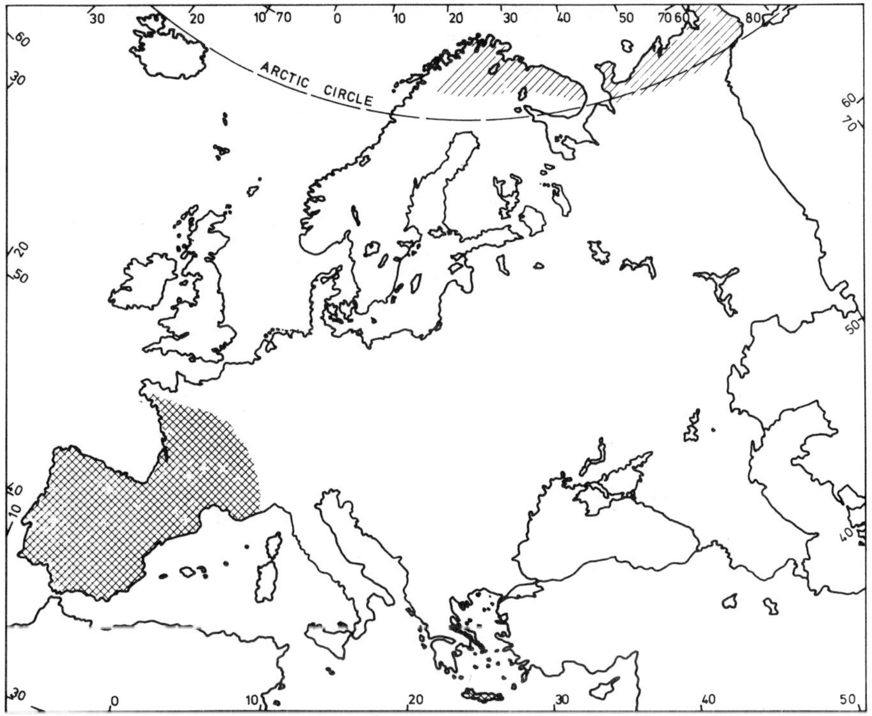

Fig. 10. Modern world distribution of *Bembidion callosum* Kust cross hatched. Modern European distribution of *Diacheila arctica* Gyll. shaded.

fuscum, *Arpedium brachypterum* and *Boreaphilus henningianus*. The increase of *Hydroporus longicornis*, a species of wet mossy flushes, and the decrease of *Ilybius angustior* suggests that there was little open water remaining at this stage. The numbers of *Plateumaris sericea* also increase markedly indicating an expansion of Cyperaceae, a conclusion probably supported by the high values of *Notaris aethiops*.

Climatically this assemblage indicates a further deterioration compared with the lower sample. The numerical increase in the representation of the northern species coupled with the evidence for opening out of the woodland, suggests that by this time the climatic limit for tree growth had been reached. The absence, however, of the characteristic species of the alpine zones, or of the true tundra, suggests that the average July temperatures were about 10°C. There is no evidence for any further change in the degree of climatic continentality and precipitation levels must have remained high enough to continue to support the bog community.

<u>Bigholm Burn</u>. Two horizons were sampled from this exposure. In Table 6 these samples are designated as follows: Bigholm A is from the folded peat bed at the base of the sequence: Bigholm B was a small (1 kg) sample from a narrow band of poorly organic silt under the blanket peat and interbedded within the uppermost part of the gravel. The faunas are so different that they will be dealt with under separate headings.

<u>Bigholm A</u>. Because of post-depositional disturbance (see Fig. 4) the basal peat was sampled as a single unit. Fortunately, because of the close proximity of the two radiocarbon dates (11,820 ± 180 B.P. and 11,580 ± 180 B.P.), it is likely that little information will have been lost by this sample lumping. In almost all respects this fauna resembles that from the Lateglacial peat at Redkirk Point and the local environmental conditions must therefore be broadly similar. There is, however, some hint from the slight increase in the representation of the phytophagous species, that the flora here was rather more varied. Thus, *Hydrothassa hanoverana* lives on *Caltha palustris* and *Phyllotreta flexuosa* feeds upon a wide variety of Cruciferae and Resedaceae. Among the weevils *Phytobius comari* lives on *Comarum palustre*, and *Gymnetron beccabungae* feeds on *Veronica beccabunga*.

Climatically this fauna indicates very similar conditions to those inferred from the assemblages of the Redkirk Point lower peat (Peat 1). There are, however, two aspects of this fauna that suggest that the climate may have been marginally less severe than at Redkirk Point. First there is the greater floral diversity hinted at by the phytophagous beetles and secondly there are two relatively southern species present in this assemblage; *Eubrychius velutus* does not extend above the conifer zone in the Scandinavian mountains and *Gymnetron beccabungae* does not range north of Latitude 64°N in Fennoscandia.* The fauna from Bigholm A thus indicates an average July temperature of about 12°C with little suggestion of any increase in climatic continentality.

<u>Bigholm B</u>. A thin organic silt of limited lateral extent was interbedded in the uppermost part of the gravel at Bigholm Burn (Fig. 4). Only about 1 kg of this material could be obtained and this yielded a very small assemblage of

*It was, however, a very common species in somewhat earlier Lateglacial deposits on the Cumbrian coast (Coope and Joachim, in preparation).

insects. In spite of the fact that only three species could be named, none of these are represented by single individuals and the commonest species *Olophrum boreale*, numbered at least five individuals. It is thus possible that this meagre assemblage is a real measure of the poverty of the fauna of the times. There can be little doubt that the local environment must have been extremely barren with little or no plant cover. The presence of *Helophorus aquaticus* suggests that the deposit accumulated in a temporary puddle. This interpretation is supported by the relatively common occurrence of the notostracan crustacean *Lepidurus arcticus* Pallas which is most usually found in shallow temporary bodies of water in the boreal parts of Europe. It does not live in the British Isles at the present day. This crustacean is a frequent member of the faunas of deposits of Jessen-Godwin pollen zone III in almost all parts of the British Isles.

It is not easy to interpret such meagre faunal evidence in terms of climate. One of the reasons for this difficulty is that, as conditions of polar desert are approached, the biological information falls to zero. However there can be little doubt that by this time the average July temperature must have been well below 10°C.

Sanquhar

This site, for which no radiocarbon dates are available, is geographically isolated from the rest of the sites, lying in Nithsdale (Fig. 1). On faunal and stratigraphical grounds there can be no doubt that the deposits here are of Lateglacial age. The samples from which the insect faunas were extracted came from two sources. First, specimens of peat with visible insect remains were obtained from this site by the Royal Scottish Museum, Edinburgh (to whom I am greatly indebted for the loan of the specimens (Sanquhar B)). After careful removal of the obvious fossils, the matrix was then processed in the laboratory and many more, often small and inconspicuous species, were thus obtained. The total weight of this material came to something less than 1 kg but since this was not a random sample of the deposit in the first place, it is not strictly comparable with the samples considered previously. The peat bed 15-20 cm in thickness outcropped in the old brickworks just north of Sanquhar railway station (NS/783107) where it was overlain by about 1 m of solifluction on which a soil profile had developed. Below the peat were about 2 m of sand and silt. The second sample (Sanquhar A) was obtained by the writer several years after the first and was taken from these underlying silts. By this time the pit had been largely filled with domestic refuse and little of the exposure was visible. At this second sampling location the peat bed was badly dried out and hard and thus unsuitable for analysis for coleopteran remains. The underlying silts were about 15 cm thick and rested upon angular blocks of Carboniferous sandstone between which the silt had been deposited. In the faunal list the sample of the lower silt is designated Sanquhar A and that from the peat Sanquhar B.

Sanquhar A. The fauna from the lower silts suggests that these deposits accumulated in a shallow pool surrounded by a sedge marsh and farther afield on the drier ground by meadow-like vegetation. Within the pool submerged plants such as *Myriophyllum* and *Potamogeton* would have provided habitats for the weevil *Litodactylus leucogaster* and for the chrysomelid *Donacia versicoloria* respectively. That the vegetation round the pool was dominated by *Carex* and *Sparganium* is indicated by *Plateumaris serricea*, *Donacia aquatica*, *Donacia obscura* and the weevil *Notaris aethiops*. The only evidence of any trees is provided by the weevil *Cryptorhynchidius lapathi* which drills into *Salix* branches of moderate size indicating at least the presence of bush willows and not the dwarf variety. On the drier places farther from the pool a light vegetation cover is indicated by *Calathus melanouphalus*, the commonest carabid species in this assemblage. This drier ground probably supported *Epilobium*, the food plant of *Adoxus obscurus*, in the northern parts of its range.

From the point of view of their geographical distributions, the species of this fossil assemblage show little that is at all unusual. With two exceptions all of them could be found living today in south-west Scotland. The first of these, *Adoxus obscurus*, is very widespread on the continent but absent from the British Isles today. This absence may well be due to the disappearance of its food plant *Epilobium*, and its inability, as yet, to re-establish itself in response to the "fire weed" explosion of modern times is probably due to the geographical isolation of these islands. The presence of this species in this assemblage thus carries little climatic significance. The second species is of much greater interest. *Bembidion callosum* is a south-west European species widely distributed in the Iberian peninsula and extending into central France (Fig. 10). There is an old record of this species from Woking, Surrey in 1851 (Fowler, 1913) but Lindroth (1974) considers the presence of this species in the modern fauna of Britain highly doubtful. Be that as it may, its presence in south-west Scotland in early Lateglacial times can now be established. In assessing the significance of this occurrence two important facts must be borne in mind. First, this is a species of open, thinly vegetated places where it is indubitably a heliophile. Secondly, it is a fine flier and hence should be a ready coloniser of newly available habitats. Its presence in this fauna might have seemed enigmatic were it not for the presence in at least two other faunas (Osborne,

1972; Coope and Brophy, 1972) of similar age in England of the related species *Bembidion grisvardi* Dew / *ibericum* Pioc which has a very similar distribution today. The presence of these thermophilous open ground species, which are by no means the only ones in the deposit, does not carry implications of Mediterranean climates since the rest of the associated species would be out of place under such conditions. Rather, it seems most probable that these exotic species were able to range farther north than their present day northern limit, because of the universality of open ground at this stage of the deglaciation of the landscape. If the presence of *Bembidion callosum* is to be explained in this way, the rest of the fauna suggests a climate very much like that of south-west Scotland today, since all the really northern species that characterise the colder episodes are absent from this assemblage.

Sanquhar B. The insect fauna from the peaty layer indicates that it was in fact a detritus mud laid down on the bottom of a pond. The relatively large number of dytisci water beetles includes several of the larger species suggesting the availability of open water. The diminution in numbers of those beetles that live on submerged vegetation may be a further indication of an increase in the open water. Surrounding the pond the vegetation was still dominated by Cyperaceae. The local environmental conditions at the time of the deposition of the peaty layer thus differ little from the conditions at the time when the underlying silts were laid down.

On the other hand the climatic implications of this fauna contrast with those inferred from the fauna of the silts. The assemblage from the peaty horizon includes a number of decidedly northern species; *Patrobus septentrionis*, *Amara torrida*, *Hydroporus arcticus*, *Agabus arcticus*, *Colymbetes dolabratus*, *Gyrinus opacus*, *Olophrum boreale*, and *Boreaphilus henningianus*. However the presence of *Bembidion doris* and *Chaetarthrea seminulum*, neither of which range above the upper limit of the conifer zone in the Scandinavian mountains, suggests that the average July temperatures were about 12°C at this time. Once again there must have been adequate precipitation available to maintain a pond with a rich biota through the summer.

In the absence of radiocarbon dates from the Sanquhar site, it seems appropriate here to mention one or two facts that are of importance in arriving at an estimate of the age of these deposits. Stratigraphically this deposit must surely post-date the retreat of the ice from this part of the Southern Uplands. The peaty horizon, though somewhat compacted, does not bear the intense compressional features characteristic of peats that have been overridden by ice sheets. Many of the insect remains still retained their three-dimensional configurations. The deposit is plainly overlain by a layer of solifluction that includes clay with stones that might well be mistaken for a till. This is not a recent land slip or dump of overburden from the old brickworks as it has on its surface a well-developed soil profile. The most appropriate time interval for this solifluction deposit is during the Loch Lomond Stadial. The insect faunas from this deposit would also support a Lateglacial age for these deposits. In general they conform to the now usual pattern of an assemblage of thermophilous species in the chiefly minerogenic deposit immediately under a peaty layer which contains many northern or even arctic beetle species. These deposits cannot be Flandrian in age because in all Flandrian deposits so far investigated in relatively lowland situations these assemblages of northern species do not reoccur. Lastly, note should be taken of the presence of *Adoxus obscurus*, a very common species in Lateglacial deposits but which is almost unknown in the numerous faunas from earlier interstadial deposits and has not yet been found in any British Flandrian deposits.

Brighouse Bay

The peaty deposits that occur well below the high tide level at Brighouse Bay have a radiocarbon date of 9,640 ± 180 B.P. (Q-398). They were sampled on two separate occasions. First, a bulk sample was obtained (5 kg) and then the peaty layer was split into a lower and upper half, again each sample weighed 5 kg. The fauna from these samples is given in Table 7 where the numerical representation of each species is provided for each sample not because these insect faunas are significantly different but to illustrate the degree of repeatability of fossil assemblage that can be obtained from such samples. In the discussion that follows these faunas will be dealt with as a single unit.

The local environment that can be inferred from this fauna is quite different from any we have so far considered. It may be summarised as follows. The peaty deposit accumulated in a marsh, richly overgrown with *Typha* and *Phragmites* and probably also by *Carex* and *Sparganium*. These conditions are indicated by the phytophagous beetles, *Plateumaris* and *Donacia* and also by the carnivorous species such as *Agonum thoreyi* and *Odacantha melanura*. The *Carex* must have been badly afflicted with smut fungus, the food of *Phalacrus caricis* which was present in all three samples. Some open water must have been available for the dytiscid water beetles and also for *Gyrinus*. Round the margins of this swamp a deciduous forest must have grown with its attendant characteristic beetle fauna. Dead and dying wood is indicated by *Melasis buprestoides*, *Dirrhagus pygmaeus*, *Grynobius excavatus* and *Anisandrus dispar*. *Cerylon histeroides* lives under dead bark. The distinctive carabid *Calosoma inquisitor* climbs

TABLE 7 Faunal list and specific abundance of coleoptera from the early Flandrian site at Brighouse Bay, Kirkcudbrightshire

	Bulk Sample	Lower Half	Upper Half
Carabidae			
Cychrus caraboides L.	1		
Calosoma inquisitor L.	1	1	
Elaphrus cupreus Duft.	1	2	1
Dyschirius globosus Hbst.	2	3	
Bembidion fumigatum Dej.		1	
B. gilvipes Sturm			1
B. guttula F.	1	3	
Trechus quadristriatus Schrk.		1	
Pterostichus minor Gyll.	1	6	2
P. oblongopunctatus F.	1	1	
P. strenuus Panz.		2	
P. vernalis Panz.			1
Agonum fuliginosum Panz.		4	
A. obscurum Hbst.	1		
A. thoreyi Dej.	5	5	3
Odacantha melanura L.		2	1
Dytiscidae			
Hydroporus palustris L.	1		
Copelatus ruficollis Schall.		1	
Agabus bipustulatus L.		1	1
Ilybius sp.	1		
Gyrinidae			
Gyrinus suffriani Scriba	2		
Gyrinus sp.	2		
Hydrophilidae			
Ochthebius minimus F.	12	14	9
Coelostoma orbiculare F.	2		
Cercyon spp.	16	12	8
Laccobius sp.		1	
Liodidae			
Liodes sp.	1		
Corylophidae			
Orthoperus sp.	10	4	5
Staphylinidae			
Olophrum piceum Gyll.		1	
Lesteva heeri Fauv.	2	2	4
Stenus juno F.	5	2	1
Stenus spp.	10	4	3
Lathrobium spp.	2	3	
Cryptobium fracticorne Payk.	1		
Philonthus sp.		1	
Quedius sp.		1	
Aleocharinae indet.	2	2	
Pselaphidae			
Brachygluta sp.	1	1	
Bryaxis sp.		1	
Eucnemidae			
Melasis buprestoides L.	1	2	1
Dirrhagus pygmaeus F.	2		
Dascillidae			
Dascillus cervinus L.	3	1	2
Helodidae			
Gen. *et* sp. indet.	12	15	8
Cryptophagidae			
Telmatophilus spp.	2		
Phalacridae			
Phalacrus caricis Sturm	2	1	2
Colydiidae			
Cerylon histeroides F.			1
Anobiidae			
Grynobius excavatus Kug.	1	3	1
Scarabaeidae			
Aphodius sp.	1		1
Cetonia aurata L.	1		
Chrysomelidae			
Donacia marginata Hoppe	3	3	2
Plateumaris affinis Kunze	1		1
P. braccata Scop.	1		1
P. sericea L.	1		1
Chalcoides cf. aurata Marsh.	2		
Chaetocnema concinna Marsh.	1		
Curculionidae			
Micrelus ericae Gyll.		1	
Rhamphus pulicarius Hbst.	5		6
Scolytidae			
Anisandrus dispar F.	1	2	

trees in search of lepidopterous larvae, usually Geometridae. On the forest floor *Cychrus caraboides* would have hunted for shell bearing snails. The insect fauna provides little evidence of the types of tree present. Only the weevil *Rhamphus pulicarius* is dependent on trees, feeding on *Salix* and *Populus*.

The climatic significance of this insect assemblage is of outstanding importance. All the northern species, so characteristic of the fauna of south-west Scotland so recently before, have disappeared and in their place is a thoroughly temperate assemblage that includes a number of species which do not range as far north as Scotland at the present time. Thus *Calosoma inquisitor* extends as far north as Cumbria (Lindroth, 1974) with an isolated outpost near Spean Bridge, Inverness-shire (R.A. Crowson, *in Litt.*). *Bembidion gilvipes* and *B. fumigatum* are distributed in Britain as far north as Yorkshire. *Agonum thoreyi* extends only as far north as north Lancashire. *Odacantha melanura* is confined in Britain to the south-east of England and the extreme south of Wales. *Melasis buprestoides* probably does not occur in modern times north of Yorkshire but, at least in part, this restriction may be due to forest clearance by man. *Dirrhagus pygmaeus* is a southern English species that has recently been found in the Wood of Cree, Kirkcudbrightshire. *Gyrinus suffriani* is also a southern species with its northern limit in Kirkcudbrightshire. The rose chafer *Cetonia aurata* is a highly conspicuous beetle and unlikely to be overlooked yet it does not occur today in Scotland apart from an old record from Ayrshire. In Fennoscandia however it does range somewhat farther north. *Phalacrus caricis* is also rather southern in Britain and does not occur in Scotland today. Finally *Anisandrus dispar* is a rare species of southern England and does not range in Europe as far north as Fennoscandia. These relatively southern species cannot owe their absence from present-day northern Europe to inadequacies in the local environments since collectively they span a wide spectrum of ecological adaptations. Rather they suggest that the climatic recovery after the Loch Lomond Stadial must have been exceedingly rapid so that by the time the Brighouse Bay deposits were laid down the climate was at least as warm or even slightly warmer than in south-west Scotland today.

SUMMARY OF LATEGLACIAL AND EARLY FLANDRIAN ENVIRONMENTS IN SOUTH-WEST SCOTLAND

Taken together the insect assemblages from all five sites investigated present a consistent picture of environmental changes. Beginning with the oldest assemblage these can be summarised as follows. At some time prior to 13,000 B.P. the climate must have been characterised by conditions of arctic severity, attested to by the existence in lowland areas of extensive ice sheets. Unfortunately no Scottish faunas are known that date from this period. Certainly by about 13,000 B.P. the climate had become as warm as present time at Roberthill and in the deposits at the base of the Lateglacial sequence at Redkirk Point, and at Sanquhar similarly temperate assemblages probably date from between 13,000 and 12,000 B.P. (Fig. 11). During this period a rich aquatic and semi-aquatic flora built up in and around the pools left behind on the ice-moulded landscape by the retreating glaciers. However, in spite of the temperate climate, there is no evidence of any tree cover except for *Salix* near the pools. The country was clothed in a thin patchy vegetation reflected in the low organic content of the sediments laid down at this time.

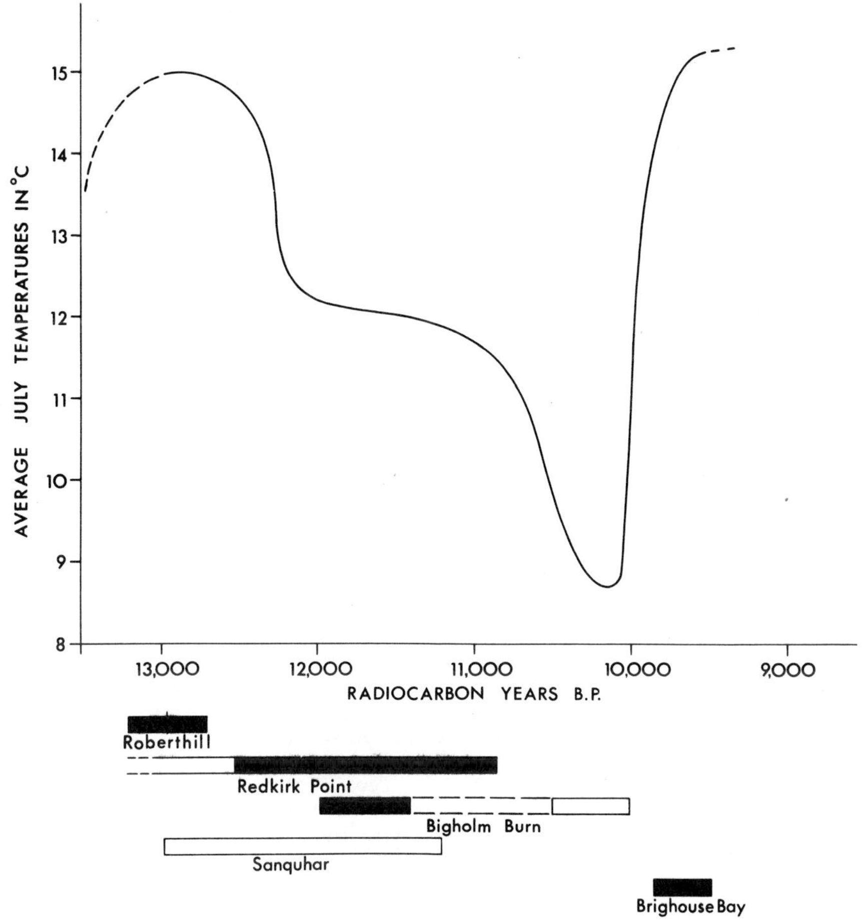

Fig. 11. Changes in the average July temperature inferred from coleoptera assemblages plotted against radiocarbon years before present. The ages of the fossil-bearing sites are shown in black where radiocarbon dates are available and inferred ages are shown by open boxes where only stratigraphical and faunal evidence exists.

The faunas that date from between 12,000 and 11,000 B.P. show abundant evidence of a considerable climatic deterioration. This was unexpected at the time this investigation was begun for two reasons. First, it was generally believed that dominantly inorganic sediments were characteristic of cold periods and those that are rich in organic matter were typical of relatively warmer episodes. In all the sites investigated here that date from this period, the reverse situation seemed to prevail. Secondly, the time between 12,000 and 11,000 B.P. was, on palynological grounds, broadly equated with the Alleröd Interstadial. The coleoptera of this period suggest a colder climate than that of the preceding period and more akin to that near to the tree line in the Scandinavian mountains where tree birches dominate the forest. The interpretation of a climatic regime so close to the critical conditions for tree growth may provide the means of reconciling these conflicting viewpoints. Any small scale fluctuation in temperature at this time may well cross and recross the threshold limiting tree growth and thus cause apparently large scale changes in pollen assemblages without any comparable large scale changes in insect assemblages.

There is little evidence of the faunas that date from between 11,000 and 10,000 B.P. The top of the Lateglacial peat at Redkirk Point dates from the very beginning of this period and provides evidence of further climatic deterioration with the resultant opening up of local woodland as the threshold for the limit of tree growth is crossed. The coarser sediments that characterise this period at Redkirk Point, Bigholm Burn and Sanquhar suggest that as the climate became still colder the vegetation cover became patchy and mass movement of the surface contributed a great deal of sediment to fill the Lateglacial hollows. At Bigholm Burn some insight may be had into the rigours of the climate at this time from the meagre fauna dominated by the low temperature stenothermic species *Olophrum boreale*.

The climatic change at the beginning of the Flandrian period must have come with extraordinary rapidity. The fauna from Brighouse Bay shows clearly that within a period of five hundred years, and the actual time involved may have been very much less than this, the climate of south-west Scotland changed from one of arctic severity enabling large ice-caps and glaciers to become established in the Highlands (Sissons, 1974c, 1976b), to a climate at least as warm or maybe rather warmer than that of the present day.

The sequence of climatic events outlined above agrees with the regional pattern of climatic changes that can be inferred from Lateglacial insect faunas elsewhere in the British Isles. Unfortunately only two sites have been investigated that present an unbroken sequence from fluvioglacial deposits associated with the melting of the Late Devensian ice-sheet through to the beginning of the Flandrian. At Glanllynnau, North Wales (Coope and Brophy, 1972) and Glen Ballyre, Isle of Man (Coope and Joachim, unpublished) deposits that date from a little prior to 13,000 B.P. contain an arctic assemblage of coleoptera. This assemblage was then replaced, with apparently dramatic suddenness, by a fauna that included many thermophilous species and lacked all those that characterised the earlier arctic assemblages. A fauna of thermophilous species of coleoptera was described from deposits of similar age from Church Stretton, Shropshire (Osborne, 1972). Thus the faunal evidence from elsewhere in the British Isles gives added support for the interpretation of temperate conditions in south-west Scotland between about 13,000 and 12,000 B.P.

At all the sites mentioned in the previous paragraph and in a number of other sites, notably Red Moss, Lancashire (Ashworth, 1972), the deposits laid down between 12,000 and 11,000 B.P. contain rich insect assemblages characterised by a return of the northern element in the fauna and the disappearance of the thermophilous species. This evidence of a marked deterioration of climate at, or slightly earlier than 12,000 B.P., followed by a continued, but slow, decline in temperature, supports the interpretation given for the coleoptera of similar age from south-west Scotland. It is likely that the faunas from Central Scotland, namely Corstorphine, Edinburgh (Coope, 1968) and Burnhead, near Airdrie (Coope, 1962) also date from this period.

The numerous sites in England that date from the Loch Lomond Stadial all carry suites of arctic species of coleoptera. In this context it is interesting to note the occurrence at Croydon in southern England of a thoroughly arctic assemblage of cool coleoptera in deposits with a radiocarbon date of 10,130 ± 120 B.P. (Birm-101) (Peake and Osborne, 1971). It seems evident that during this period in south-west Scotland the climate must have been very harsh indeed judging from the paucity of the fauna from the Bigholm Burn gravels.

The rapid temperature rise at the beginning of the Flandrian period has long been recognised (Iversen, 1954) though it has been difficult to quantify. Assemblages of coleoptera from the English Midlands that date from this period suggest that within the space of five hundred years, or even less, the climate changed from arctic severity to something like that of the present day (Ashworth, 1973; Osborne, 1974). The thermophilous assemblage of beetles from the early Flandrian deposits of Brighouse Bay thus fits into a regional picture of climatic change and provides the justification for the sharp upturning of the inferred Lateglacial/early Flandrian temperature curves already published (Coope, 1970, 1975).

The climatic implications based on evidence from coleoptera, shown in graphical form in Fig. 11, should be compared with

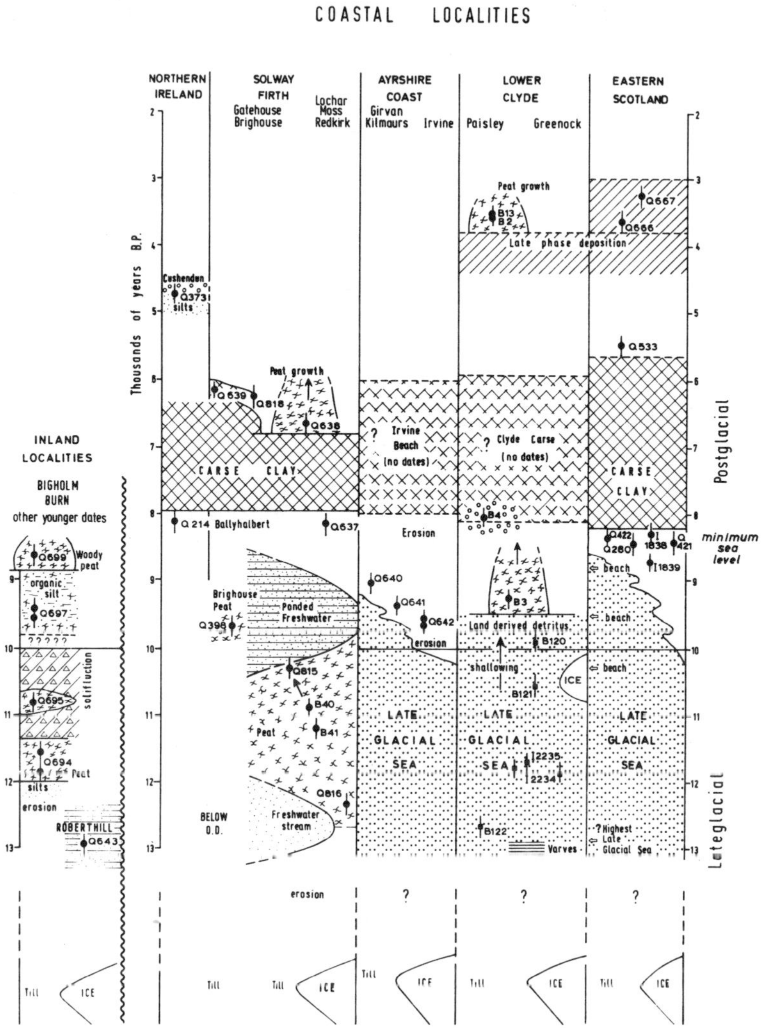

Fig. 12. Time-locality diagram illustrating environmental conditions established in three areas of south-west Scotland plotted against a radiocarbon time-scale and compared with east central Scotland. Radiocarbon dates from Godwin and Willis (1960, 1961, 1962, 1964); Godwin *et al.* (1965); Godwin and Switsur (1966); Shotton *et al.* (1967, 1968); Sissons (1966, 1967c).
B = Birmingham University Dating Lab.; Q = Cambridge University Dating Lab.; I = Isotopes Dating Lab.

the summary of environmental and chronological data given in Fig. 12. This reveals differences in timing of environmental changes in the three study areas and particularly between the Solway Firth and the Clyde-Forth region.

There is no evidence of the Lateglacial sea having inundated the present land areas adjacent to the Solway Firth. Indeed in this area there is evidence of freshwater streams below O.D. contemporary with dates in excess of 12,000 years for marine faunas implying a sea-level well above O.D. in the Firth of Clyde. Peat growth and ponded freshwater at Redkirk Point on the Solway are contemporaneous with marine shells in growth position in silts receiving land-derived detritus in the Greenock-Paisley area spanning a period from over 10,000 years to about 9,500 years ago.

The onset of carse clay deposition shows remarkably accordant ages for the replacement of peat growth by marine and estuarine conditions about 8,000 years ago. There is evidence of synchronous transgression in areas as widely spaced as Ballyhalbert in Northern Ireland, Redkirk on the Solway and several localities on the Firth of Forth.

The Ayrshire coast seems to represent intermediate conditions throughout the period between 12 - 13,000 and 8,000 years ago. Lateglacial clays were mapped by the Geological Survey (Richey *et al.*, 1930) as occurring round the base of drumlins and hummocky drift composed of tills beneath which the Kilmaurs "elephant bed" outcrops. At some stage in the Lateglacial, in excess of 10,000 years ago, sea-level fell below O.D. but returned early in Flandrian times as shown by dates on peats at Girvan and Irvine.

The area east of Irvine does not contain carse clay deposits because it was open to strong wave action from the west, but the compacted peat was partially eroded by sands comprising beach deposits which culminate at about 12 m O.D. at the landward side of the bay. At Girvan the peats occur beneath finer grained "raised beach" deposits that accumulated in more sheltered conditions. In the Linwood area erosion of earlier peats and other deposits between 8,000 and 4,000 years ago probably represents an event equivalent to the carse clay transgression.

The underlying cause of all the differences in timing between 13,000 and 8,000 B.P. is to be sought in local differences in the behaviour of the crust as it recovered from differential ice loading in the Late Devensian.

ACKNOWLEDGEMENTS

We are grateful to many students and colleagues who assisted with field work and particularly to Mr. Robert Little of East Hayrigg, Lockerbie, Dumfriesshire, whose infectious enthusiasm, liaison with local landowners and nose for an intriguing locality triggered both sides of these investigations. W. W. B. benefitted from the local knowledge of his extra-mural class in Lockerbie, from the assistance of members of the Dumfriesshire and Galloway Natural History Society and from the observations of the 'Drift' Group of the Glasgow Geological Society. Dr. R. H. Cummings and Mr. A. Howie provided help at Kilmaurs, including the arrangement of mechanical aids, and Mr. H. A. Moisley was a never-failing source of archival material and a tireless sleuth in the search for temporary exposures in the Linwood area. Dr. Ian Rolfe made available records of collections in the Hunterian Museum, and Dr. N. T. Moar allowed access to relevant sections of his Ph.D. thesis and gave other helpful advice. Organisations who replied to our enquiries by providing additional assistance included Cementation Ltd. and Kilmarnock Corporation. Dr. R. A. Crowson and Dr. R. B. Angus assisted G.R.C. with identification and interpretation of the beetle faunas. Mr. and Mrs. G. Trafford provided accommodation and helpful discussion, and Ruth was happy to be pressganged as a peat digger when the turning tide threatened operations at Brighouse Bay. Finally, we thank our wives and families for pretending not to notice when so many ostensible holiday outings returned unerringly to where the peat lay dinted.

Evolution and Chronology of Lateglacial Marine Environments at Lochgilphead, Scotland

J. D. PEACOCK (I.G.S. Edinburgh)
D. K. GRAHAM (I.G.S. Edinburgh)
J. E. ROBINSON (University College, London)
and
I. WILKINSON (I.G.S. Leeds)

ABSTRACT

A section in Lateglacial marine clays ('Clyde beds') on the foreshore of Loch Gilp yields a history of events covering the latter part of the Lateglacial Interstadial. Prior to approximately 12,500 B.P. Loch Gilp was connected by a strait to the Sound of Jura, and strong tidal currents led either to erosion or to the deposition of gravel only. As relative sea-level fell to the level of the cols between Lochgilphead and Crinan current strength was reduced and offshore marine sediments were laid down under quiet conditions. Further reduction of sea-level led to the initiation of estuarine conditions about 12,100 B.P. which continued to at least 11,000 B.P. During this time-interval sea-level was probably between 4 and 10 m O.D. Sea temperatures were lower throughout than those obtaining at present and conditions apparently became even more severe after 12,000 B.P. It is suggested that, following deglaciation, sea temperatures never reached present day values near western and northern Scottish coasts during the Lateglacial Interstadial. The inferred change, if valid, from a condition of rapidly falling sea-level to one where the variations in this parameter were relatively slight may be due to a greatly increased rise in eustatic sea-level beginning about 12,000 B.P.

INTRODUCTION

Lateglacial marine deposits described from the Lochgilphead area some 100 years ago (Crosskey and Robertson, 1871, pps. 18-25; Brady *et al.*, 1874, p.61) belong to the 'Clyde beds' (Jamieson, 1865), a series of often richly fossiliferous clays and silts which are widespread in this part of western Scotland. Good exposures of these beds are rare, but a temporary excavation made in such deposits on the foreshore a short distance south of Lochgilphead on the east side of Loch Gilp (Pit A, Fig. 2) allowed a few samples to be collected, including material for a radiocarbon date. These results were supplemented by screw auger samples collected nearby (Fig. 2). The mollusca and ostracoda from these two sites were determined as part of a private research programme by Peacock and Robinson. Although the base of the marine sequence was not reached the results clearly showed that in the 2 m or so penetrated there was an upward transition from a marine to an estuarine environment and that the possibility of dating the sequence by marine shell radiocarbon dates was good.

Following this preliminary work, a second pit (Pit B, Fig. 2) immediately adjacent to the auger hole was excavated to the base of the marine sequence by the Institute of Geological Sciences and a larger number of samples was retrieved for faunal analysis and radiocarbon age determination. Because the sediments have a dip, the depths of samples collected from the floor of the pit during excavation do not coincide with those subsequently collected at the northwest corner of the pit. The correlation is shown in Fig. 3. For Pit B the mollusca were identified by Graham, the foraminifera by Wilkinson and the ostracoda by Robinson. The fossils from Pit A and the auger hole are stored in the private collections of Peacock and Robinson and the remainder in the Palaeontology Department of the Institute of Geological Sciences in Edinburgh and Leeds.

GENERAL CONSIDERATIONS

Lochgilphead lies at the head of a small sea loch which opens southwards onto the middle reaches of Loch Fyne. To the

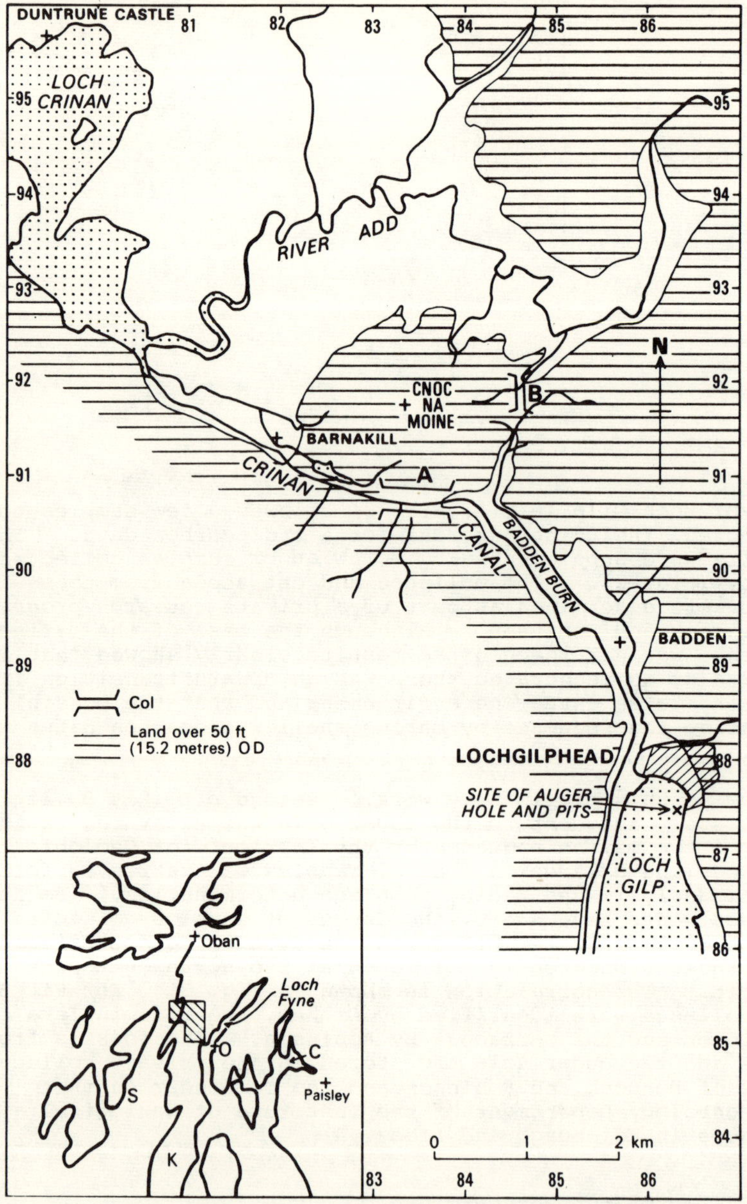

Fig. 1. Location and surroundings of Lochgilphead. A and B, cols; C = Clyde estuary; K = Kintyre; O = Otter Ferry; S = Sound of Jura.

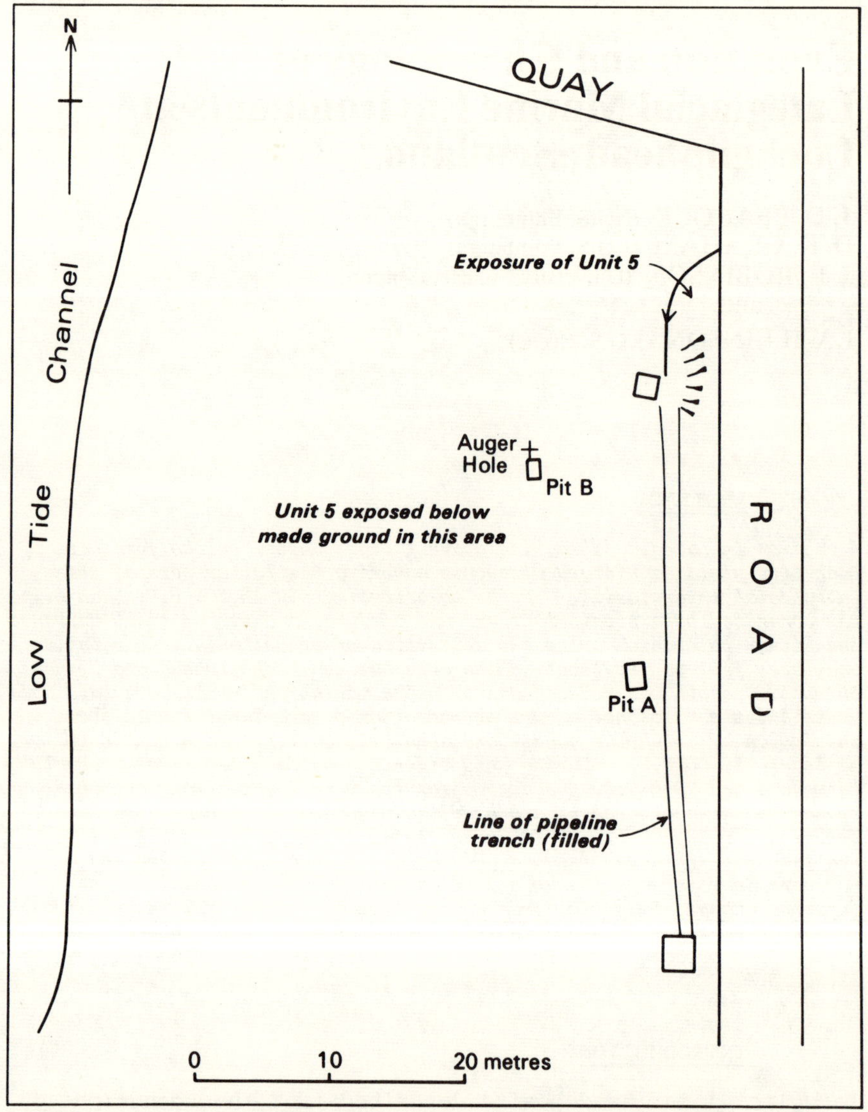

Fig. 2. Sketch map of 'Clyde bed' site on foreshore near Lochgilphead. For location see Fig. 1.

northwest, the route followed by the Crinan Canal, which connects Loch Fyne with the Sound of Jura (Fig. 1), now rises to 19 m O.D., the highest level of the Crinan Canal (but probably originally above 20 m), and the road from Lochgilphead into the valley of the River Add crosses another low col at about 20 m O.D. Both cols appear to have solid rock near the surface. Thus when the sea was at its highest levels during Lateglacial times (about 36 m - see below) these low cols were submerged, and Kintyre was a series of islands.

During the last glacial maximum ice flowed west or north-west across the area as a whole, depositing a medium grey or brownish-grey till derived mainly from the local Dalradian phyllite, quartz-schist and hornblende-schist. In the vininity of the site described below, the till is thin and solid rock is exposed on the foreshore and at localities inland.

Lateglacial marine deposits are known to occur on the low ground in the vicinity of Lochgilphead, near Barnakill (NR/819913) on the Crinan Canal, and on the foreshore near Duntrune Castle (NR/794956). At Barnakill they reach up to 9 m O.D. (Crosskey and Robertson, 1871, p.327), and at Lochgilphead the height is given as 15 - 20 ft (4.6 - 6.1 m) above sea-level (Brady *et al.*, 1874, p.62). South of Lochgilphead, the Lateglacial deposits can be traced along the foreshore for some 500 m on the east side of Loch Gilp and underlie parts of the town. They are also known to extend locally inland below the Flandrian raised beach deposits on the east side of the loch, and, from borehole evidence, appear to underlie estuarine deposits in Loch Gilp itself.

From the past and present investigations it is clear that the Lateglacial marine deposits found on the foreshore and on the neighbouring low ground occupy hollows in the till or rock floor, and that the thin successions have been subsequently truncated. However, as shown below, the succession near Lochgilphead was laid down over a time span of about 1500 years and indicates conditions over a significant proportion of Lateglacial time.

LITHOSTRATIGRAPHY

The strata seen in Pit B can be subdivided into five units, four of which are marine. There is an original dip of about 10° to the north.

Unit 1. Soft-to-hard, brownish- to bluish-grey, sandy, silty till was encountered at the base of Pit B (Fig. 3). In this type of material, which occurs elsewhere on the foreshore below the thin modern beach deposits, the

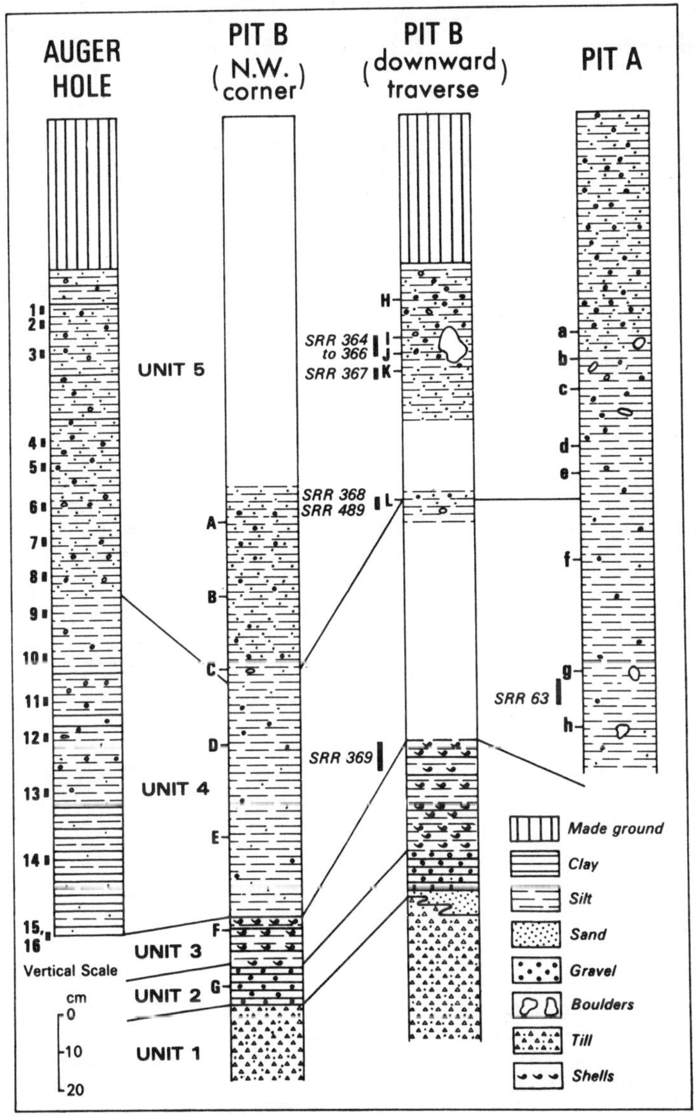

Fig. 3. Sections in Lateglacial deposits at Lochgilphead. Surface levels approximately 1.5 m O.D. at Pit A, and 1 m O.D. at Pit B and the auger hole. Letters a-h, A-L and numbers 1-16 are sample localities in Pit A, Pit B and the auger hole respectively. SRR-63 *etc.* are C14 dated samples.

fragments are mainly of local rocks (quartz-schist, phyllite, and hornblende-schist). The top few centimetres of till are sandy and locally overlain by up to 10 cm of pale bluish-grey sand and silty sand with pebbles and cobbles.

Unit 2. Above a sharp contact with Unit 1, the marine succession begins with a bed of permeable, clayey gravel some 10 cm thick containing numerous echinoid spines. The subangular to angular fragments are generally 2 - 10 mm across, with a few ranging up to 4 cm, and are composed of quartz-schist, psammite and phyllite together with numerous fragments of barnacles and the spines of echinoids.

Unit 3. This unit, 12 - 25 cm thick, is formed predominantly of large specimens of *Modiolus* in a matrix of clayey silt with scattered, angular fragments of phyllite and quartz-schist. A substantial proportion of the debris upwards of sand-size is of echinoid fragments (spines and plates) and barnacle plates.

Unit 4. Above the *Modiolus* bed are some 60 - 90 cm of grey silt and silty clay, well-sorted for the most part, but containing scattered fragments of gravel size and rare cobbles. Much of the material is bioturbated. This unit is clearly visible in the records of Pit A and the auger hole near Pit B (Fig. 3).

Unit 5. This is the thickest of the five units, varying from 60 to 90 cm thick in Pit B (Fig. 3) and reaching a thickness of 105 cm in Pit A. It consists of silt and silty sand which is less well-sorted than the strata in Unit 4. There are numerous fragments of phyllite and quartz-schist, commonly angular and in places shard-like. Many of the coarser sand-size quartz grains are broken although some are well-rounded. Larger cobbles and boulders occur less commonly, together with wisps and lenses of gravel. Bedding is poorly developed. Lithologies similar to that of Unit 5 are exposed at the sewage intake adjacent to the road (Fig. 2) and also occur below thin modern gravels on the lower foreshore west of Pit B.

FAUNA

The faunal identifications from the two pits and the auger hole are summarized in Appendices 1-6. In general the fauna is similar to that recorded by Crosskey and Robertson (1871) and Brady *et al.* (1874), but significant differences are apparent among the lithological units. No fauna was recovered from Unit 1. The ostracoda recorded from Pit A were similar in species and distribution to those in the more detailed record from the auger hole, and are not shown in the Appendices.

The methods of presenting the faunal data in Appendices 1-6 is a compromise because of different techniques adopted during the course of the investigations. For the mollusca a semi-quantitative approach has been used because of the great size range in the phylum (2 to 100 mm in the taxa studied) and the limited weight of the sediment samples which could be collected and processed. A somewhat more quantitative approach has been used for the foraminifera and ostracoda, based on numbers per 100 g of sediment. No allowance has been made for juveniles in the count of ostracoda where up to 5 moult stages with both valves present may be preserved. The latter, if present, are of environmental significance, but might give a misleading importance to the percentage of an individual species in a particular table. Interpretations based on percentages or absolute figures of microfauna must allow for these qualifications as well as for environmental aspects such as the effects of bioturbation and sorting by bottom currents.

Unit 2. The molluscan species in the gravel are either adults of small forms, or juveniles and young adults of larger animals. Most are eurythermal species to be found in a variety of water depths at the present day in British seas, but *Trophon clathratus** and *Thyasira gouldi* are cold water forms. The foraminifera are dominated by *Cibicides lobatulus*, a cosmopolitan form distributed where shelter can be found in weeds, though apparently not under estuarine conditions (Haynes, 1973). Although generally found in shallow water, this species has been recorded at depths of up to 80 m on the Grand Banks of Newfoundland (Sen Gupta and McMullen, 1969; Sen Gupta, 1971). Other foraminiferal species present are epineritic forms with a boreal to subarctic distribution. There is only a sparse fauna of ostracods, but low water temperatures may be inferred from the occurrence of *Cytheropteron arcticum* (Neale and Howe, 1973).

Unit 3. This, the *Modiolus* bed, is characterized by a greater diversity of molluscan species than Unit 2. It is dominated by *Modiolus modiolus*, an epifaunal species in position of growth, accompanied by burrowing forms such as *Macoma calcarea*, *Nuculana pernula* and *Thyasira gouldi*. The last three are northern bivalves absent from the British list though locally present in the eastern North Sea. *Abra alba* however reaches its northernmost station near Lofoten. The relatively high Miliolina content of the foraminifera indicates open marine conditions with salinities in excess of 30‰. Shallow water (less than 20 m on comparison with modern boreal assemblages) is suggested by the high proportion of epineritic species such as *Cibicides lobatulus*, *Elphidium clavatum*, *Quinqueloculina lata* and *Q. seminulum*. The dominance of the arctic to subarctic form *E. clavatum* indicates low water temperatures. Open sea conditions and low water temperatures are the conclusions to be drawn from

---

*Authors of species are given in the Appendices.

the ostracod fauna which includes *Acanthocythereis dunelmensis*, *Elofsonella concinna* and *Eucytheridea punctillata*, all species which have a northern or deep water distribution at the present day, and normally make up an important part of the 'Clyde bed' fauna. Diversity of species of ostracods is greatest for this unit of the succession.

Unit 4. The greater variety of mollusca collected from this unit compared with that from the underlying beds is probably partly a function of the relatively large number of samples examined and partly a function of bottom conditions (see below). Infaunal *Macoma* and *Thyasira* are numerous and *Arctica islandica* is characteristic, occurring either as isolated individuals or in horizons with numerous fully grown individuals. The *Portlandia* spp. are cold water bivalves generally living at depths greater than 20 m offshore in Northern Norway, but *P. fraterna* (as *P. frigida* Torell) may extend southwards along the Norwegian coast (Sars, 1878; Ockelman, 1958). The number of species and individuals declines upwards, and the *Portlandia* spp. tend to be restricted more to the lower part of the unit, though locally occurring at the Unit 4/5 boundary (Sample L, Pit B). An environment similar to that of Unit 3 in terms of temperature and depth is suggested by the foraminifera at the base of this unit, although salinities appear to be somewhat reduced. The assemblage is dominated by *Elphidium clavatum*, *E. incertum* and *E. Williamsoni*. Miliolina are greatly reduced, a feature compatible with salinities below 30‰ (Murray, 1968). More estuarine, hyposaline conditions are therefore indicated especially in the upper part. The ostracod fauna likewise indicates cool-to-cold conditions, without any particularly cold, stenothermal species. In the auger hole samples, the ostracod fauna of the *Modiolus* bed (Unit 3) apparently continues into the lower part of this unit, at which point the species diversity declines. This coincides with the appearance of *Patagonacythere dubia* (southern limit Faroes-Shetlands today). Generally the changes in the ostracod fauna occur a little higher in the succession than those observed in the foraminifera and mollusca, the last mentioned changes coinciding with the onset of sandier deposits.

Unit 5. In Pit A there is a sharp drop both in diversity and numbers of macrofauna above Unit 4. *Hiatella arctica*, *Macoma calcarea* and *Mya truncata* are known from field observation to continue to the top of the succession, though too sparsely to be represented in all the samples. A better representation of the mollusca is given by those recovered in Pit B, where, omitting Sample C, the dominant species are infaunal *Macoma calcarea*, *Mya truncata* and *Thyasira gouldi*. *Arctica islandica* is also well represented at one horizon. The foraminifera suggest that the shallowing and the tendency towards estuarine conditions, apparent throughout Unit 4 is continued in this stratum. *Elphidium williamsoni*, an estuarine species (Haynes, 1973) dominates, with subordinate *E. clavatum* and *E. incertum*, and the assemblage as a whole suggests hyposaline conditions with depths of less than 20 m. The sharp reduction in numbers and diversity of species in the ostracod fauna, coupled with a high proportion of juveniles is characteristic of periods of reduced salinity or increased cold. Taken as a whole the faunal evidence suggests the onset of more severe conditions and a shallow-water estuarine environment with a water depth of only a few metres below low tide level.

CONDITIONS OF DEPOSITION

The gravel (Unit 2) at the base of the marine succession appears to lie unconformably on the underlying glacial deposit (Unit 1). It was clearly laid down at a period of relatively strong bottom currents, and largely for that reason, proved an unsuitable substratum for the kind of infauna and epifauna which was to characterize later deposits. Most of the shells which do occur appear to have been derived, but as there is little evidence of abrasion, the distances travelled appear to have been small.

The *Modiolus* bed, Unit 3, is a fossil equivalent of the offshore epifaunal *Modiolus* communities, widespread around the British Isles at the present day (*e.g.* Tebble, 1966, p.43) and Iceland (Spärck, 1937). Such beds commonly rest on a foundation of gravel, and occur where sediment accumulation is minimal. Currents of a strength sufficient only to keep the bottom clear of fine sediment are implied.

The silty clays of Unit 4 contain the most diverse and abundant fauna. Many of the species are either in or near the interpreted position in life, and have been disturbed only by general organic bioturbation. This impression of an *in situ* fauna established in quiet water conditions is reinforced by the completeness of the record of ostracod moult stages in the samples. With any measure of current action, the small juvenile moult stages of ostracods are quickly separated from the full-grown adult valves. Most samples from Unit 4 yield full growth series of valves. Sedimentation was probably intermittent, being slow enough at times to allow the growth of *Arctica islandica*, which,as a surface living bivalve virtually incapable of locomotion, can only live where sedimentation rates are low (Schäfer, 1972, p.376). Shallowing of the water is indicated by the upward disappearance of *Portlandia*, which here were in any case probably close to their minimum depth limit. The decrease in diversity of all species suggests a harsher, more stringent environment, for which additional evidence may lie in the scattered gravel-size clasts being rafted erratics, brought in by winter sea-ice (see below).

The transition to the estuarine environment of Unit 5 is particularly clear in the lithological log of the auger hole, being marked by the increase in sand between 140 and 120 cm. A constraint in the hyposaline conditions suggested by the faunal evidence is provided by the stable carbon isotope ratios of the dated shells (see below) which indicate that salinities did not diverge greatly from those of a normal marine environment (*cf.* Keith *et al.*, 1964). The intermittent nature of the sedimentation in this unit is indicated by the occasional lenses of gravel which break the sequence and by the presence at other times of *Arctica* with the implications set out above. Although there is no clear incoming of cold stenothermal species, the low diversity of animal life, together with the physical evidence of fractured quartz grains and shard-like clasts, suggests a cold environment with much winter frost. The numerous scattered clasts set in the silty sandy matrix were probably dropped by sea-ice which was sufficiently thick to transport boulders and pebbles from the adjacent shores. However, it is possible that this unfavourable environment resulted in part from the shallowing of the water in Loch Gilp, and it is interesting in this connection to note the presence in Unit 5 of *Ammonia batavus*, a 'warm' indicator which could reflect summer warming near the shore.

Throughout the marine sequence the faunal evidence combines to suggest water temperatures considerably lower than those prevailing at present. Similar fauna could well occur on the Lofoten coast where several arctic species reach their southernmost limit, to overlap with boreal species reaching their northernmost stations. In this connection, however, it is of interest that *Portlandia (Y.) lenticula* occurs at Lofoten (its southern limit) at a depth of 40 m (Brögger, 1900-01), whereas at Lochgilphead its postulated depth would have been about 20 m or less (see below), a depth in keeping with its arctic stations today.

No typical tidal flat sedimentary sequences have been identified in the upper parts of the succession (there being no laminated silts and muds which would be the prevailing conditions at the present day) and the identification of the estuarine environment is principally based on the fauna. Such sequences, however, would have been vulnerable to erosion during the events which led to the development of the modern beaches.

INTERPRETATION OF THE SUCCESSION

Within the area under consideration, the highest raised beaches formed on the retreat of the ice from the Late Devensian limit are thought to reach 36 m O.D. at nearby Otter Ferry (D. G. Sutherland *in* Sissons, 1974c, p.329) and 35 - 38 m O.D. in the area east of Loch Crinan (Gray and Sutherland, this volume). At this level the two cols between Lochgilphead and the Sound of Jura would be submerged by 10 - 15 m of water. Thereafter, on the assumption of a general fall in relative sea-level during the Lateglacial Interstadial, now partly documented by radiocarbon dates (Bishop and Dickson, 1970), the cols would emerge and Kintyre would be joined to the mainland. At the present day there are significant differences in the time and height of tides between the Sound of Jura and the middle reaches of Loch Fyne on the other side of the peninsula, and at the highest sea-levels the cols and areas nearby would have been subjected to strong tidal currents. Under these conditions little or no sediment would have been deposited other than gravel, and erosion would have made unconformable relationships highly probable. Thereafter, with falling sea-levels, there would have been first, a reduced, and second, no tidal flow across the cols. In this last situation there would be little to disturb the bottom with sea-levels at about 20 m O.D. Truly shallow water conditions would lead to a return of tidal scour. Overall, the evidence preserved in the Lochgilphead succession is consistent with this interpretation of events.

It is clear from the present study and from Crosskey and Robertson (1871) that there is a sharp break at the base of the Lateglacial marine strata in the Lochgilphead area, and that the laminated silts and clays so often preserved between the till and the overlying shelly beds are absent. From a study of exposures of the laminated beds elsewhere in the Clyde it seems likely that these are the deposits of bottom traction currents and of turbid surface meltwater plumes in the neighbourhood of glacier fronts (*cf.* Crosskey, 1867, p.51), and their absence at Lochgilphead suggests that they were either not laid down or were removed before the deposition of the faunally rich marine beds. It can therefore be concluded that marine deposition began after, perhaps well after glacier ice had disappeared from the immediate neighbourhood. In Pit B the only record of this period of highest sea-level is the shelly gravel at the base of the succession (Unit 2), which certainly dates from well before 12,360 B.P., the first reliable radiocarbon date (Pit A).

The decrease in current strength through the Sound of Jura to Loch Gilp gap is indicated by the *Modiolus* bed (Unit 3) which on geological grounds was probably laid down when sea-level was a little above the height of the cols, *i.e.* about 25 - 30 m O.D. Unit 4, laid down in quiet water of moderate depth, clearly illustrates the conditions to be expected at Lochgilphead once tidal flow across the cols had diminished and finally ceased. Sea-level would have been 20 - 25 m O.D. when the lower part of the bed was being deposited. As sea-level fell towards 6 - 8 m O.D. a large area of ground then underlain by till and marine deposits would have been exposed to wave and subaerial erosion in the valley of the Baden Burn north of Lochgilphead, and sediment would have been carried seawards by stream and tidal currents. This phase is presumably recorded by the coarsening-upward sequence at the junction of Units 4 and 5 and by the accompanying faunal change from an offshore to an estuarine assemblage. Sea-level at this boundary was therefore probably near 10 m O.D. There is no evidence of a change to littoral conditions in Unit 5, and since the top of the deposit (in Pit A) is at approximately 1.5 m O.D. it seems unlikely that mean sea-level ever fell below 4 m O.D. over the time-interval represented by Unit 5, assuming that the tidal range in Loch Fyne was the same during Lateglacial times as it is at present (about 3 m).

DATING OF THE DEPOSITS

One sample (SRR-63) of marine shells was dated from Pit A, and a further seven from Pit B. The positions from which these samples were taken are shown against the sections of Pits A and B in Fig. 3 and are listed in Table 1 (data from Harkness and Wilson, 1974, p.240, and in preparation).

In Table 1 SRR-364 and 366 refer to the same horizon. Some difficulty was experienced in obtaining a correct level for SRR-365, which was in fact taken from a slightly lower horizon than 364 and 366. There is, however, good agreement between the dates for the three species and the 0.5 m level in Pit B is therefore dated with confidence at about 11,100 B.P. SRR-367, though only 10 - 15 cm lower in the sequence, yields a date some 1,000 years older. Although there is nothing to indicate that it is incorrect in laboratory terms, the value appears to be too high.

Both its general position in the section and the discrepancy between the 'outer' and 'inner' values indicate that the date obtained for SRR-368 from the Unit 4/5 boundary is too high. In contrast, the second sample dated from the Unit 4/5 boundary (SRR-489) yields virtually identical 'inner' and 'outer' dates and generally accords with the other dates from the sequence. For this reason the latter is accepted as reliable, and the Unit 4/5 boundary is accordingly dated at 12,100 B.P. The *Arctica* sample SRR-63 was composed of well-preserved uneroded, entire valves with the periostracum preserved, and the date, which is in keeping with its position in the section, is regarded as reliable.

The dates from the *Modiolus* sample from Unit 3 (SRR-369) show a discrepancy between the 'outer' and 'inner' values and this material may have been contaminated by groundwater

TABLE 1 Marine shell radiocarbon dates from Lochgilphead

Lab No.	Sample and depth	Date	Stable isotope ratio (δC13‰)
SRR-63	*Arctica islandica*	12,360 ± 85 B.P.	- 0.8
	1.6 m depth		
SRR-364	*Modiolus modiolus*	11,279 ± 125 B.P. (outer)	0.0
	0.6 m depth	10,974 ± 110 B.P. (inner)	+ 0.3
SRR-365	*Mya truncata*	11,694 ± 130 B.P. (outer)	+ 0.5
	0.55 to 0.60 m	11,313 ± 140 B.P. (inner)	+ 0.5
SRR-366	*Arctica islandica*	10,962 ± 115 B.P. (outer)	+ 1.0
	0.6 m depth	11,196 ± 130 B.P. (inner)	+ 0.6
SRR-367	*Macoma calcarea*	12,092 ± 190 B.P.	- 0.7
	0.75 m depth		
SRR-368	*Arctica islandica*	13,187 ± 85 B.P. (outer)	+ 1.1
	1.05 m depth	12,765 ± 85 B.P. (inner)	+ 1.0
SRR-369	*Modiolus modiolus*	12,729 ± 85 B.P. (outer)	+ 0.1
	1.7 m depth	12,210 ± 100 B.P. (inner)	+ 1.8
SRR-489	*Arctica islandica*	12,151 ± 60 B.P. (outer)	+ 1.6
	(as SRR-368)	12,066 ± 70 B.P. (inner)	+ 1.8

All dates are calculated to δC14. No subtraction is required for apparent age.

known to be present in Unit 2. The δC13 values are also less consistent than for the other shells. For these reasons no reliable date for the *Modiolus* bed can be given, but by extrapolation this could well have been laid down about 12,500 B.P.

No dates are available for Unit 2, which spans the time between deglaciation and the formation of the *Modiolus* bed, but on the basis of data summarised by Sissons (1974c) the former event probably took place about 13,000 B.P., at which time the highest raised beach in the area (about 36 m O.D.) was formed. These conclusions are in agreement with marine shell radiocarbon dates of about 12,600 B.P. on samples collected from about 15 m O.D. near Paisley (Welin *et al.*, 1972, p.141).

To summarize, Unit 2 is considered to have been laid down between 13,000 and 12,500 B.P., Unit 3 about 12,500 B.P., Unit 4 from roughly 12,500 B.P. to 12,100 B.P., and Unit 5 from 12,100 B.P. until some time after 11,100 B.P.

DISCUSSION AND CONCLUSIONS

Relative Sea-level Curve for the Lochgilphead Area

Combining the evidence for relative sea-level with the dates of deposition of Units 2 to 5 it is inferred that sea-level was about 25 - 30 m O.D. at 12,500 B.P. falling to about 10 m O.D. at 12,100 B.P. and remaining between this level and 4 m O.D. until some time after 11,100 B.P. (Fig. 4). This evidence is consistent with that at Garscadden Mains a short distance west of Glasgow, where it is known from stratigraphy and pollen analytical data that sea-level was below 22 m at the Jessen-Godwin pollen zone I/II boundary (Mitchell, 1952), usually dated about 11,950 B.P. (West, 1968). It apparently disagrees with that from Oban, however, where plates of the sub-littoral barnacle *Balanus balanus* found up to an altitude of 20.5 m O.D. by Gray (1974b) yielded much lower ages (Shotton and Williams, 1973, p.459) but with a considerable divergence between 'inner', 'middle' and 'outer' values. Such apparent disagreement could be caused either by a genuine difference in height reflecting the trend of the isobases or, more likely, by errors in the radiocarbon dates from Oban (Gray, 1974b). The abrupt flattening of the curve at 12,000 B.P., if valid, suggests the possibility of a major eustatic transgression beginning at this date (*cf.* Emiliani *et al.*, 1975) for areas little affected by isostatic uplift.

Palaeotemperature

The depositional history of Lateglacial marine deposits, temporarily exposed on the foreshore near Lochgilphead, records changes in the local marine environment over a time span of about 1,500 radiocarbon years. During this period, which ended at about 11,000 B.P., sea-water temperatures were consistently lower than at present, and for the 12,000 - 11,000 B.P. interval, high boreal to low-arctic conditions are inferred.

The Lochgilphead sequence is the first Lateglacial marine succession in Scotland to be adequately dated, and as such cannot be directly compared with other areas. Undated, or inadequately dated thicker sequences of Lateglacial marine strata have been reported in the Colonsay area (Binns *et al.*,

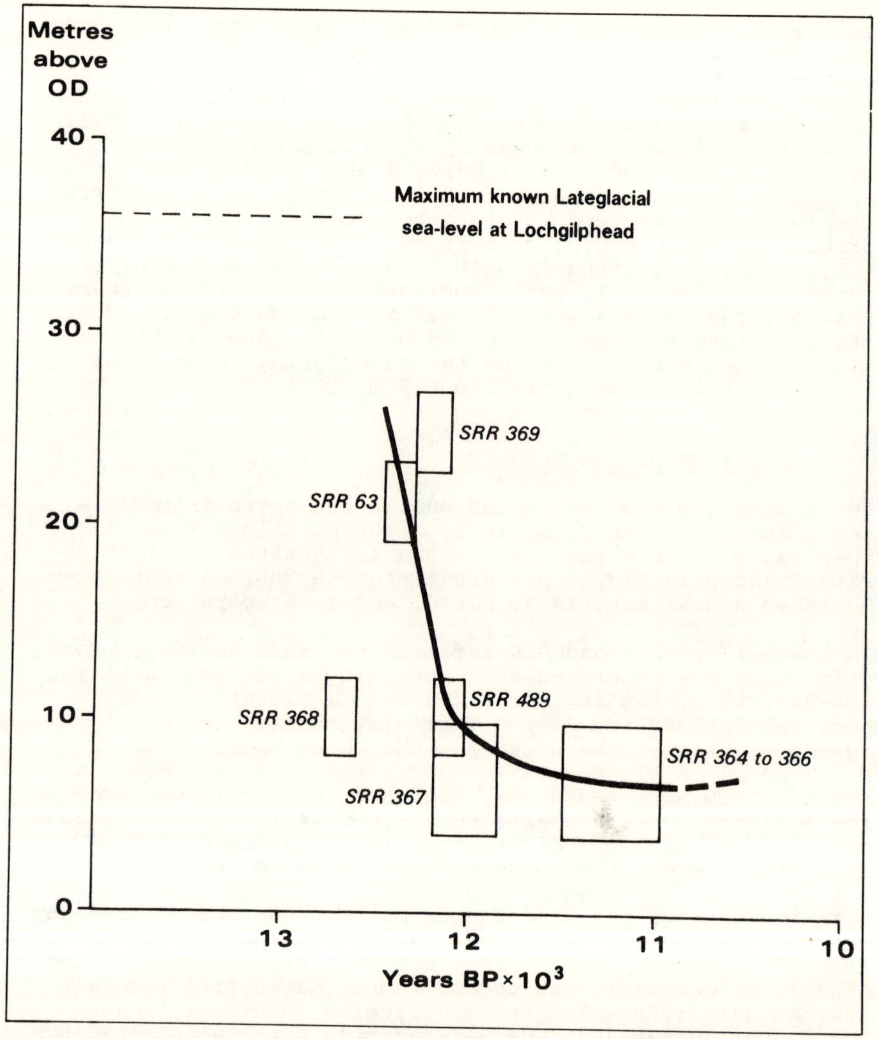

Fig. 4. Lateglacial sea-level curve for the Lochgilphead area. Notional dates for the curve shown by boxes. SRR-63 *etc.* refer to C14 dated samples (Fig. 3).

1974), and in the Cromarty Firth (Peacock, 1974). At Cromarty, the evidence from the Lateglacial part of a borehole sequence indicates that water temperatures were lower than at present throughout the period following the disappearance of glacier ice. From the more complete unpublished faunal list now available for the Colonsay Borehole (I.G.S. No. 71/9), it is now clear that water temperatures were likewise below present values for the whole of the Lateglacial part of the sequence (*cf.* Ruddiman and McIntyre, 1973).

Comparison of the Lochgilphead and other Lateglacial marine data with the palaeontologically derived climatic curves from Lateglacial sites in Shropshire and North Wales (Osborne, 1972; Coope and Brophy, 1972) shows some interesting similarities and dissimilarities. From these studies it is thought that temperate conditions with July air temperatures at or above present values obtained in parts of southern Britain from roughly 13,000 B.P. to 12,000 B.P., that temperatures fell in the 12,000 to 11,000 B.P. interval, and that they were much lower thereafter. The climatic curve for the period post-dating 12,000 B.P. fits the Lochgilphead data well. However, the faunas from sites in the Clyde estuary for which dates as early as 12,600 B.P. have been obtained, such as those from Paisley (Welin *et al.*, 1972), are cool water faunas similar to those at the base of the Lochgilphead strata, and thus water temperatures lower than at present must be inferred for the period 13,000 to 12,000 B.P. on the available evidence for the Clyde sea-area, i.e. for the period following retreat of glacier ice from the district about 13,000 B.P. This agrees with the data from the Colonsay and Cromarty boreholes mentioned above, which provide a seemingly continuous record of Lateglacial events following ice retreat. Since many marine animals are temporary components of the plankton in the early larval stages of their lives they may be expected to be capable of rapid dispersal and migration following temperature changes. The record of temperature inferred from their remains, particularly the planktotrophic forms, thus probably reflects actual conditions fairly closely. It can therefore be concluded that offshore summer and winter sea-temperatures around the Scottish coast between the Clyde and Cromarty were at best several degrees lower during the Lateglacial Interstadial than at present. This suggests the climate of Scotland, particularly of Highland areas, was probably more severe than at present even during the early part of the Lateglacial Interstadial (*ca.* 13,000 to 12,000 B.P.). The contrast between early Lateglacial Interstadial temperatures in England and Scotland implied by the two lines of evidence may be real but could on the other hand be the result of insufficient knowledge of Lateglacial environments.

ACKNOWLEDGEMENTS

The writers are indebted to colleagues in Edinburgh, Leeds and London for critical discussion of the manuscript through several drafts. The paper is published with the permission of the Director of the Institute of Geological Sciences.

APPENDIX 1. LOCHGILPHEAD AUGER HOLE

MOLLUSCA	Sample positions (see Fig. 3)														
	1	2	3	4	5	6	7	8	9	10	11	12	13	14	15
	no molluscs found	no molluscs found	no molluscs found	no molluscs found	not sampled for molluscs									not sampled for molluscs	
Littorina littoralis (Linné)											□		▣		
Littorina sp.															□
Natica pallida Broderip & Sowerby															
Natica sp.										□					
Omalogyra atomus (Philippi)												□			
Retusa obtusa (Montagu)												□			□
Rissoa parva (da Costa)												■			
Skeneopsis planorbis (Fabricius)												□			
Abra sp.						□									
Astarte montagui (Dillwyn)															▣
Macoma calcarea (Chemnitz)						■		▣	▣	■	▣	■	▣		▣
Mytilus sp.							□	□		□					
Nucula sp.															□
Nuculana pernula (Müller)															▣
Parvicardium sp.						□	■	□							
Portlandia (Yoldiella) lenticula (Möller)												□			▣
Thyasira gouldi (Philippi)												▣			▣

□ present

▣ several

■ abundant

APPENDIX 2. LOCHGILPHEAD AUGER HOLE

OSTRACODA — Sample positions (see Fig. 3)

OSTRACODA	1	2	3	4	5	6	7	8	9	10	11	12	13	14	15
Acanthocythereis dunelmensis (Norman)	no ostracods found										+	+	not sampled for ostracods	+	◎
Cythere lutea (Müller)		◎	+	+	○	○		+	○	○	◎	◎		⊙	⊙
Cytheropteron latissimum Norman														◎	●
Cytherura undata Sars														○	○
Elofsonella concinna (Jones)		○				○			●	●	●	●		●	●
Eucytheridea bradii (Norman)			+		+	◎	○	○	⊙	⊙	●	●		●	●
E. punctillata (Brady)							○	+	◎	●	⊙	●		⊙	●
Hemicythere villosa (Sars)						○		+				+		+	
Hirschmannia tamarindus (Jones)														+	○
H. viridis (Müller)														○	
Jonesia simplex (Norman)															+
Leptocythere pellucida (Baird)						◎	◎	◎	◎	⊙	⊙	○			
Loxoconcha granulata (Sars)														○	○
Patagonacythere dubia (Brady)						◎	◎	◎	●	●	⊙	●			◎
Robertsonites tuberculata (Sars)						⊙	⊙	◎	⊙	○	○	◎		○	○

Numbers of ostracods in each sample. (All samples *c.* 100 g)

- + 1
- ○ 2 - 5
- ◎ 6 - 10
- ⊙ 11 - 20
- ● > 20

APPENDIX 3. LOCHGILPHEAD PIT A

MOLLUSCA — Sample positions (see Fig. 3)

MOLLUSCA	a	b	c	d	e	f	g†	h
Aporrhais pespelecani (Linné)			no molluscs found	no molluscs found			□	
Cingula semicostata (Montagu)								□
Littorina littorea (Linné)						□		
L. littoralis (Linné)							□	
L. saxatilis (Olivi)							□	
Natica pallida Broderip & Sowerby						□	▣	
Natica sp.								□
Oenopota turricula (Montagu)							▣	
Omalogyra atomus (Philippi)						▣	□	
Retusa obtusa (Montagu)						▣		▣
Skeneopsis planorbis (Fabricius)						▣	▣	
Abra sp.						□		
Acanthocardia echinata (Linné)								□
Arctica islandica (Linné)							▣	□
Chlamys varia (Linné)							□	
Crenella sp.							□	
Hiatella arctica (Linné)	□							
Macoma calcarea (Chemnitz)					▣	■	■	■
Mya Truncata (Linné)	□	□				□	■	
Mytilus edulis (Linné)						□		□
Nucula sp.							▣	
Nuculana pernula (Müller)							▣	
Parvicardium scabrum (Philippi)							□	
Portlandia (Yoldiella) fraterna (Verrill & Bush)							□	
P. (Y.) lenticula (Möller)							▣	
Thyasira gouldi (Philippi)						■	□	

- □ present
- ▣ several
- ■ abundant

† Large sample

APPENDIX 4. LOCHGILPHEAD PIT B

MOLLUSCA

MOLLUSCA	Sample positions (see Fig. 3)											
	H	I	J	K	A	B	L	C	D	E	F	G
Aporrhais pespelecani (Linné)									□			
Buccinum undatum Linné							□			□		
Cingula semicostata (Montagu)							⊡			■		□
Gibbula sp.							□					
Lacuna vincta (Montagu)							□			□		□
Lacuna sp.										□		
Littorina littoralis (Linné)		□										
L. littorea (Linné)		□						⊡				
Natica pallida (Broderip & Sowerby)		□		□		□	□					
Natica sp.								⊡				
Neptunea ?							□					
Odostomia unidentata (Montagu)							□					
Oenopota turricula (Montagu)												□
Omalogyra atomus (Philippi)										⊡		
Puncturella noachina (Linné)												□
Retusa obtusa (Montagu)							□			■		
Rissoa parva (da Costa)								□				
Trophon clathratus (Linné)											□	■
Abra alba (Wood)										■	□	
Arctica islandica (Linné)							⊡					
Astarte elliptica (Brown)	⊡											
A. montagui (Dillwyn)						⊡						
A. sulcata (da Costa)		□										
Astarte sp.						□	□			□		
Chlamys islandica (Müller)		⊡									□	
Heteranomia squamula squamula (Linné)											□	⊡
H. squamula aculeata (Müller)											□	⊡
Hiatella arctica (Linné)						□					□	■
Macoma calcarea (Chemnitz)	⊡	■			□	■	■	⊡	⊡		□	
Macoma sp.	■											
Modiolus modiolus (Linné)							□				□	
Monia patelliformis (Linné)												⊡
M. squama (Gmelin)				□								
Mya arenaria Linné							□					
M. truncata Linné	□	⊡	⊡								□	
Mya sp.	□	⊡					□			□		
Mytilus edulis Linné										□	□	
Nucula tenuis (Montagu)							□			⊡	□	
Nucula sp.										□		
Nuculana minuta (Müller)							⊡			⊡	⊡	
N. pernula (Müller)							■			⊡		
Ostrea edulis Linné												
Parvicardium ovale (Sowerby)		⊡			□							
Portlandia (Yoldiella) fraterna (Verrill & Bush)							⊡			⊡		
P. (Y.) lenticula (Möller)							⊡			■		
Portlandia sp.							□					
Spisula elliptica (Brown)	□											
Thyasira cf. gouldi (Philippi)	⊡	■				■	■			■	⊡	
Thyasira sp.					⊡	□				■	⊡	

□ present

⊡ several

■ abundant

APPENDIX 5. LOCHGILPHEAD PIT B

FORAMINIFERA	H	I	J	K	A	B	L	C	D	E	F	G
Ammonia batavus (Hofker)	○		not sampled for foraminifera	not sampled for foraminifera					not sampled for foraminifera			
Bucella frigida (Cushman)										○		○
Bulimina marginata d'Orbigny										○		
Cassidulina crassa d'Orbigny											○	
Cibicides lobatulus (Walker & Jacob)							⦿			◎	⦿	⦿
Elphidium askluni Brotzen	○											
E. bartletti Cushman	◎				◎			◎				
E. clavatum Cushman	○	○			⦿	⦿	●	●		⦿	⦿	○
E. incertum Williamson	⦿	⦿			⦿	⦿	●	⦿		◎		
E. subarcticum Cushman							◎					
E. williamsoni Haynes	⦿	⦿			⦿	⦿	⦿	⦿		◎	○	
Fissurina cf. cucurbitasema Loeblich & Tappan										○	○	
F. marginata (Montagu)							◎					
Guttulina lactea (Walker & Jacob)		○										
Lagena mollis Cushman										◎		
L. semilineata Wright		○										
L. substriata Williamson							◎					
Laryngosigma hyalascida Loeblich & Tappan										○		
Miliammina fusca (Brady)	○											
Pateoris hauerinoides (Rhumbler)											◎	○
Protelphidium anglicum Murray	○	◎			⦿			◎				
P. orbiculare (Brady)							⦿	⦿				
Pyrgo depressa (d'Orbigny)										⦿	○	
P. williamsoni (Silvestri)											⦿	○
Quinqueloculina agglutinata Cushman								◎		◎	◎	
Q. lata Terquem										◎	⦿	+
Q. seminulum (Linné)		○				○	⦿	⦿		⦿	◎	○
Quinqueloculina sp.										○		+
Rosalina sp.												+

Sample positions (see Fig. 3)

Numbers of foraminifera in 100 g: + 1 - 5; ○ 6 - 20; ◎ 21 - 100; ⦿ 101 - 1000; ● > 1000

APPENDIX 6. LOCHGILPHEAD PIT B

OSTRACODA	H	I	J	K	A	B	L	C	D	E	F	G
Acanthocythereis dunelmensis (Norman)			not sampled for ostracods	not sampled for ostracods			⦿		not sampled for ostracods	◎		
Argilloecia conoidea Sars											+	
Cythere lutea (Müller)	+				+		⦿	◎		◎	○	+
Cytheropteron cf. arcticum Neale & Howe												
C. latissimum Norman							◎			●	+	
Cytheropteron sp.							◎					+
Elofsonella concinna (Jones)							⦿	⦿		●	+	
Eucytheridea bradii (Norman)					+	+	●	◎		●	○	+
E. punctillata (Brady)	+	◎				○	⦿	⦿		⦿		
Finmarchinella finmarchia (Sars)	+	◎										
Finmarchinella ?					+							
Hemicythere villosa (Sars)						+	◎			⦿	+	
Hirschmannia tamarindus (Jones)							◎			⦿	○	
H. viridis (Müller)							◎				+	
Jonesia simplex (Norman)										⦿		
Leptocythere castanea Sars	+											
L. pellucida (Baird)		◎			○							
Leptocythere sp.										◎	+	
Paradoxostoma sp.											+	
Patagonacythere dubia (Brady)	+	⦿			+	◎	◎	◎		⦿	○	
Pseudocythere caudata Sars												+
Robertsonites tuberculata (Sars)		⦿			+	○					+	
Sclerochilus contortus Norman						+					○	+
Semicytherura striata (Sars)		◎										
S. nigrescens (Baird)										◎	+	+

Sample positions (see Fig. 3)

Numbers of ostracods in 100 g: + 1 - 5; ○ 6 - 20; ◎ 21 - 100; ⦿ 101 - 1000; ● > 1000

The Reconstruction of the Lateglacial Environment in the Southern and Eastern Grampian Highlands

J. J. LOWE (City of London Polytechnic)

and

M. J. C. WALKER (St. David's University College, Lampeter)

ABSTRACT

This paper describes the results of pollen-stratigraphic investigations at six Lateglacial sites in the southern and eastern Grampian Highlands. Detailed pollen diagrams are presented, along with associated litho-stratigraphic data, results of analyses of deteriorated pollen, and radiocarbon dates from critical horizons. Following a discussion of problems encountered in the interpretation of these records the palynological, sedimentological and geomorphological evidence is used to reconstruct environmental conditions in this part of Scotland during the Lateglacial period.

INTRODUCTION

The post-war years have witnessed a rapidly growing interest in the Lateglacial landscape history of the Scottish Highlands with the publication of a large number of pollen diagrams from upland sites (Gray and Lowe, this volume, Fig. 2). The distribution of these sites, however, is very uneven, and there are still extensive areas of the Highlands where vegetational and environmental changes during the Lateglacial period are poorly documented. This paper presents the results from such a region, for prior to this study only one Lateglacial pollen diagram had been published from the area (Donner, 1958), and no data was available from Loch Lomond Stadial (Jessen-Godwin zone III) deposits.

The area under consideration lies on the south-east facing slopes of the Grampian Highlands. It is dominated in the north and west by an undulating plateau between 600 and 800 m in altitude which discordantly truncates outcrops of metamorphic and igneous strata. In the south and east the landscape becomes less rugged as the valleys broaden out beyond the Highland Boundary Fault towards the wide plain of Strathmore and the lowland valleys of Strathearn, the Teith and the Forth. All the major valleys have been glacially-overdeepened and cut back sharply into the uplands to depths of 700 m in places. The upper slopes of the glens are characterised by large areas of ice-scoured bedrock, while the valley floors are masked by variable thicknesses of till and fluvioglacial sediments.

The present vegetation pattern of the Grampians has been described by McVean and Ratcliffe (1962) and Burnett (1964), and is composed largely of acid grassland and heathland on the middle and lower slopes, with occasional copses of willow and tree birch on the valley floors. Heather moor, blanket bog, and moss heath cover the upper mountain slopes.

Within this region six Lateglacial pollen sites were investigated (Fig. 1). Two sites are located in the north-east of the area, Blackness in Glen Esk (NO/463786), and Roineach Mhor in Glen Clova (NO/331728). Both of these sites have been described in detail elsewhere (Walker, 1975a). A further two sites were found in the Teith Valley near Callander, these being Tynaspirit (NN/667047) and Cambusbeg (NN/658051). A fifth site, high on the south-facing slopes of the Grampian massif is at Tirinie in Glen Fender (NN/889678), while the sixth site is located on the bleak moorland to the south of Aberfeldy at Amulree (NN/895338).

The paper is essentially in two parts. The first describes the lithostratigraphic and biostratigraphic characteristics of each profile, and assesses the difficulties encountered in the interpretation of these records. The results of

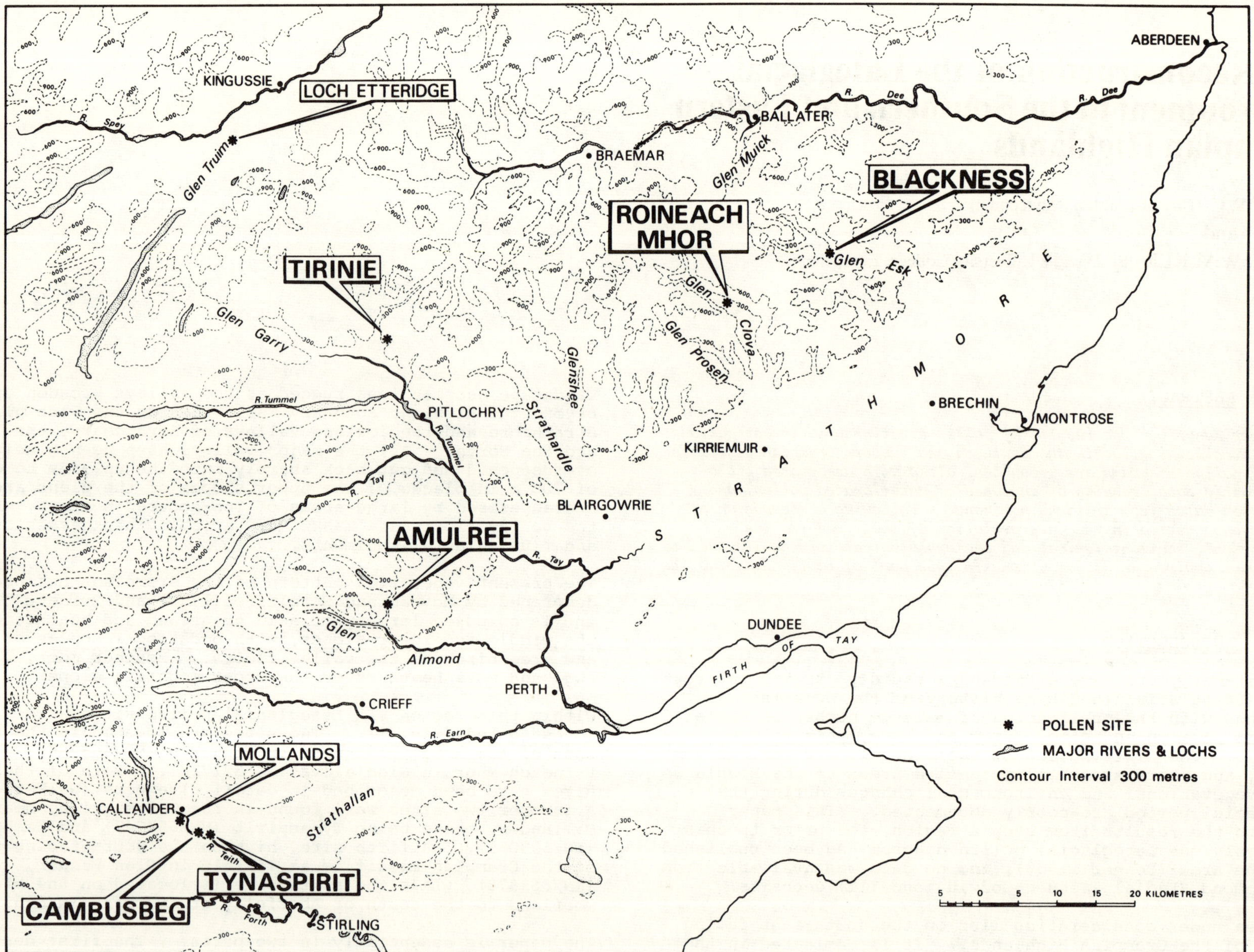

Fig. 1. Location Map

deteriorated pollen analyses and of radiocarbon dating of critical horizons are also discussed. The second part draws together the relevant information from the six sites and presents an overall impression of environmental change and landscape evolution in the southern and eastern Grampian Highlands during the Lateglacial period.

LITHOSTRATIGRAPHY

The stratigraphy at the deepest point in each of the six basins is shown in Fig. 2. Samples from Blackness, Cambusbeg, Amulree 1 and Tynaspirit 1 were obtained by Hiller borer, from Roineach Mhor, Amulree 2 and Tirinie by Abbey piston sampler, and from Tynaspirit 2 by modified Livingstone piston sampler. At each site a number of test bores were first made to establish the deepest point, and in most cases downward progress of the sampling chamber was eventually arrested by a layer of gravel. Only at Amulree is a basal gravel layer not recorded; further penetration there is prevented by an extremely stiff laminated clay.

In most basins in Scotland where sediments of Lateglacial age are present the infill consists of a considerable thickness of Flandrian peats and lake muds underlain by a three-fold sequence of minerogenic, organic, and basal minerogenic sediments (Sissons *et al.*, 1973). This lithostratigraphic sequence has in the past been equated with the Jessen-Godwin pollen zones I, II and III (see below). A three-fold sequence of Lateglacial deposits is present at Blackness, Tirinie and Tynaspirit. At Blackness a basal layer of coarse gravel is overlain by sands, silts and eventually by a brown organic mud (gyttja) which is over 60 cm thick at the sampling point (Walker, 1975a). The organic sediments are in turn overlain by silty-clays and sands. At Tirinie a basal gravel layer is overlain by 17 cm of shell marl, above which are 46 cm of clays and silts. The basal sediments at Tynaspirit consist of a thin organic-rich layer overlain and underlain by compact grey clays.

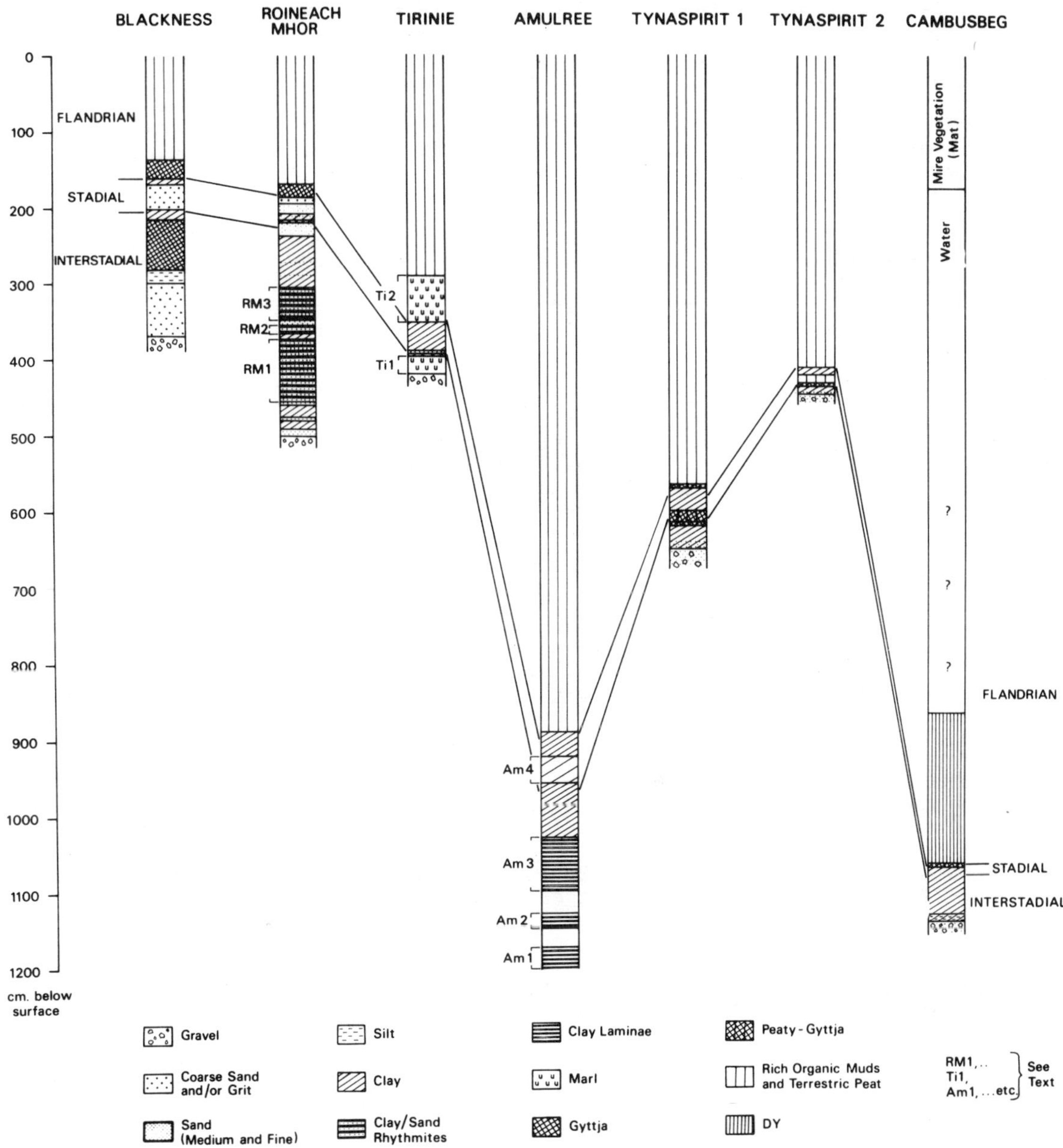

Fig. 2. Sedimentary records for the sampling points at the six sites.

By contrast, a three-fold sedimentary sequence is not present in the Roineach Mhor, Amulree and Cambusbeg basins. Above the basal gravels at Roineach Mhor alternating layers of sand and silt grade upwards into three separate but quite distinct series of rhythmically-bedded units of sand and silty-clay (RM1, RM2, RM3; Fig. 2). A total of 79 rhythmite pairs was counted over a vertical distance of 82 cm. Overlying the rhythmite series is a further complex of sands and clays, indicating greater variations in sedimentation rates (Walker, 1975a).

At Amulree the pre-Flandrian sediments consist entirely of clay, which is occasionally laminated (Am1, Am2, Am3; Fig. 2). The laminations are revealed by a change of colour in the clays either from a dark to a light grey, or sometimes from grey to pink. Individual laminae can be extremely thin (less than 1 mm), although the average thickness is between 2 and 3 mm. An estimate of the total number of paired laminations for lithostratigraphic unit Am3 is 250. Units Am1 and Am2 could not be sampled by piston corer, and thus the total number of laminations at the site is unknown. Immediately above unit Am3 is a stiff, dark grey clay which is thought to be of Lateglacial Interstadial age, while a further change to a much softer light grey clay (Am4) marks the transition to Loch Lomond Stadial sediments at Amulree.

The Lateglacial deposits at Cambusbeg are relatively shallow by comparison with other sites in the eastern Grampians, and are composed of soft grey clay. These deposits therefore contrast markedly with the basin infill at Tynaspirit only 1 km to the south-east where the characteristic three-fold sequence of Lateglacial deposits is present (see above). The grey clays at Cambusbeg appear to reflect a very slow rate of sedimentation throughout the Lateglacial period. Organic detritus of Flandrian age overlies the clays, and the evidence suggests that these sediments too accumulated relatively slowly. The Flandrian sediments consist of an extremely soft, homogeneous organic mud which is watery in nature between 870 and 950 cm (Fig. 2), and gel-like between 950 and 1070 cm. The latter deposit has been termed dy (Faegri and Iversen, 1964). Between 150 and 870 cm an extremely watery organic mud, or in some places only water, is enclosed by a floating mat of mire vegetation supporting a number of birch trees. This basin thus corresponds to the "schwingmoore" (Moore and Bellamy, 1974).

The nature and thickness of both Lateglacial Interstadial and Loch Lomond Stadial deposits therefore vary considerably between the six sites. Organic sediments accumulated during the Lateglacial Interstadial at Blackness, Tirinie and Tynaspirit. By contrast, the two largest basins lack Interstadial organic deposits. At Amulree, the Lateglacial minerogenic infill is almost 4 m in thickness, while at Cambusbeg only about 50 cm of sediment accumulated during the same time interval. Over 3 m of sand, silt and clay were deposited at Roineach Mhor, yet this basin is considerably smaller than either Amulree or Cambusbeg. Where organic Lateglacial Interstadial deposits are present, minerogenic sediments of Lateglacial Interstadial age are relatively thin, and the organic sediments themselves are quite different at each site. An organic detrital layer so rich in plant remains that it has here been termed "peat" is found at Tynaspirit, while a much finer organic deposit (gyttja) occurs at Blackness. Of particular interest are the marl deposits at Tirinie (Ti1 and Ti2; Fig. 2). Ti1 consists of a cream-coloured shell marl, and Ti2 of a light grey to white marl. Both units contain abundant shells of *Lymnaea peregra* and species of *Pisidium* (Walker, 1974).

BIOSTRATIGRAPHY

Recent pollen-stratigraphic investigations in Scotland have shown that the Jessen-Godwin zonation scheme for the Lateglacial period (zones I, II and III) and the corresponding chronostratigraphic terms (Older Dryas, Alleröd, Younger Dryas) can no longer be applied (Moar, 1969a; Pennington *et al.*, 1972; Birks, 1973; Walker, 1974; Pennington and Sackin, 1975). The problem is further compounded by the results of studies in Highland Britain as a whole which have revealed marked discrepancies between lithostratigraphic, biostratigraphic and climatostratigraphic units hitherto considered to be broadly synchronous (Coope, 1970; Pennington, 1970; Pennington and Bonny, 1970; Coope and Brophy, 1972; Pennington, 1975b; Bishop and Coope, this volume). These investigations underline the need to adopt an inductive approach to biostratigraphic subdivision by defining individual pollen assemblage zones for each site, before attempting to draw up a broader scheme with climatic and/or chronological implications.

The six pollen diagrams presented here (Figs. 3-8) have therefore been zoned independently on the basis of local pollen (and certain spore) assemblages alone. In other words, subdivisions of each profile are defined by stable or constant relative values of different pollen or spore types, irrespective of their ecological or climatic affinities. These local pollen assemblage zones are then used in the interpretation of regional vegetational history, and serve as the basis for comparison between neighbouring sites in the southern and eastern Grampians. Wider correlations will remain questionable until precise chronozone criteria (West, 1970) have been established (see Radiocarbon Dating, below).

The diagrams (Figs. 3-8) have a number of features in common. First, it is possible to recognise a sequence of zones at

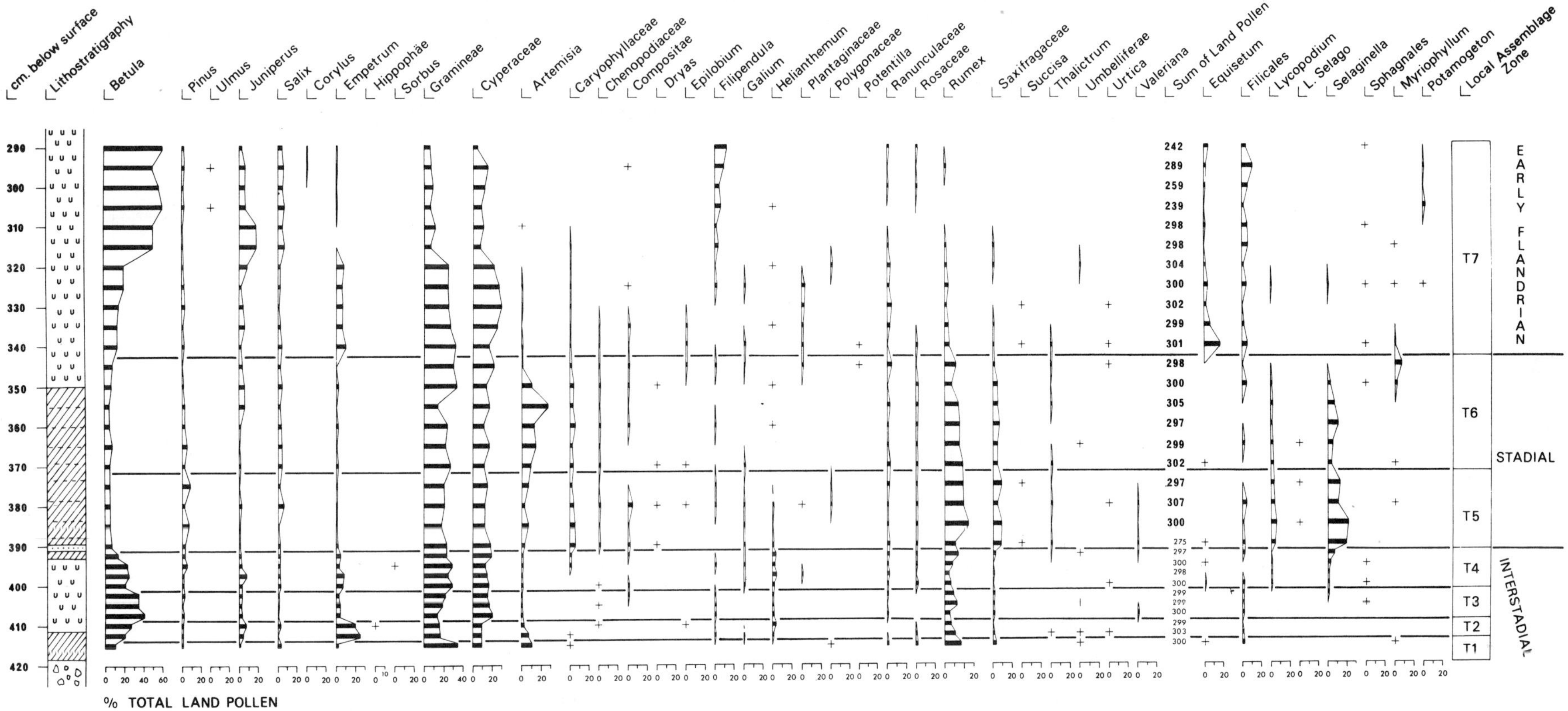

Fig. 3. Tirinie pollen diagram. Percentages based on total pollen excluding aquatic pollen.

the base of each of the profiles representing a developing vegetation cover. The number of subdivisions of this sequence (*e.g.* RM1 and RM2, B1 - B4; Figs. 4 and 5) depends on local site factors, such as the time of initial sedimentation in the basin, the rate of sedimentation, and the nature and rate of vegetation succession. The sequence may correspond approximately in some of the diagrams with the Jessen-Godwin zones I and II, but the age of the basal assemblage zone is unknown, and is likely to be variable from site to site.

Secondly, a period of vegetation revertence is recognised, where a return to herbaceous-, sedge-, and grass-dominated communities, often associated with increased *Lycopodium* and *Selaginella* percentages, indicates a major set-back in the local seral successions. At Tirinie (Ti5 and Ti6, Fig. 3) and Amulree (Am2d and Am2e, Fig. 7) this phase can be divided into two separate assemblage zones. Thirdly, the revertence phase is terminated at all sites by a pronounced increase in woody plant pollen, and particularly by a marked upward trend in the curve for *Juniperus*.

The phase of vegetational revertence is correlated with the Loch Lomond Stadial (Sissons *et al.*, 1973), which equates approximately with Jessen-Godwin zone III. The Lateglacial Interstadial encompasses all the pre-Stadial pollen zones. It is recognised, however, that the pre-Stadial zones at any one site may encompass the whole of the period between

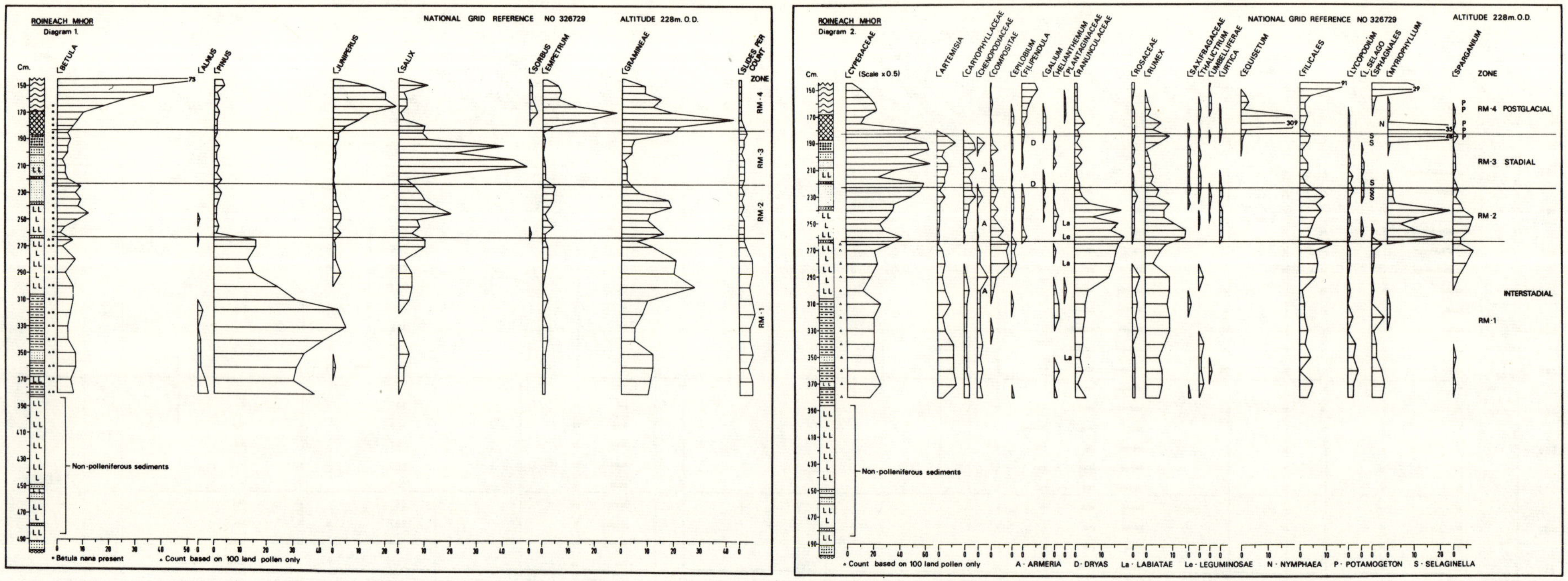

Fig. 4. Roineach Mhor pollen diagrams. Percentages based on total pollen excluding aquatic pollen. (Reproduced from *Pollen et Spores* 17, 1975).

regional deglaciation (from the Late Devensian ice sheet) and the Loch Lomond Stadial (Sissons and Walker, 1974), or that they may represent a later part of that period only. There is no evidence to support a major subdivision of the Lateglacial Interstadial into, for example, Jessen-Godwin zones I and II. Moreover, there is no indication at any of the sites within the present study area of a climatic fluctuation during the early Lateglacial Interstadial which may equate with the Bölling Oscillation of continental north-west Europe, and which has been recognised at certain sites in northern Britain (Bartley, 1962; Pennington, 1970, 1975b; Pennington and Sackin, 1975). The Lateglacial/Flandrian boundary is characterised at all sites in the eastern Grampian Highlands by significant increases in the percentages of *Juniperus* and *Betula*.

The characteristic vegetational elements of the local pollen assemblage zones are summarised in Fig. 9. No strict chronostratigraphic correlation is implied in the figure, neither in the local assemblage zone boundaries, nor in the major Lateglacial Interstadial/Loch Lomond Stadial and Loch Lomond Stadial/Flandrian boundaries. However, it is assumed that both of these important horizons are virtually synchronous within the region of the southern and eastern Grampians, and that the pollen diagrams can therefore be broadly correlated on the basis of these boundaries. The remarkable similarity between the general vegetation sequences outlined in Fig. 9 (with the single exception of Amulree) supports this assumption. Although the pollen assemblage zones at Amulree do not accord with those at other sites, the developing vegetational phase, the revertence phase, and the period of *Juniperus*-dominated woody plant expansion during the early Flandrian can all be recognised. Since only one distinctive revertence phase is recognised at the other sites, and this has been shown to be Loch Lomond Stadial in age, then it is assumed that the comparable vegetational phase at Amulree corresponds with the same period.

Several problems were encountered in the interpretation of the biostratigraphic records at the six sites. First, there is clear evidence for contamination through using the

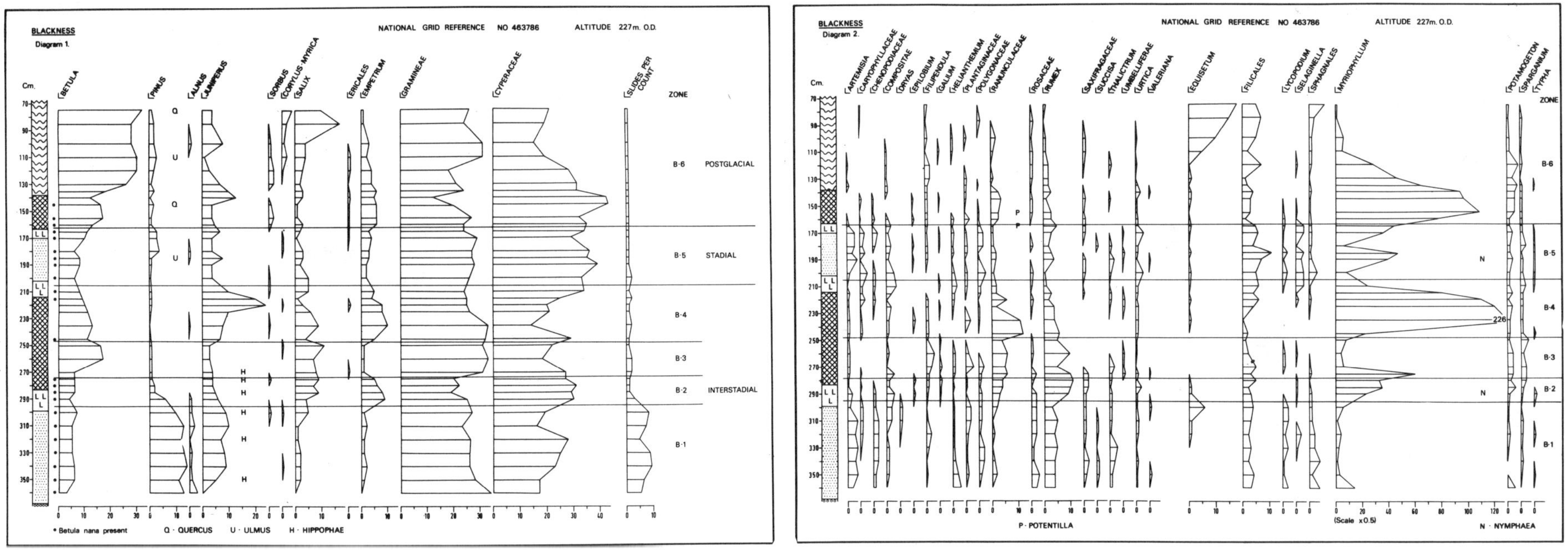

Fig. 5. Blackness pollen diagrams. Percentages based on total pollen excluding aquatic pollen. (Reproduced from *Pollen et Spores* 17, 1975).

Hiller borer. The distinctive clay laminations in the Lateglacial sediments at Amulree (see above) were only discovered after a second visit had been made to the site and cores removed by piston sampler (Amulree 2). Previous sampling (Amulree 1) at the site using the Hiller borer produced a stratigraphic record in which no clear laminae were visible, and which was initially described in field notes as "pink mottles within grey clay". Clearly the fragile laminations had been destroyed by the action of the auger mechanism on the Hiller sampling chamber. Further, clay samples obtained by Hiller from Amulree and Cambusbeg were found to be stained with organic material, presumably carried down from higher levels by the sampler. These problems of sediment disruption and contamination from overlying organic deposits were not encountered when the piston corer was employed. Thus, where two pollen diagrams are available from the same site (Tynaspirit 1 and 2 and Amulree 1 and 2) only those based on samples obtained by piston corer (Tynaspirit 2 and Amulree 2) are presented here. Contamination was not found to be serious at all the sites investigated, but was mostly associated with sampling in the minerogenic deposits at Amulree and Cambusbeg.

Additional evidence of contamination is revealed by the pollen records, and particularly by the presence of Coryloid grains in sediments of Lateglacial age. Pollen diagrams based on samples collected by piston corer (Roineach Mhor, Tirinie, Tynaspirit 2 and Amulree 2) show that Coryloid pollen is absent from pre-Flandrian sediments, yet Coryloid grains are found in deposits of Lateglacial age at Blackness, Tynaspirit 1, Amulree 1 and Cambusbeg, where sampling was carried out using a Hiller borer. Contamination by Coryloid pollen was found to be particularly severe in the soft clays at Cambusbeg. In each case, these grains appear to have been carried down through the profiles by the Hiller chamber from the early and mid-Flandrian organic muds and peats in which percentages of *Corylus* pollen are particularly high. If this was so, then the possibility cannot be discounted that at least some of the *Betula*, *Pinus*, *Alnus* pollen *etc.*, which seem anomalous in Lateglacial sediments, may also have been derived in a similar fashion. This applies especially to the minerogenic Loch Lomond Stadial deposits in which pollen is often extremely sparse, and where even minimal contamination can seriously distort the pollen spectra.

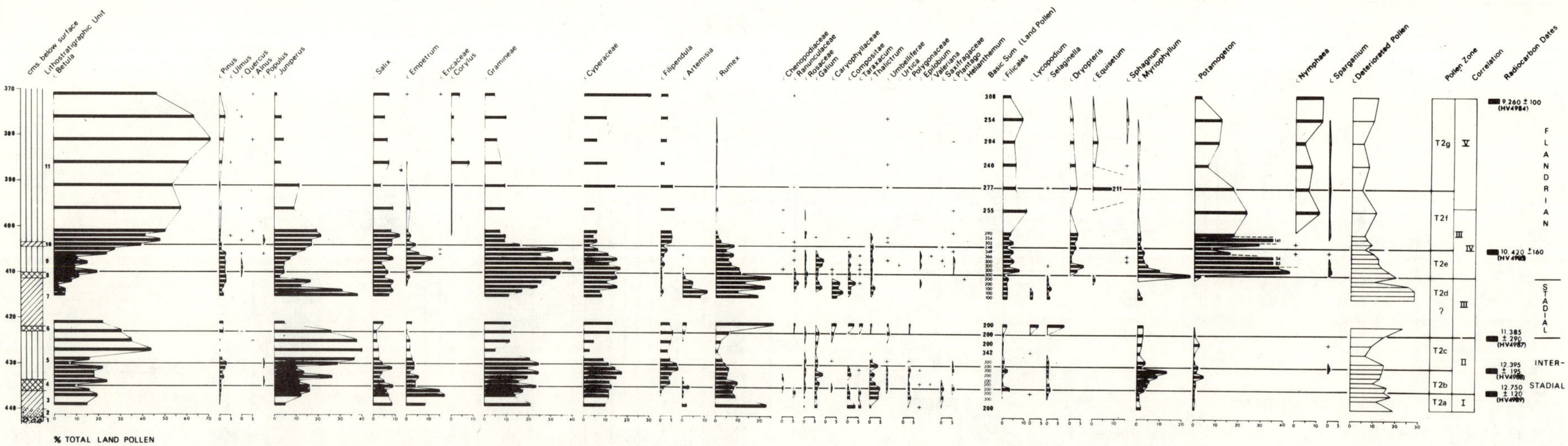

Fig. 6. Tynaspirit 2 pollen diagram. Percentages based on total pollen excluding aquatic pollen.

Secondly, a large proportion of the woody plant pollen recorded during the early Lateglacial Interstadial is unlikely to be of local origin. The low pollen influx at that time, due to the impoverished nature of the local vegetation cover in association with rapid minerogenic sedimentation in a number of the basins, tended to produce pollen spectra that emphasise grains derived from long-distance transfer. The problems of over-representation of tree pollen in sediments accumulating in virtually treeless areas are well documented (Faegri and Iversen, 1964; Nichols, 1970; Tyldesley, 1973a, 1973b, 1973c). Moreover, Pennington (1970) is of the opinion that much of the birch pollen recorded in Lateglacial pollen profiles from north and west Britain is of the species *Betula nana*, and that much of the tree birch and all of the *Pinus* pollen are the product of long-distance transport. Hence, the vastly inflated values particularly for *Pinus*, but also for *Betula* and to a lesser extent for *Alnus* at Amulree, Roineach Mhor and Blackness, are most probably attributable to the effect of long-distance transfer.

Thirdly, there is the problem of over- and under-representation of local taxa. This is best reflected in the pollen records of Tynaspirit 1 and Tynaspirit 2. These two basins are less than 100 m apart and yet a dominant *Juniperus* phase recorded in the upper Lateglacial Interstadial deposits at Tynaspirit 2 is not found in sediments of a similar age at Tynaspirit 1. This suggests that *Juniperus* development was extremely localised at that time, and that the high values recorded in the profile from Tynaspirit 2 do not necessarily reflect a regional *Juniperus* expansion in the area at the end of the Lateglacial Interstadial.

ANALYSIS OF DETERIORATED POLLEN

Data are available from three of the sites on percentages of deteriorated pollen grains. Following the classification introduced by Cushing (1964) and modified by H. J. B. Birks (1970), deteriorated pollen grains were classified into corroded, amorphous, broken and crumpled grains. Corrosion effects are attributed to oxidation processes (Sangster and Dale, 1961, 1964; Faegri and Iversen, 1964; Havinga, 1964; Königsson, 1969), while broken and crumpled grains reflect physical processes of corrasion and compaction. The interpretation of amorphous grains, however, is much more problematic (Cushing, 1964, 1967; H. J. B. Birks, 1970).

The counts for deteriorated pollen (unclassified) are given in the pollen diagrams as percentages of total land pollen. Figure 10 is a schematic diagram of the curves for total deteriorated pollen from five Lateglacial profiles. The highest percentages of deteriorated pollen are found in

Fig. 7. Amulree 2 pollen diagram. Percentages based on total pollen excluding aquatic pollen.

sediments of Lateglacial age, with maximum totals at Tynaspirit and Cambusbeg occurring in early Lateglacial Interstadial and Loch Lomond Stadial deposits (Figs. 6 and 8). At Amulree the curve for total deteriorated pollen differs from that at the other two sites, in that percentages do not decline as markedly during the Lateglacial Interstadial period. The relative percentages for the individual deterioration classes (Fig. 11) show that in general broken and crumpled pollen grains account for most of the deterioration, although there is a small but significant increase in the proportion of corroded pollen at all three sites in sediments of Loch Lomond Stadial age. In addition, there is a significant decline in total deteriorated pollen at the Loch Lomond Stadial/Flandrian boundary in all of the

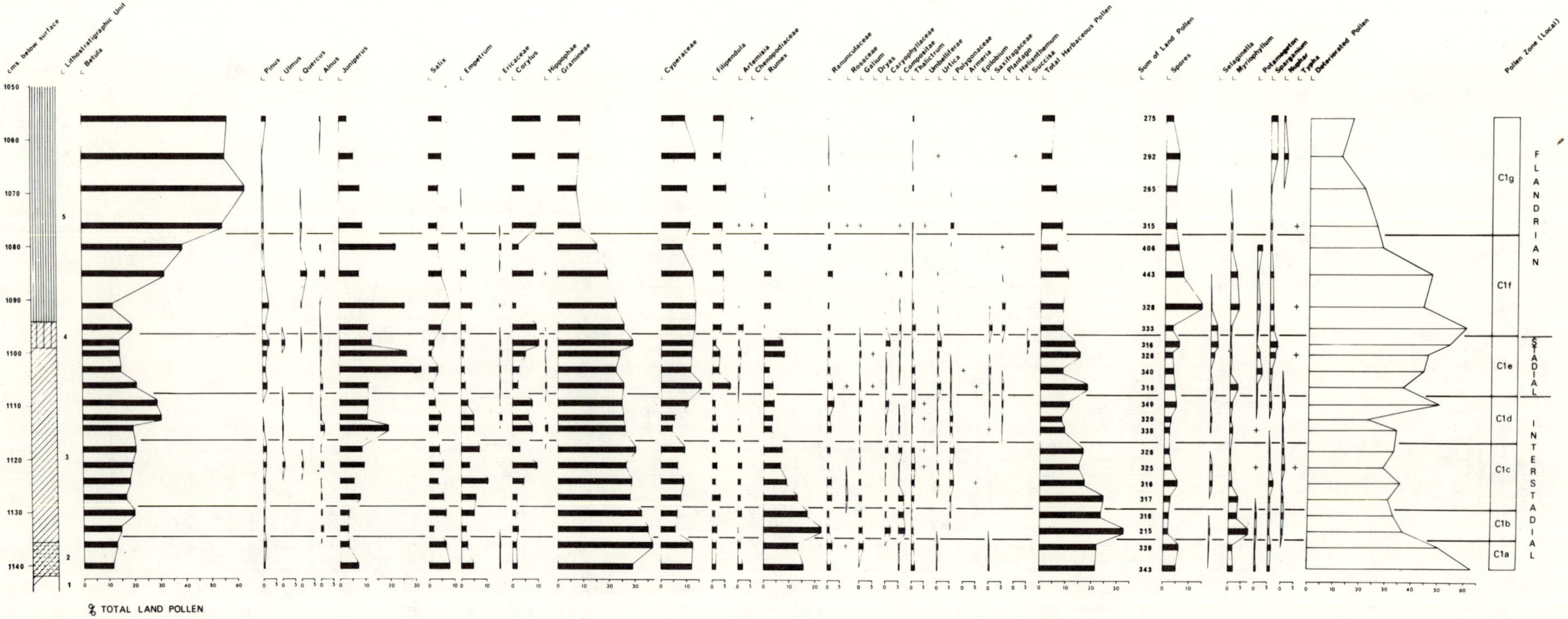

Fig. 8. Cambusbeg pollen diagram. Percentages based on total pollen excluding aquatic pollen.

diagrams (Fig. 10).

As yet the causes of pollen deterioration are not well understood, and the problem is compounded by the fact that certain pollen grains have a higher resistance to deterioration than others (Faegri and Iversen, 1964). In general it would seem that much of the deterioration results from recycling of pollen grains within the basins, particularly from exposed basin-edge deposits. In the present study area, analyses suggest that deterioration was mainly contemporaneous, or penecontemporaneous, affecting pollen grains in the soils and streams on surrounding slopes prior to, or during, transport into the basin (Lowe, unpublished). The high percentages of broken and crumpled pollen in early Lateglacial Interstadial and Loch Lomond Stadial deposits are therefore a reflection of unstable soil conditions around the basins during those time periods. The increased proportion of corroded pollen in the Loch Lomond Stadial deposits is probably a result of redeposition of pollen from the more mature soils that developed during the milder Lateglacial Interstadial, and which were subsequently disrupted by cryoturbation and increased erosion during the colder Loch Lomond Stadial period.

RADIOCARBON DATING

Samples were obtained by enlarged piston corer from three of the sites for the purpose of radiocarbon dating. Four cores were taken from Amulree and Blackness, and three cores from Tynaspirit 2. On each occasion the cores were removed from the corners of a *c*.30 cm square described on the surface of the bog. Pollen diagrams were constructed to identify critical horizons, and then slices of sediment of average

SITE	INTERSTADIAL	STADIAL	EARLY FLANDRIAN
BLACKNESS	B 1 Gramineae – *Rumex* B 2 *Empetrum* – *Rumex* B 3 *Betula* – *Salix* – Gramineae B 4 *Betula* – *Juniperus*	B 5 Gramineae – Cyperaceae	B 6 *Betula* – *Juniperus* – *Empetrum*
ROINEACH MHOR	RM 1 *Rumex* – *Artemisia* – Gramineae RM 2 *Betula* – *Salix* – *Rumex*	RM 3 *Salix* – Cyperaceae	RM 4 *Juniperus* – *Empetrum* – *Betula*
TIRINIE	T 1 *Rumex* – *Artemisia* T 2 *Empetrum* – *Betula* T 3 *Betula* T 4 *Betula* – *Juniperus* – *Empetrum*	T 5 *Rumex* – *Selaginella* T 6 *Artemisia* – *Rumex*	T 7 *Betula* – *Juniperus* – *Empetrum*
AMULREE 2	Am 2 a *Lycopodium* – *Selaginella* Am 2 b Cyperaceae – *Rumex* Am 2 c *Rumex* – Cyperaceae – *Juniperus* – *Empetrum*	Am 2 d Cyperaceae – *Rumex* – *Lycopodium* Am 2 e *Artemisia* – *Rumex* – *Lycopodium*	Am 2 f *Juniperus* – *Empetrum* – *Rumex* – Gramineae Am 2 g *Betula* – *Juniperus*
TYNASPIRIT 2	T 2a *Rumex* – *Empetrum* – *Salix* T 2b *Betula* – *Juniperus* – Gramineae T 2c *Betula* – *Juniperus*	T 2d *Rumex* – *Artemisia* – *Selaginella*	T 2e *Empetrum* – Gramineae – *Betula* T 2f *Betula* – *Juniperus*
CAMBUSBEG	C 1a Gramineae – *Salix* C 1b *Rumex* – Gramineae C 1c *Empetrum* – *Betula* – Gramineae C 1d *Betula* – *Juniperus*	C 1e *Rumex* – Cyperaceae – *Selaginella*	C 1f *Betula* – *Juniperus*

Fig. 9. Comparison of Lateglacial and early Flandrian local pollen assemblage zones for the six sites.

thickness 1.5 cm were taken from the cores and bulked to provide sufficient material for radiocarbon assay. Figure 12 shows how these dates relate to the Lateglacial and early Flandrian boundaries as determined by pollen analysis. All dates were assayed from gyttja or peaty-gyttja material.

Comparisons with other radiocarbon age determinations on these boundaries from sites in northern Britain (Godwin and Willis, 1959a, 1964; Shotton *et al.*, 1970; Sissons and Walker, 1974; Pennington, this volume; Vasari, this volume) suggest that a number of the dates from the present study area are in error,

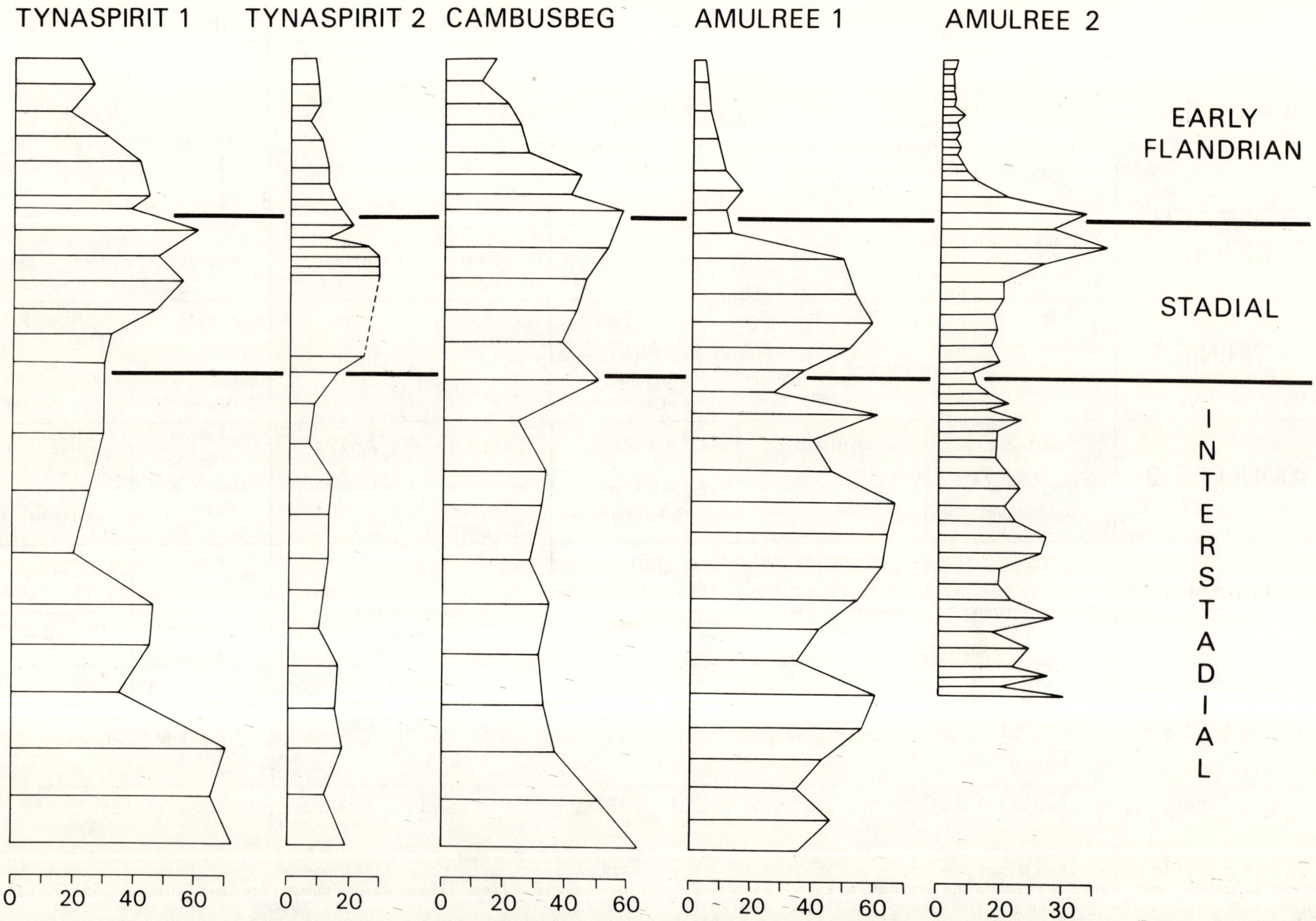

Fig. 10. Total deteriorated pollen curves for Lateglacial sediments at Tynaspirit 1, Tynaspirit 2, Amulree 1, Amulree 2 and Cambusbeg.

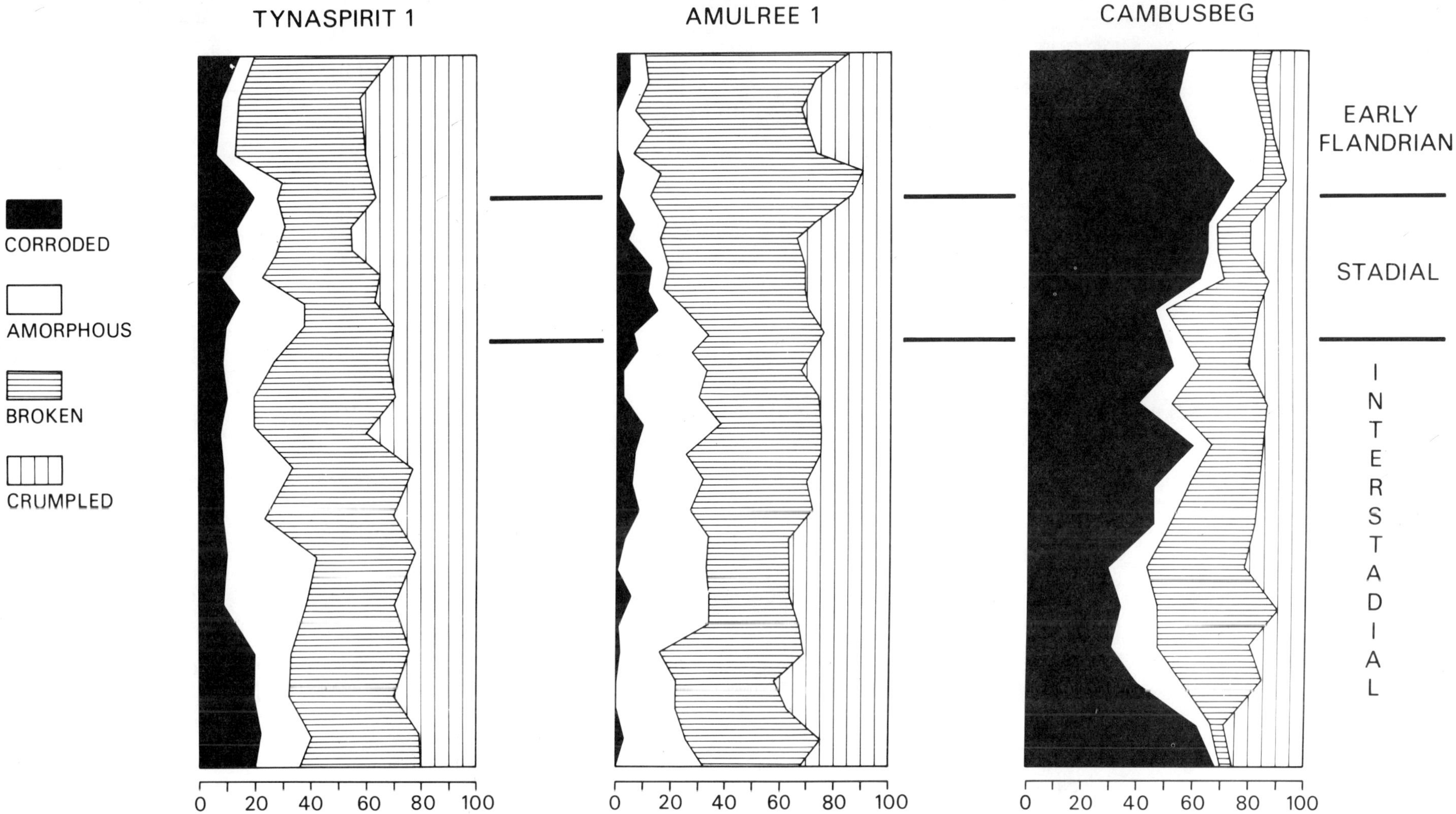

Fig. 11. Proportion of corroded, amorphous, broken and crumpled pollen grains in Lateglacial sediments at Tynaspirit 1, Amulree 1 and Cambusbeg.

and thus it is clear that these dates cannot be accepted for chronozone definitions.

It has already been suggested (Walker, 1975a) that those dates which are markedly younger than radiocarbon age determinations on comparable horizons at other sites may be a result of contamination of the dated sediments by the infiltration of humus-rich ground water. However, many of the dates in Fig. 12 are considerably older than those already published from sites in northern Britain. The reasons for these discrepancies are not immediately apparent, but it is significant that all the dates which are thought to

		BLACKNESS	TYNASPIRIT 2	AMULREE 2	MOLLANDS	LOCH ETTERIDGE
FLANDRIAN	*CORYLUS* RISE		9,260 ± 100	9,115 ± 120	9,365 ± 120	
	JUNIPER MAXIMUM	9,140 ± 105	10,420 ± 160	10,270 ± 100	10,480 ± 150	
					10,670 ± 85	
	EARLIEST ORGANIC DEPOSITS	9,490 ± 160		10,770 ± 90		9,405 ± 260
STADIAL						
INTERSTADIAL	YOUNGEST ORGANIC DEPOSITS	10,100 ± 135	11,385 ± 290			10,764 ± 120
			12,395 ± 195			11,290 ± 165
	OLDEST ORGANIC DEPOSITS		12,750 ± 120			13,151 ± 390

1. BLACKNESS (HV 5648 - 50) Walker, 1975a
2. TYNASPIRIT 2 (HV 4984 - 89) Previously unpublished
3. MOLLANDS (HV 5645 - 47) Previously unpublished
4. AMULREE 2 (HV 5642 - 44) Previously unpublished
5. LOCH ETTERIDGE (SRR 301 - 304) Sissons & Walker, 1974

Fig. 12. Lateglacial and early Flandrian radiocarbon dates from the southern and eastern Grampians. All dates in radiocarbon years B.P.

be suspect were obtained from gyttja material, and serious doubts are now being expressed about the reliability of gyttja as a medium for radiocarbon dating (R. E. G. Williams, pers. comm.). Thus until a type site with more suitable material for radiocarbon assay is found, chronozones cannot be established for this area of the Grampian Highlands.

THE ENVIRONMENT DURING THE LATEGLACIAL INTERSTADIAL

The palynological, sedimentological and geomorphological record preserved in and around these six Lateglacial sites makes possible the reconstruction of the landscape and environment over a large area of the southern and eastern Grampian Highlands from the time of decay of the Late Devensian ice sheet to the disappearance of the Loch Lomond Readvance glaciers.

At the maximum of the Late Devensian glaciation between 17,000 and 20,000 years ago, the Grampian Highlands were submerged beneath an ice sheet at least 1000 m in thickness which is known to have extended as far south as Holderness in Yorkshire (Penny *et al.*, 1969). If the radiocarbon dates are correct, the Lockerbie area of the Southern Uplands was ice-free shortly after 13,000 B.P. (Bishop, 1963), while ice had retreated from Drymen near the southern end of Loch Lomond before 12,500 B.P. (Vasari, this volume). Within the present study area only one date was obtained which has any bearing on ice-sheet wastage; the basal organic deposits in the Tynaspirit 2 basin gave a radiocarbon age of 12,750 ± 120 B.P. This date, if correct, is minimal for deglaciation of the Callander area, for up to 10 cm of minerogenic lacustrine sediment underlies the dated horizon. The Tynaspirit date is thus in broad agreement with the Drymen and Lockerbie dates, and together with the recently obtained dates of 13,151 ± 390 B.P. (Sissons and Walker, 1974) and 12,700 ± 270 B.P. (Vasari, this volume) from the Spey Valley lends credence to the hypothesis that shortly after 13,000 years ago large areas, if not all, of the Grampian Highlands were free from glacier ice.

The thinning and retreating ice exposed extensive areas of bare ground to rapid colonisation by pioneer vegetation. Many of the herbaceous plants that were characteristic of that period, *e.g.* *Rumex* and *Artemisia*, have steppe or tundra affinities. However, the fresh substrate following deglaciation exercised a selective influence on diaspores landing on it, such that only those species that today occur on open and generally humus-free soils (*e.g.* soils typical of steppe and tundra regions) became established (Faegri, 1963). The presence of these open-habitat taxa in the basal pollen assemblage zones is therefore not necessarily indicative of a severe climatic regime in the Grampians during and immediately after ice-sheet decay, but is more likely a reflection of competition, migration rates and edaphic conditions. Indeed, in view of the high summer temperatures postulated for parts of England and Wales at that time, based on the occurrence of thermophilous coleoptera (Coope *et al.*, 1971; Coope and Brophy, 1972), it would seem that the vegetation of large parts of upland Britain bore little relation to climate during the early part of the Lateglacial Interstadial.

About 13,000 years ago therefore, in spite of apparently mild climatic conditions, the landscape of the eastern and southern Grampians was one of open habitats, with a predominantly herbaceous vegetation cover interspersed with large areas of relatively unstable bare ground, and here and there small remnants of decaying glacier ice in sheltered localities. Only in more southerly areas around Callander were there any significant numbers of shrub birch and willow. High values for arboreal pollen are recorded in the basal assemblage zones at Amulree, Blackness and Roineach Mhor, but these are considered to be the product of long-distance transfer of tree pollen to these upland sites (see above).

Following the initial period of vegetational development, the trend towards soil stability is marked in all the profiles by an increase in percentages of locally-derived woody plant pollen, and, with the exception of Amulree, a significant decrease in numbers of open-habitat taxa. A rising birch curve is a common element of all the pollen diagrams and, while it is evident that a large proportion of the grains are of *Betula nana*, many of the grains recorded at the sites near Callander and at Tirinie appear to be of tree birch. Birch development was restricted at the exposed upland site of Amulree, while few birch trees seem to have penetrated into the more northerly valleys of Glen Esk and Glen Clova. Thus if tree birch was present in the study area during the Lateglacial Interstadial, the main concentration would seem to have been in the more southerly and westerly valleys, with absolute numbers declining north-eastwards. At none of the sites, however, are birch values sufficiently high to postulate the establishment of anything other than a limited birch woodland cover during the Lateglacial period.

The relatively sparse nature of the tree cover is further emphasised by the presence of shade-intolerant woody taxa in the Lateglacial Interstadial spectra, notably *Juniperus* and *Empetrum*. *Juniperus* is especially common at the lowland sites near Callander, and is also present in sediments that accumulated towards the end of the Lateglacial Interstadial at Blackness. However, as was discussed earlier, it would seem that *Juniperus* expansion may have been very localised during the Lateglacial, suggesting edaphic or possibly climatic restrictions to widespread colonisation. *Empetrum* occurs in Lateglacial Interstadial spectra at all of the sites, but is a less significant element than in diagrams from sites to the north and west of the main Highland watershed (Pennington *et al.*, 1972; Walker, 1975b). As the distribution of *Empetrum* during the Lateglacial has been closely connected with former moisture levels (Brown, 1971), the lower *Empetrum* percentages within the present study area are most probably a reflection of lower precipitation values in the south-east Grampians by comparison with the uplands to the north and west.

Other woody taxa occurring in the pollen spectra are *Salix*, *Pinus* and *Corylus/Myrica*. *Salix* is recorded in significant quantities at all of the sites and the presence of this genus probably reflects the diffusion of shrub willows along the floors of the glens. This is in marked contrast to the early Lateglacial Interstadial records of *Salix*, for a large number of the grains found in the basal assemblage zones resembled those of the least willow, *Salix herbacea*. It is unlikely, however, that the *Pinus* records are indicative of local pine growth and, as in the early Lateglacial Interstadial spectra, these grains are probably due to the effects of long-distance transport. It is possible that some of the Coryloid pollen can be attributed to the same source, for it is almost certain that the relatively warmth-demanding *Corylus avellana* was not growing in the Grampian Highlands during the Lateglacial. However, it is considered more likely that the occurrence of these grains in sediments of Lateglacial Interstadial age are the product of contamination during the sampling process (see above).

Characteristic herbaceous taxa of the Lateglacial Interstadial period include Cyperaceae, Gramineae, and *Rumex*. The constant record for Cyperaceae, particularly at the more exposed sites such as Amulree, is unlikely to represent reedswamp vegetation only, for studies of modern pollen rain in upland areas of Scotland (Birks, 1973) and in tundra regions (Lichti-Federovich and Ritchie, 1968; Rampton, 1971) have shown high percentages of sedge pollen where local reedswamp vegetation is minimal. It is suggested that the presence of both Cyperaceae and Gramineae in considerable quantities at all of the sites during the Lateglacial Interstadial is a reflection of the former widespread occurrence of grassland communities, not dissimilar to those found over large areas of upland Britain at the present time. In general, taxa indicative of bare soils and open ground are less frequent than during the early part of the Lateglacial Interstadial, but *Rumex* remains a common element in the pollen spectra at all of the sites. Values for this genus are particularly high during the initial expansion of woody plants, but percentages decline progressively throughout the Lateglacial Interstadial. At Amulree, however, *Rumex* maintains a constant record during this time period. The continued presence of this genus in significant quantities reflects the incomplete nature of the vegetation cover over large areas of the eastern Grampians and the existence of areas of bare ground and disturbed soils which provided suitable habitats for *Rumex*, and also for other open-habitat taxa such as species of Caryophyllaceae, Chenopodiaceae and Compositae, as well as for the clubmosses *Lycopodium* and *Selaginella*.

The landscape of the eastern Grampians during the Lateglacial Interstadial therefore ranged from floristically-rich closed or semi-closed grassland, with juniper, willow heath, dwarf birch and occasional copses of tree birch in sheltered localities, to grass and moss heath communities interspersed with patches of bare and unstable soils on the upper slopes and plateau surfaces. Tree birch appear to have been more common in the more southerly and westerly valleys and, if the radiocarbon dates are reliable, may have been established as early as 12,500 B.P. in the Teith Valley near Callander. Tree birch may also have spread into Glen Fender, but on the exposed uplands around Amulree more severe environmental conditions allowed the development of a poor montane grassland only, dominated by arctic-alpine species of sedge and moss. The spread of woody taxa was very restricted in the vicinity of this site.

The trend towards a closed vegetation cover and eventual soil stability is reflected not only in the pollen record, but also in the stratigraphy and composition of the basin sediments, and in the degree of pollen preservation. The presence of minerogenic sediment at the base of all the profiles is due to the inwash of inorganic material resulting from the erosion of humus-free skeletal soils. The trend towards organic deposition at Tynaspirit, Tirinie and Blackness reflects the cessation of minerogenic inwash and the development of a closed vegetation cover. The gradual increase in organic carbon content of the basin sediments is a result of the progressive leaching of humic acids from gradually maturing soils around the basins. The sediments would have been further enriched by the addition of organic material from decaying macrophytic vegetation, phytoplankton and associated fauna within the lake basins.

At Roineach Mhor, Amulree and Cambusbeg however, minerogenic deposition continued throughout the Lateglacial Interstadial period. This is thought to have been due mainly to soil instability and a sparse vegetation cover on the very steep slopes around Roineach Mhor (Walker, 1975a). The situation at Amulree is rather different. There, the size of the basin, the virtual absence of macrophytic vegetation within the lake, and the impoverished nature of the local vegetation cover around the site, all appear to have been contributory factors leading to the uninterrupted deposition of minerogenic sediments during the Lateglacial Interstadial. At Cambusbeg, the Flandrian gel-like muds that are poor in macrofossil content and the present "schwingmoore" development within the basin suggest base-poor and possibly dystrophic conditions (Moore and Bellamy, 1974). Why such conditions should have prevailed at this site, while only 1 km distant at Tynaspirit an apparently rich basin ecosystem developed during the Lateglacial Interstadial, is difficult to explain, and will be discussed more fully elsewhere.

The degree of pollen preservation further serves to emphasise the overall trend towards soil stability in the eastern Grampians during the Lateglacial Interstadial. At Cambusbeg and Tynaspirit the decline in percentages of broken and crumpled grains from high values in early Lateglacial Interstadial sediments to markedly lower percentages in sediments of late Lateglacial Interstadial age reflects a significant decrease in the amount of sediment inwash into the basins from the surrounding catchments. However, at Amulree, where minerogenic inwash persisted, more constant values for deteriorated pollen are recorded than at lowland sites near Callander, emphasising the harsher Lateglacial environment at the more exposed upland site.

Little direct evidence can be gained from the study area of former temperature levels in the eastern Grampians during the Lateglacial Interstadial, but if the curves derived by Coope and others are correct, then it would appear that the thermal maximum was reached at, or slightly before, the establishment of heathland or woody plant communities, *i.e.* 12,500 to 13,000 B.P. According to Coope *et al.* (1971), the mean July temperatures in the English Midlands at around 13,000 B.P. was *c.*17°C, after which temperatures declined to about 12-13°C at the Lateglacial Interstadial/Loch Lomond Stadial boundary. The present sea-level temperature difference between the eastern Grampians and the English Midlands is *c.*1.4°C and so, in more sheltered localities, Lateglacial Interstadial July temperatures of 1-2°C cooler than the estimates quoted above would not seem to be unreasonable. On the upper slopes and plateau surfaces however, conditions would have been relatively more severe, and temperatures may not have been much in excess of 10°C in these more exposed localities.

THE ENVIRONMENT DURING THE LOCH LOMOND STADIAL

At about or possibly even before 11,000 B.P. a marked change took place in the landscape and environment of the eastern and southern Grampians (Sissons, 1974c; Vasari, this volume). In response to falling temperatures ice caps developed on the Gaick Plateau and on the uplands above Glen Esk and Glen Clova, from which outlet glaciers descended to within 7 km and 3 km respectively of the sites at Tirinie and Blackness (Sissons, 1972a, 1974b). Small glaciers also formed in Corrie Brandy and Corrie Clova only 1.5 km from the site at Roineach Mhor. A large glacier moved south-eastwards down the Teith Valley to its maximal extent at Callander, some 2 km north-west of Cambusbeg and Tynaspirit, while the site at Amulree was only 3 km to the north of a major valley glacier moving eastwards along Glen Almond (Thompson, 1972).

At each of these sites close to the limits of the Loch Lomond Readvance, the change from interstadial to stadial conditions is reflected in the renewed inwash of coarse minerogenic material, and in the marked upward trends in the curves for deteriorated pollen. The harsher environmental conditions led to a rapid decline in numbers of shrub and arboreal pollen, to a break-up of heath communities, and to a sharp increase in open-habitat taxa such as *Artemisia*, *Rumex*, and species of Caryophyllaceae, Chenopodiaceae and Compositae. Present-day montane and arctic species of moss and sedge were also much more common. Particularly severe environmental conditions seem to have prevailed in the north-easterly glens where the former extent of solifluction activity is reflected in the extremely coarse sediments in the basins at Blackness and Roineach Mhor. The presence of *Salix herbacea* and other markedly chionophilous taxa in the pollen records are indicative of the widespread distribution of snowpatches in that area. This factor may well have acted to restrict the spread of *Artemisia* in the more easterly glens, for percentages are significantly lower than at Tirinie, Amulree and at sites to the west of the main Highland watershed (Walker, 1975b), and present-day records for this genus indicate that while it is characteristic of disturbed ground, it has marked xerophilous affinities and is intolerant of snow cover (Andersen, 1961).

At the lowland sites around Callander values for Artemisa are very low indeed, and the Loch Lomond Stadial spectra are dominated by *Rumex* with small percentages of locally-derived woody plant pollen such as *Juniperus* and *Betula nana*. This is in marked contrast to the upland sites where values for woody taxa are significantly lower, and the majority of such pollen in the spectra are thought to be due almost exclusively to the effects of long-distance transport. In the lowlands, in spite of the prevailing periglacial climate, areas of moving soils appear to have been more restricted, and minerogenic inwash into the basins seems to have been due more to the processes of normal run-off over bare ground than to cryoturbation and solifluction, contrasting with the evidence from the more exposed upland sites. Nevertheless, the increased percentages of corroded pollen at Tynaspirit and Cambusbeg, as well as at Amulree, suggests that even in the lowland areas soils which developed during the Lateglacial Interstadial were at least partially destroyed during the succeeding Loch Lomond Stadial period.

Using data from estimated firn lines and former precipitation levels, Sissons (1974b) has calculated that the June - September temperatures in the Gaick area of the Grampians at the time of the Loch Lomond Readvance were around 1.5°C at the firn line, while the equivalent sea-level temperatures were inferred to be about 6.3°C for the four summer months, or 7.6°C for July. Taking altitudinal and latitudinal differences into account, there is good agreement between these figures and those suggested for localities in England. Manley (1964) has estimated that the July temperatures in the Windermere area were 7.5 - 8.0°C and in the north Midlands were *c.*10°C, while Coope *et al.* (1971), on the basis of coleoptera remains, suggested a July temperature of "at, or just below 10°C" for the Loch Lomond Stadial in the English Midlands. In the present study area, therefore, summer temperatures of 5 - 6°C would seem to be reasonable estimates, while in more exposed localities and on upper slopes mean July temperatures may have been as low as 1.5 - 2.0°C. The mildest conditions within the study area would have been experienced in the more sheltered valleys of the south and west, while a particularly severe climatic regime appears to have prevailed in the glens of the north-east. The latter area would have been directly exposed to bitterly cold winds blowing off the frozen North Sea, and conditions would have been further exacerbated by the effects of local glacier winds blowing downvalley from the ice-caps above Glen Esk and Glen Clova, and from the plateau of the Gaick Forest.

The sudden termination of harsh stadial conditions at around 10,000 B.P. is reflected in all of the profiles by an abrupt change from minerogenic to organic sediment accumulation and a sharp fall in percentages of deteriorated pollen, indicating the cessation of minerogenic inwash and the beginning of a trend towards vegetation and soil stability. The early Flandrian pollen records show a widespread colonisation initially by grassland species, followed by *Salix-Empetrum* heath, *Juniperus* heath and finally by birch woodland. The rate of vegetation succession was extremely rapid, for if the radiocarbon dates from northern Scotland are correct (Pennington *et al.*, 1972; Birks, 1972; Harkness and Wilson, 1973; Pennington, this volume; Vasari, this volume), large areas of the Scottish Highlands supported tree birch before 9,000 B.P.

CONCLUSIONS

The lithostratigraphic and biostratigraphic data from the six sites discussed in this paper indicate that in the southern and eastern Grampian Highlands vegetational succession and soil development progressed without interruption from the decay of the Late Devensian ice sheet (*c.*13,000 years ago) until the beginning of the Loch Lomond Stadial (*c.*11,000 years ago). The early Lateglacial pollen spectra reflect an initial period of colonisation by open-habitat taxa. Throughout the Lateglacial Interstadial period a vegetation of closed grassland with juniper, dwarf birch, willow, and occasional copses of tree birch gradually developed over the valley floors, while moss heath and poor grassland communities were more characteristic on the upper slopes. In exposed localities the vegetation cover was never fully developed, and conditions of soil instability persisted throughout the Lateglacial Interstadial. At around 11,000 B.P. declining temperatures heralded the onset of harsher environmental conditions, with the recrudescence of glacier ice of the Loch Lomond Readvance, and the break-up of existing plant communities as a result of widespread cryoturbation and solifluction. As conditions in the Loch Lomond Stadial became increasingly more severe open-habitat and chionophilous vegetation proliferated and a tundra landscape developed. This phase was relatively short-lived, however, for rapid amelioration about 10,000 years ago resulted in the final disappearance of glacier ice from Scotland, the cessation of solifluction, and the initiation of a plant succession that led to the establishment of birch woodland over much of the area less than 1,000 years later.

ACKNOWLEDGEMENTS

We are grateful to the University of Edinburgh and the Natural Environment Research Council for financial assistance, and to the Niedersächsisches Landesamt für Bodenforschung, Hanover, for the radiocarbon dates. We thank all who helped in our investigations, especially Dr. J. B. Sissons, Dr. W. W. Newey, Dr. M. A. Geyh, and Professor Y. Vasari.

Lake Sediments and the Lateglacial Environment in Northern Scotland

W. PENNINGTON (MRS. T.G. TUTIN)
(Freshwater Biological Association and University of Leicester)

ABSTRACT

Analysis for biological (pollen and diatoms) and chemical variables in the sediment profiles from open water in nine Scottish lakes, from the Great Glen northwards, has provided evidence on the Lateglacial environment of northern mainland Scotland since well before 13,000 B.P. A series of C14 dates from Cam Loch, Sutherland, provided a time-scale from which the boundaries of pollen assemblage zones were related to chronozones, correlated with the subdivisions of Late Weichselian time used by Mangerud et al. (1974). Evidence suggests the following major environmental changes: 1. a widespread climatic amelioration at c. 13,000 B.P. (opening of Bölling chronozone) which initiated accumulation of organic interstadial sediments in lakes and led to a change from discontinuous vegetation of pioneer and snowbed communities to a more complete vegetation cover of dwarf-shrub heath, with acceleration of processes of soil maturation as soils became stabilised; 2. a minor recession from interstadial equilibrium at c. 12,000, coinciding with the Older Dryas chronozone, involving only reduced pollen production by woody plants and a slightly increased rate of soil erosion; 3. a return to interstadial conditions, c. 11,800 - 11,000 (Alleröd time), with maximum Lateglacial rates of annual pollen deposition but erosion of organic soils; and 4. a period of low temperatures reflected in break-up of continuous plant communities and universal frost-disturbance of soils, c. 11,000 - 10,400, (roughly equivalent to the Loch Lomond Stadial), followed by rapid climatic amelioration between 10,400 and 10,000 B.P. In seven other lakes, within or near the limits of the Loch Lomond Readvance, the minerogenic sediment below Flandrian muds is coarse or thick and was not penetrated by our samplers.

INTRODUCTION

During the last ten years the palaeolimnology group of the Freshwater Biological Association has worked on the sediments of 18 lakes of the mainland of Scotland, from the Great Glen northwards. Information on the Lateglacial environment provided by lithostratigraphy and, at one site, chronostratigraphy, has been supplemented by biological analysis (pollen and diatoms) and geochemical analysis (inorganic and organic) of sediment (Pennington and Lishman, 1971; Pennington *et al.*, 1972; Pennington and Sackin, 1975; Pennington, 1975a, 1975b; Haworth, 1974; Cranwell, in press). The present paper will review these integrated studies of lake sediments and their contribution to an understanding of the environmental changes in northern Scotland between *c.* 13,000 and 10,000 radiocarbon years ago.

All this work has been carried out on open-water sediments from deep and predominantly rocky-shored lakes, and the results cannot therefore be compared directly with results of other authors from analysis of the deposits of shallow lakes with marginal hydroseres, where sampling was carried out from the surface of lacustrine peat (*e.g.* Vasari and Vasari, 1968; Birks, 1973). It has been shown in the English Lake District that the deep-water muds of sizeable lakes which lie on the courses of vigorous streams are characteristic of each lake and its catchment, and record an integrated history of both (Mackereth, 1965, 1966). A major result of this palaeolimnological project has been to demonstrate how the close relationship found in Lake District lakes between changes in pollen spectra, indicative of the type of vegetation, and changes in sediment composition, indicative of soil history, was found also in northern Scotland in lakes

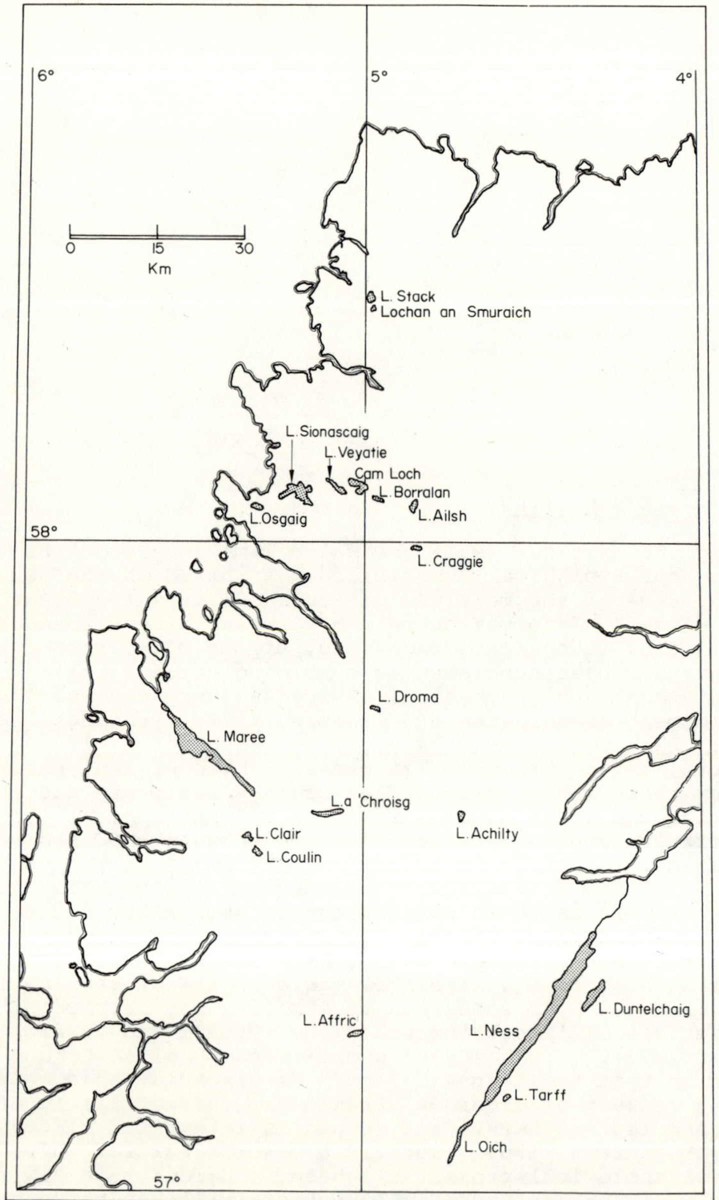

Fig. 1. Map of northern Scotland, showing positions of lakes.

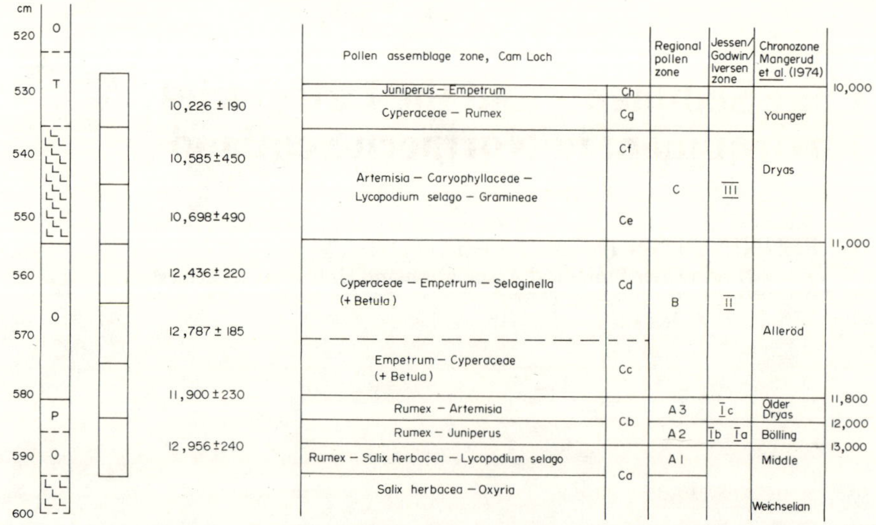

Fig. 2. Correlation of chronostratigraphy and pollen assemblage zones at Cam Loch, Sutherland, with Late Devensian and Late Weichselian subdivisions. Radiocarbon dates are SRR-247 to 253.

which, though generally similar in morphometry and their rather low biological productivity, differ from Lake District lakes in geology and vegetation (present and past) of their catchments, as well as in the history of deglaciation.

Figure 1 shows the positions of sites, and Fig. 2 tabulates chronology and pollen assemblage zones at the radiocarbon-dated site, Cam Loch in Sutherland, correlating these with the pollen zonation used by Pennington *et al.* (1972), with the Jessen-Godwin zonation of pollen zones I, II and III recommended by Mitchell *et al.* (1973), and with the chronostratigraphic subdivisions of the Late Weichselian proposed by Mangerud *et al.* (1974) and applied to the Late Devensian by Pennington (1975b). In this paper, the term "Interstadial" will be used for the period between 13,000 and 11,000 B.P., including the Bölling, Older Dryas and Alleröd chronozones of Mangerud *et al.*, and the term "Loch Lomond Stadial" for the period between *c.*11,000 and 10,400 B.P. which represents the first part of the Younger Dryas chronozone of Mangerud *et al.* The term "interstadial" was used differently in Pennington *et al.* (1972) where it referred to the period equivalent to pollen zone B, which has subsequently been correlated by radiocarbon dating with the

TABLE 1 Data on sites

Loch	Altitude		Surface Area		Maximum Depth		Mean Depth		Catchment		Site (L or R)	Zone C deposit	Loch Lomond ice in Catchment
	m	(ft)	km^2	($miles^2$)	m	(ft)	m	(ft)	km^2	($miles^2$)			
Stack	36	(118)	2.56	(0.99)	33	(108)	11	(36)	103.6	(40)	L	Laminated Clay	Yes
Lochan an Smuraich	37	(120)	0.39	(0.015)	7	(22)	-		Very small		L	S. Clay	No
Cam	123	(405)	2.62	(1.01)	37	(122)	12	(38)	86.8	(33.5)	L	S. Clay	No
Veyatie	112	(366)	2.41	(0.93)	38	(126)	12	(41)	120.4	(46.5)	L	S. Clay	No
Sionascaig	74	(243)	5.41	(2.09)	66	(216)	18	(60)	40.1	(15.5)	L	S. Clay	No
Borralan	140	(460)	0.47	(0.18)	6	(21)	3	(10)	16.3	(6.3)	L	S. Clay	No
Craggie	154	(506)	0.18	(0.07)	12	(40)	5	(15)	-		L	S. Clay	No
Tarff	291	(956)	0.54	(0.21)	27	(89)	7	(24)	2.6	(1.0)	L	S. Clay	No
Duntelchaig	214	(702)	5.05	(1.95)	62	(205)	26	(84)	22.0	(8.5)	L	S. Clay	No
Clair	92	(303)	0.62	(0.24)	28	(93)	13	(42)	52.6	(20.3)	R	S/G	Yes
a'Chroisg	155	(508)	2.59	(1.0)	51	(168)	23	(74)	49.2	(19)	R	S/G	Yes
Affric	228	(747)	2.12	(0.82)	67	(221)	29	(94)	121.7	(47)	R	S/G	Yes
Achilty	36	(117)	0.60	(0.23)	36	(119)	16	(52)	5.2	(2)	R	S/G	?
Maree	10	(32)	28.57	(11.03)	112	(367)	38	(125)	442.9	(171)	R	Laminated Clay	Yes
Ness	16	(52)	56.41	(21.78)	230	(754)	132	(433)	1776.7	(686)	R	Laminated Clay	Yes
Oich	32	(105)	1.97	(0.76)	47	(154)	13	(42)	440.3	(170)	R	Sandy Silt	Yes
Ailsh	152	(498)	0.98	(0.38)	7	(24)	2	(8)	44.0	(17)	?*		
Coulin	93	(305)	0.47	(0.18)	15	(49)	5	(18)	36.3	(14)	?*		

Metric conversions of original data published by Murray and Pullar (1910)

S.Clay = Solifluction Clay
S/G = Sand and Gravel present
* = Flandrian muds not completely penetrated

Alleröd chronozone. In the present paper "Interstadial" includes pollen zones A 2, A 3 and B of Pennington *et al.* (see Fig. 2).

The carbon content of open-water lake sediments is low, so only at one site has it been possible to obtain a series of radiocarbon dates from a Lateglacial profile (Figs. 2 and 3). From the consistent agreement from site to site between pollen zone boundaries and horizons of maximum change in sediment composition (an agreement supported statistically by Principal Components analysis of the Cam Loch data (Pennington and Sackin, 1975)) it is likely that these boundaries represent the results of approximately synchronous changes in vegetation and soils consequent on major climatic shifts. It is postulated that corresponding pollen zones at other sites can be correlated with the equivalent chronozone at Cam Loch. This agrees with the opinion of Mangerud *et al.* (1974, p.118) that "chronozone boundaries seem to be climatically conditioned within southern Scandinavia" during Late Weichselian time.

METHODS

Full descriptions of methods of both sampling and analysis appear in Pennington *et al.* (1972).

Sampling

This was carried out from a boat in open water, using either a Mackereth sampler 6 m in length and 5 cm in sample diameter (Mackereth, 1958) or the smaller 4 m sampler operated by an aqualung diver (Walker, 1967), the choice of sampler being governed by depths of water and sediment. Both samplers take a continuous and uninterrupted core.

Analysis

Pollen and spore analysis. This followed in general the methods of Faegri and Iversen (1964) but was supplemented by absolute pollen analysis using the exotic pollen method described by Benninghoff (1962) and Bonny (1972). Criteria for recognition of pollen taxa are given in Pennington *et al.* (1972); at later sites *Salix herbacea* type pollen was separated from *Salix* undiff. on the criteria discussed by Fredskild (1973).

Diatom analysis. This was carried out as described by Haworth in Pennington *et al.* (1972) and more fully in Haworth (1974).

Chemical analysis. Inorganic methods for sediments of lochs Sionascaig and Tarff are described in Pennington *et al.* (1972), as are the more automated methods originally used at Loch Clair and subsequently used at Cam Loch and Lochan an Smuraich. Organic methods are given by Cranwell (1973, 1974 and in press).

LITHOSTRATIGRAPHY

In seven lakes ("R" sites) no organic Lateglacial sediment was found. At the base of their Flandrian muds is a thick and coarse minerogenic deposit which our samplers were unable to penetrate (Table 1). The thickness of this included in our cores varied from 50 cm (Loch Clair) to 6 m (Loch Ness). We attribute this minerogenic deposit to deposition from active ice during the Loch Lomond Readvance, the postulated limit of which (Sissons, 1974c, Fig. 1) enters the catchments of all R sites except Loch Achilty. In nine lakes ("L" sites, Table 1) which all lie well outside the postulated limit of the Loch Lomond Readvance, there is a complete and conformable sequence of Lateglacial sediments, passing downwards with no evidence of unconformity into barren, fluvioglacially redeposited clays, silts and sands.

The lithostratigraphy for lochs Clair, Tarff, Sionascaig, Borralan, Craggie and Cam has been described (Pennington *et al.*, 1972; Pennington and Sackin, 1975; Pennington, 1975a). These Scottish profiles agree lithostratigraphically with those from Lake District lakes to indicate a weakly divided Late Devensian interstadial, followed by a stadial (corresponding to Younger Dryas time) which saw recrudescence of mountain glaciation in some catchments, with consistent differences from lake to lake in the lithostratigraphy of stadial deposits. In Loch Stack (Fig. 1), where the sediments have not yet been analysed, the lithostratigraphic sequence is similar to that found in Windermere (Pennington, 1970, 1973).

BIOSTRATIGRAPHY

Pollen Analysis

Zonation of pollen diagrams. Comparison of four percentage pollen diagrams (Fig. 4) led to definition of three regional pollen assemblage zones for northern mainland Scotland - A, B and C - each named for the taxon whose maximum Lateglacial percentages delimit that zone (Pennington *et al.*, 1972). Zones A and B were subdivided on the basis of associated taxa (Figs. 2 and 4). At Cam Loch (Fig. 4b) an additional pollen zone was recognised below A 1, and the statistical validity of zone boundaries was established by Principal Components analysis of both percentage and absolute data.

The following pollen assemblage zones are recognised:
1. *Rumex-Salix herbacea-Oxyria* (below A 1) is defined by the presence of grains of *Oxyria* type and grains which cannot be referred with certainty either to *Oxyria* or *Rumex* (Cam Loch only).
2. Zone A is defined by maximum percentages for the profile of *Rumex* type *acetosa* (up to 50 per cent) and is divided into three sub-zones by a temporary rise in percentages of woody plants, giving a middle sub-zone, A 2, which is a *Rumex-Empetrum* zone in which *Juniperus* is present at some sites (*e.g.* Loch Sionascaig). At most sites the zone boundary A 1/A 2 is marked by a rapid fall in *Lycopodium selago* spores from their maximum frequencies in A 1. Sub-zone A 3 is defined by minima in *Empetrum* and (where present) *Juniperus*, and either a continuous curve or a small maximum of *Artemisia* pollen which is not all referable to *A.norvegica*, though pollen of this species is present at some sites.
3. Zone B is defined by maximum values for the Lateglacial profile of the woody plants *Empetrum* and *Juniperus*, associated with high percentages of Cyperaceae pollen, and is subdivided into three sub-zones by changes in their relative values.
4. Zone C is defined by maximum percentages for the profile of total *Artemisia* pollen, and at all sites corresponds closely with a lithostratigraphic unit of low organic content.
5. Two pollen assemblage zones lie between the top of the *Artemisia* zone (C) and the horizon dated to 10,000 radiocarbon years ago which by definition forms the base of the Flandrian. The lower, a herb zone (see Cg, Fig. 2) is usually dominated by Cyperaceae, with some *Rumex*, and the upper is a woody plants zone with *Empetrum* and varying amounts of *Juniperus* (NWS I at Loch Sionascaig (Fig. 4a) and Ch at Cam Loch (Figs. 2 and 4b)).

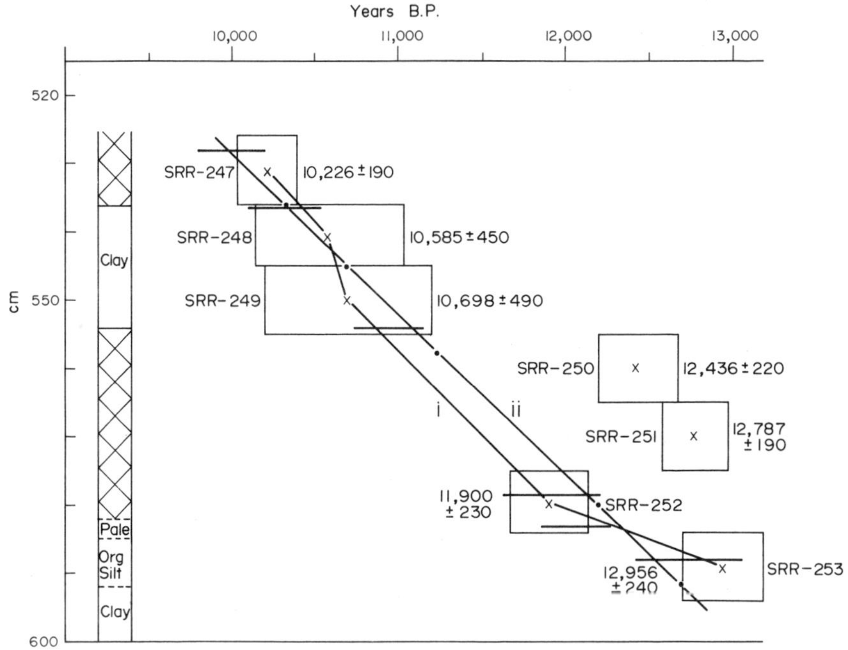

Fig. 3. Depth-time-scales for lake sediments:

3a. Cam Loch, C14 samples. i. and ii. represent alternative accumulation rates calculated from the C14 dates.

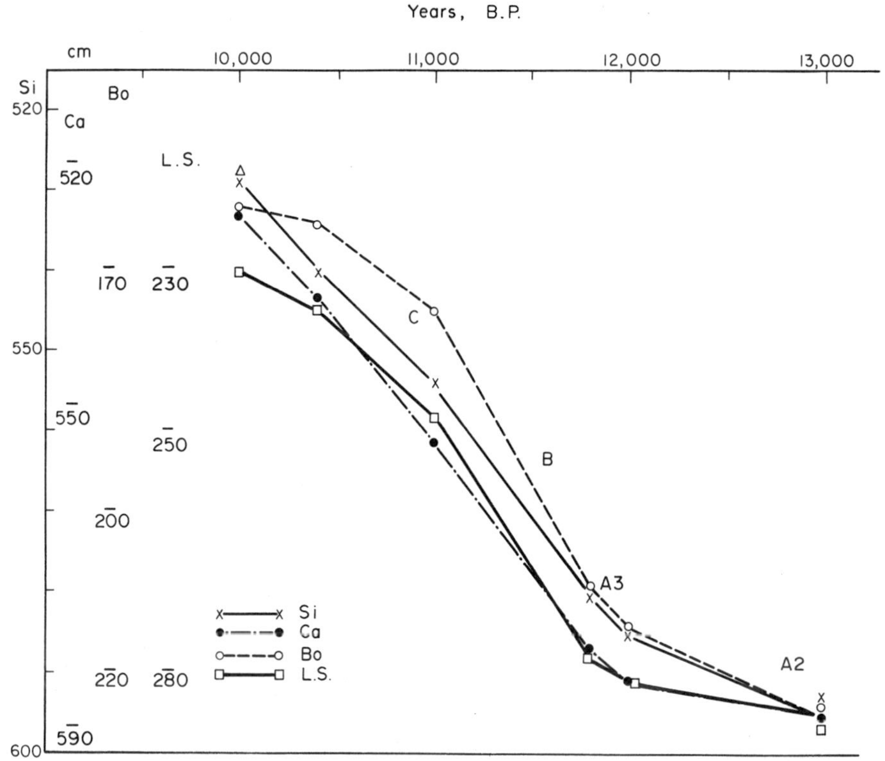

3b. Accumulation rates calculated from chronozone boundaries; Si = Loch Sionascaig, Ca = Cam Loch, Bo = Loch Borralan, L.S. = Lochan an Smuraich. A 2, A 3, B and C = pollen zones.

Chronozones based on the chronostratigraphic subdivision of the Late Weichselian proposed by Mangerud *et al.* (1974) are shown in Fig. 2 and are referred to the radiocarbon chronology at Cam Loch. The pollen sub-zone boundary A 1/ A 2, which is drawn at the first rise in percentage frequencies of the pollen of woody plants (*Empetrum* with or without juniper), coincides with the lower boundary of the Bölling chronozone at *c.*13,000 B.P., so that pollen sub-zone A 1 and the lower zone at Cam Loch are referred to the Middle Weichselian of Mangerud *et al.* (1974) (see Fig. 2). Pollen sub-zone A 3 represents the Older Dryas chronozone, and pollen zone B coincides with Alleröd time (chronozone). Pollen zone C, *Artemisia*, coinciding with a lithostratigraphic unit, represents the period from *c.*11,000 to 10,400 B.P., the first part of the Younger Dryas chronozone of Mangerud *et al.* (1974). The herb pollen zone, Cg at Cam Loch and Transitional at Loch Sionascaig (Figs. 2 and 4) and the overlying zone of *Empetrum* with varying amounts of juniper (Ch and NWS I), coincide with the second part of the Younger Dryas chronozone (the period between *c.*10,400 and 10,000 B.P.) on radiocarbon dates from Loch Sionascaig (Fig. 4a) and Cam Loch (Fig. 4b).

Numerical analysis of percentage pollen data. Principal Components analysis of the percentage pollen data from Cam Loch showed an oscillation pattern giving a threefold division of the Lateglacial profile (Pennington and Sackin, 1975, Fig. 2, Curve b). Radiocarbon dating has identified

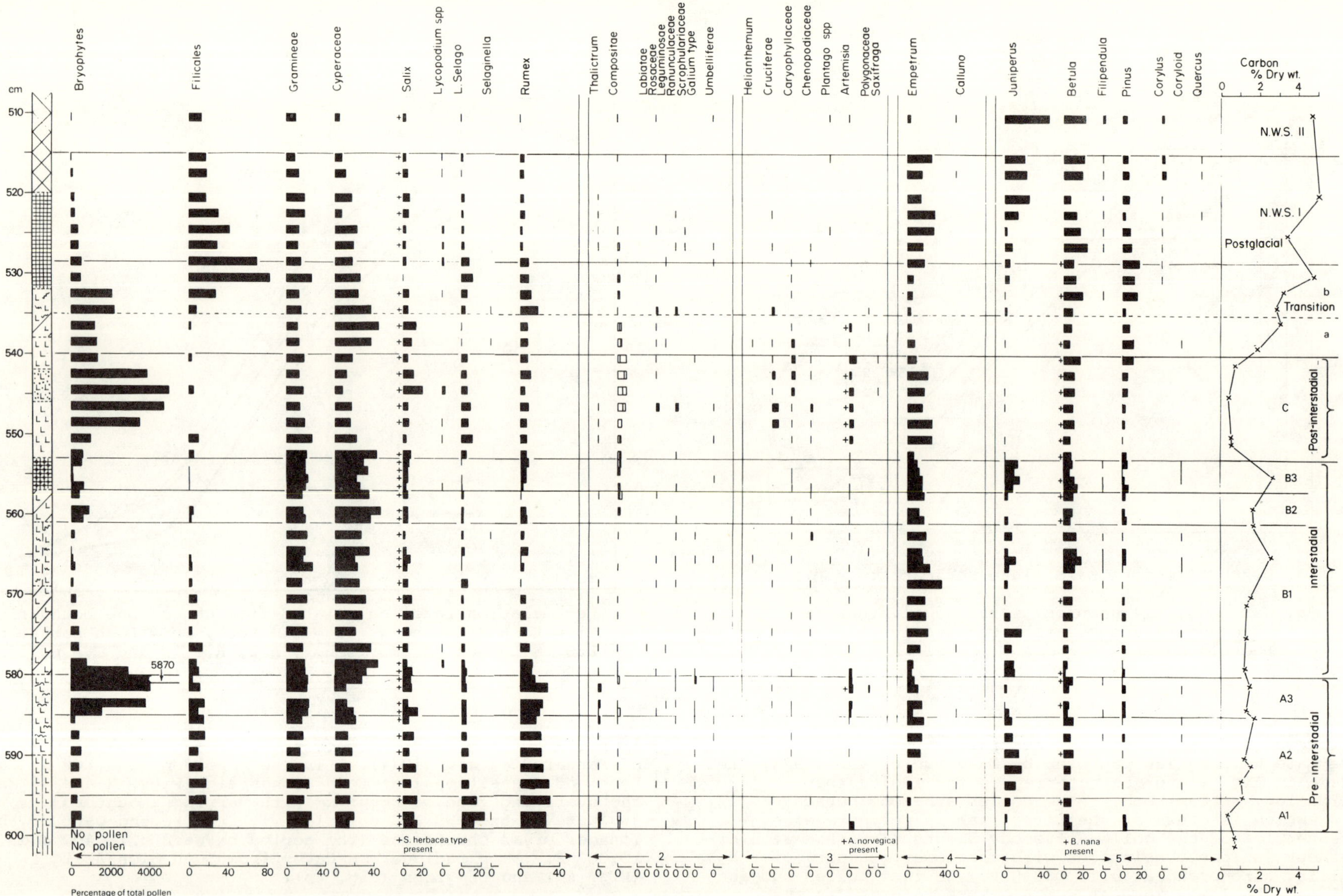

Fig. 4. Percentage pollen diagrams; all percentages represent percentage of total pollen. For stratigraphic symbols, see Fig. 4d.

4a. *Loch Sionascaig*: reproduced from *Phil. Trans.R.Soc.* B264, 1972, by permission of the Royal Society. The sample (500-510 cm) immediately above this profile, gave a C14 date of 9,474 ± 160 B.P. (SRR-15).

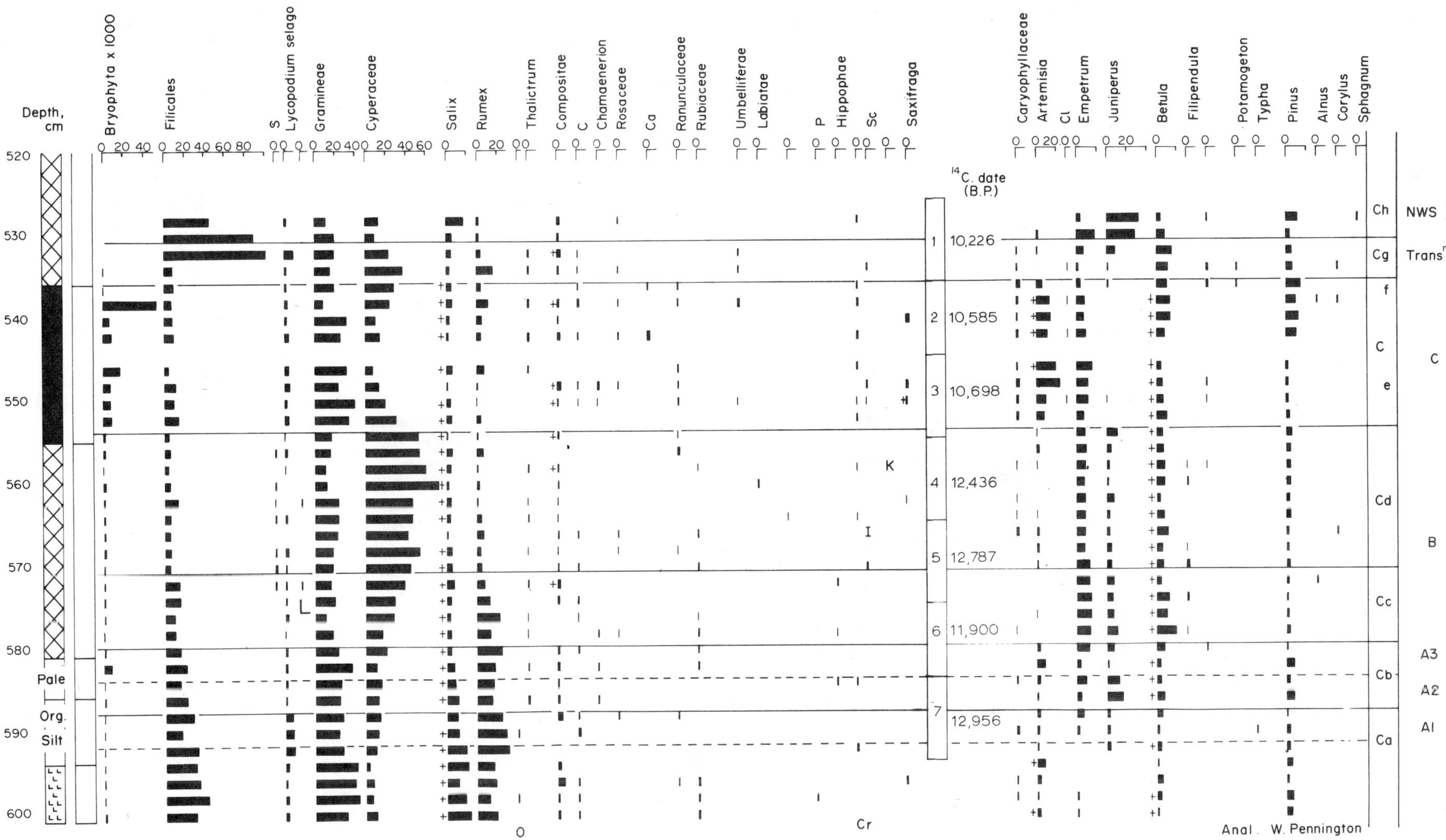

4b. *Cam Loch*: reproduced from *Boreas* 4, 1975. C = Chenopodiaceae, Ca = *Caltha* type, Cl = *Calluna*, Cr = Cruciferae, I = *Iris*, K = *Koenigia*, L = *Lycopodium* sp., O = *Oxyria* type, P = *Plantago* spp., S = *Selaginella*, Sc = Scrophulariaceae. Stratigraphic symbols as in Fig. 4d + black = faintly banded clay.

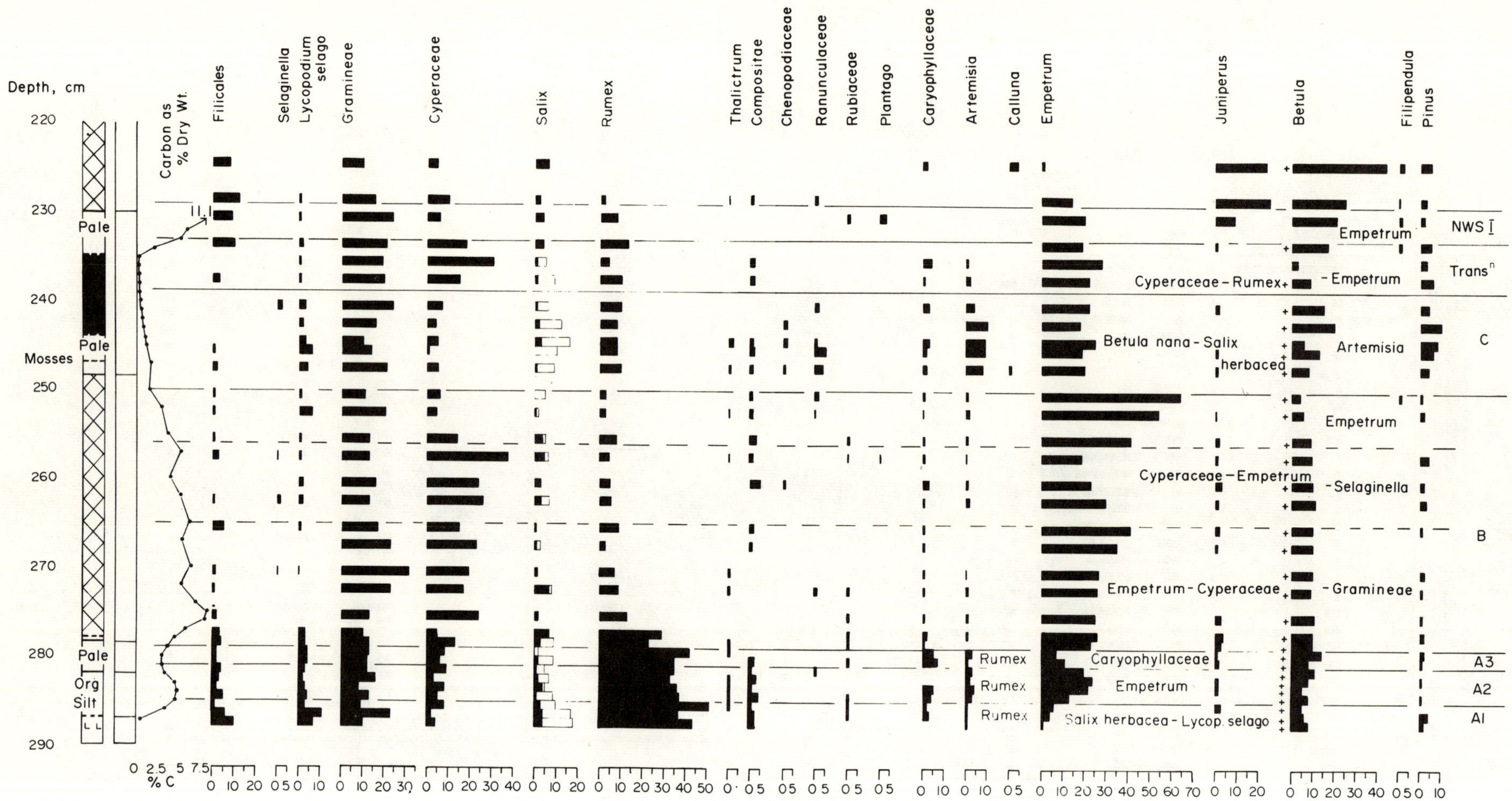

4c. *Lochan an Smuraich*: The pollen assemblage zones are named (*cf*. Cam Loch, Fig. 2). Stratigraphic symbols as in Fig. 4d + black = smooth grey clay.

this with the classical Late Weichselian pattern (Mangerud *et al.*, 1974) of a weakly subdivided interstadial followed by a stadial, and has justified the hypothesis of Pennington *et al.* (1972) that the environmental fluctuations of Late Weichselian time can be recognised in these pollen diagrams, although they do not show any expansion of the total *Betula* percentages to a figure from which the presence of Lateglacial birch woodland could be supposed.

Absolute pollen analysis. Figures 5a and 5b compare Lateglacial concentrations (grains/cm^3) of pollen of those taxa of major importance at (a) Cam Loch, a pollen-poor site (Pennington, 1973) and (b) Lochan an Smuraich, a pollen-rich site. Figures 5c and 5d present this data as annual pollen deposition rates (grains/cm^2/yr) using the time-scales shown in Fig. 3 (Cam Loch in Fig. 3a and Lochan an Smuraich in Fig. 3b). These absolute pollen diagrams will be discussed

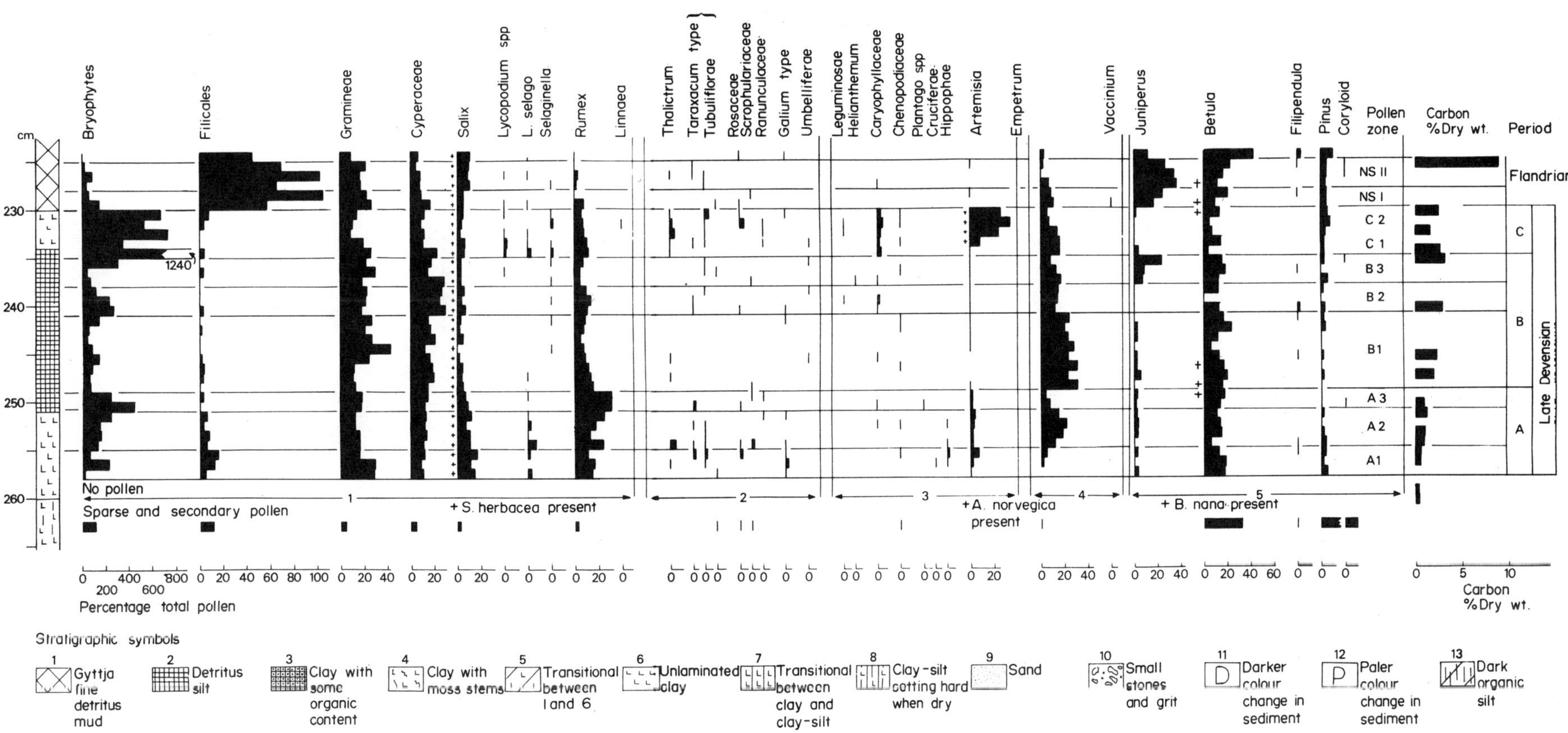

4d. *Loch Tarff*: reproduced from *Phil. Trans.R.Soc.* B264, 1972, by permission of the Royal Society.

more fully elsewhere (Pennington, in preparation), but their immediate contribution to an understanding of the Lateglacial environment in northern Scotland is to show that at no time during the Interstadial did annual deposition rates of tree birch pollen exceed *c.* 150 grains/cm^2/year, contrasting with rates of up to 1,500 grains/cm^2/year at Blelham Bog in the English Lake District (Pennington and Bonny, 1970). No local presence of birch woodland near these sites in Sutherland can be supposed. Deposition rates for total pollen at Cam Loch during the Interstadial did not exceed the figure of 762 grains/cm^2/year collected in dwarf-shrub tundra in the Canadian Arctic by Ritchie and Lichti-Federovich (1967). At the pollen-rich site Lochan an Smuraich total deposition rates during the first part of Alleröd time exceeded 1,500 grains/cm^2/year, but of this significantly less than 10 per cent was contributed by tree birches.

Interpretation of the pollen data in terms of vegetation and environment. The percentage pollen spectra from the Lateglacial sediments of these lakes cannot be matched with any published surface pollen spectra from Highland Britain today (*e.g.* Birks, 1973) because of:

i. High percentages of *Rumex* type *acetosa* in pollen zone A
ii. High percentages of *Empetrum* without *Calluna* in pollen zone B

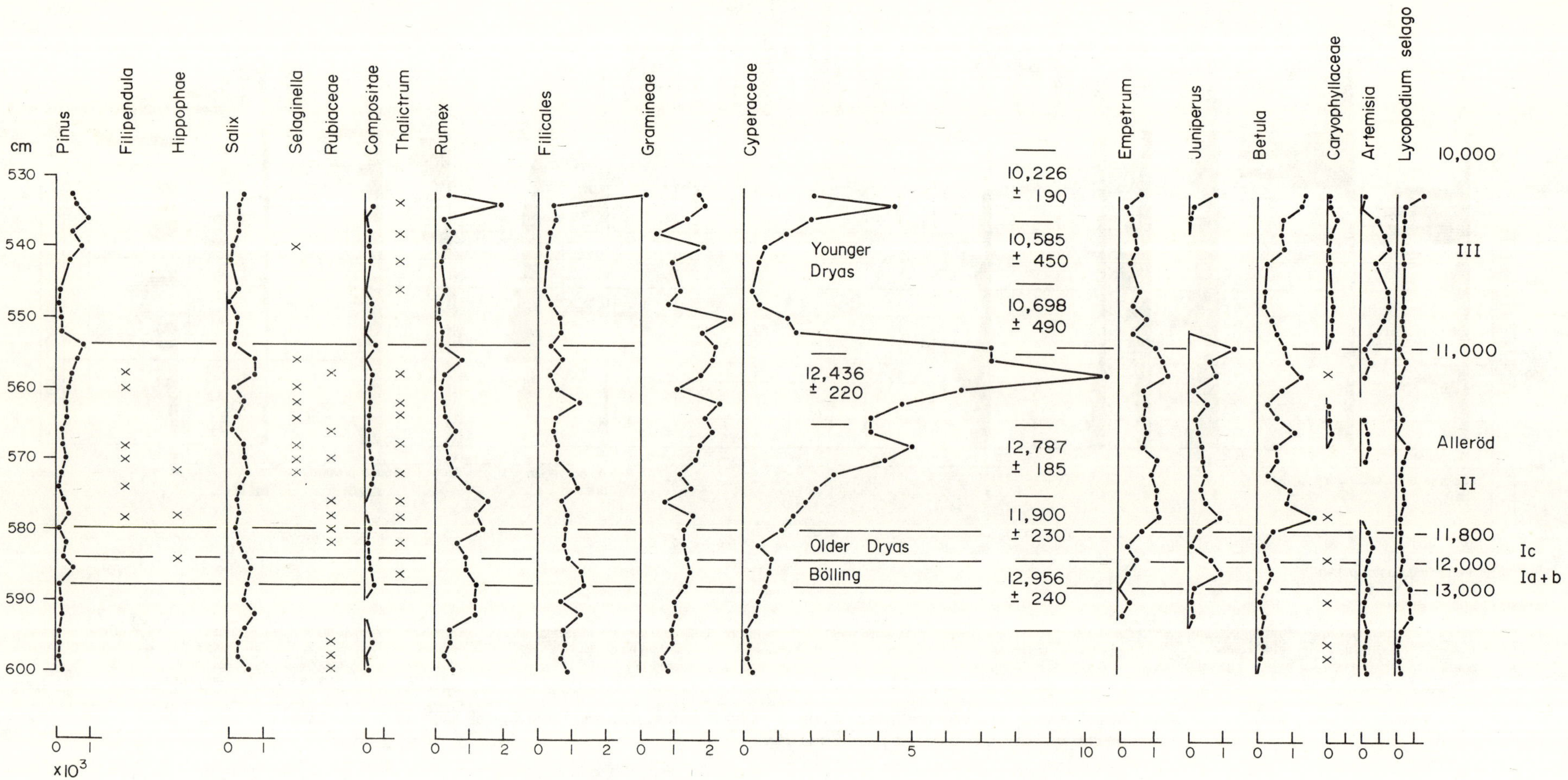

Fig. 5. Absolute pollen diagrams.

5a. *Cam Loch*: concentration of selected taxa, grains/cm³. Reproduced by permission from *Boreas* 4, 1975.

iii. Moderately high percentages of *Artemisia* (not all *A.norvegica*) in pollen zone C.

It is therefore deduced that on the mainland of northern Scotland the Lateglacial vegetation was different from the vegetation found at any altitude today; in this the mainland differs from the Isle of Skye (Birks, 1973).

Table 2 compares pollen spectra from zones A and B with some of the author's data on contemporary pollen spectra from moss polsters and lake sediments in Norway, and with pollen data from Greenland lakes given by Fredskild (1973); these sites are at higher latitudes and in Norway include higher altitudes than the Scottish lakes. No modern pollen spectra from northern Europe, of which the author is aware, contain the high percentages of *Artemisia* pollen found in zone C, and it is concluded that some of these Lateglacial pollen spectra

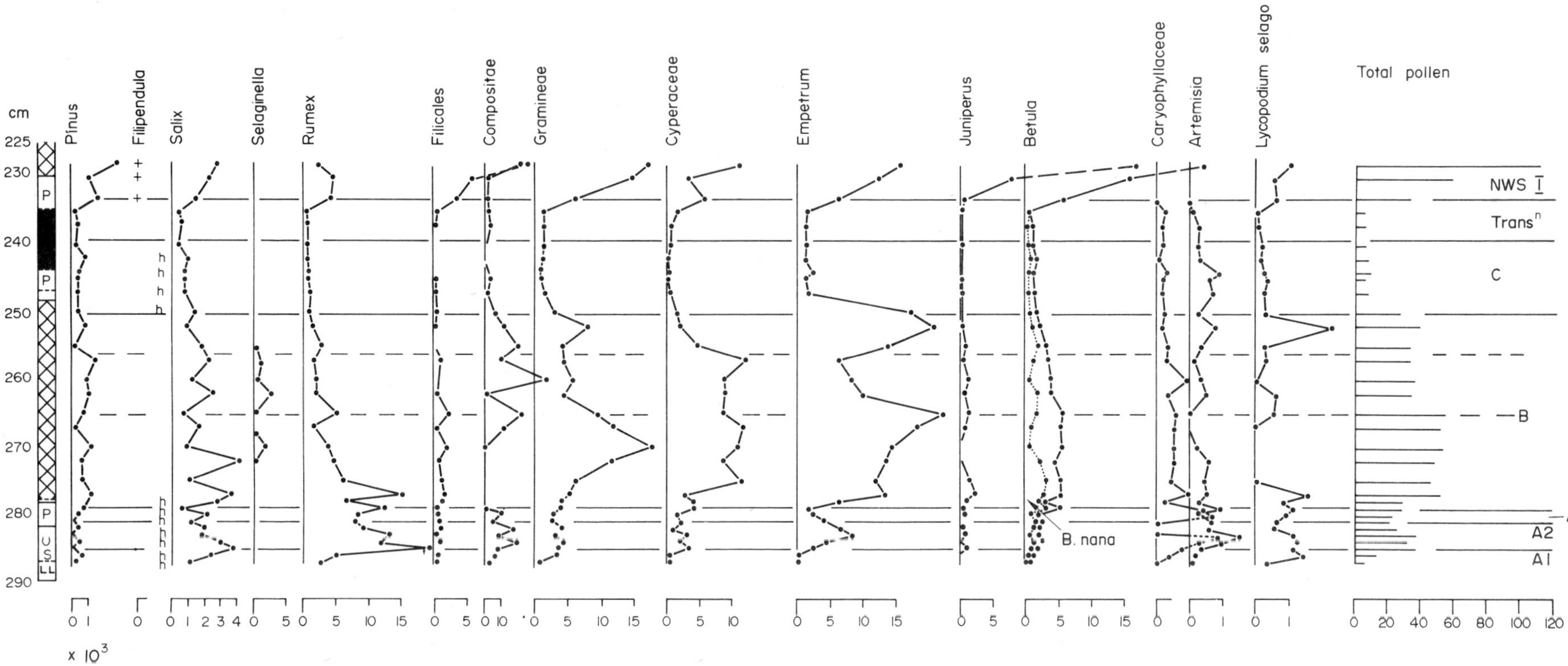

5b. *Lochan an Smuraich*: concentration of selected taxa, grains/cm^3. h signifies *Salix herbacea*.

from the northern Scottish mainland represent types of vegetation which are not found in northern Europe today.

1. A 1, before 13,000 B.P. (*Rumex-Salix herbacea-Lycopodium selago*) and the lower zone with *Oxyria* at Cam Loch. The high percentages of *Rumex*, type *acetosa*, have been found by the author in only one type of contemporary vegetation, *viz.* that colonising ground left bare during the last two centuries by the retreat of the snouts of valley glaciers originating in the Jostedalsbre ice-cap of S.W. Norway. Table 2a compares taxa lists and pollen percentages from moss polsters collected immediately down valley from the snouts of glaciers in two Norwegian valleys, with samples from zone A 1 in four of the Scottish lakes; moss polster percentages are calculated using total non-arboreal pollen as the sum, since these samples contain much non-local tree pollen. High percentages of *Salix* type *herbacea* pollen are found in the surface sediments of the small lake Leirvatn at 1400 m altitude in the Jotunheim mountains of S. W. Norway (Pennington, 1973). This lake is surrounded by snow-beds dominated by *S. herbacea*, *Ranunculus glacialis* and *Anthelia* and small glaciers are present on the surrounding mountains. Concentrations of non-arboreal pollen in the surface mud of Leirvatn have been shown to resemble those in the sediments of the basal *Salix herbacea* pollen zone in the Lateglacial profile of Blea Tarn, in the Lake District uplands, and Table 2b compares pollen percentages and concentrations in zone A 1 at Cam Loch and Lochan an Smuraich with similar data from Leirvatn surface sediment.

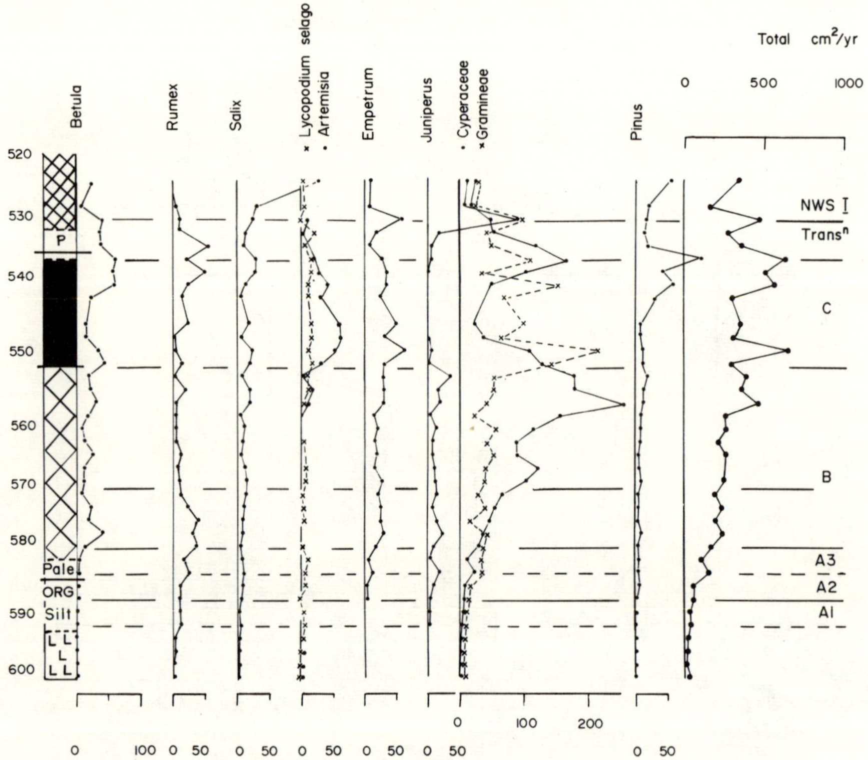

5c. *Cam Loch*: annual deposition rate (influx) of grains of selected taxa, using accumulation rates (i) of Fig. 3a.

From this comparison it is postulated that the pollen spectra of zone A 1 represent a vegetation which was a mosaic of pioneer communities of chionophilous character. *Salix herbacea* would dominate areas of long snow-lie, and *Rumex acetosa* and *Oxyria digyna* would be pioneer plants on well-drained coarse drift. Iversen (1954, p.103) cited these three species as chionophilous, pointing out how scarce is their pollen in Danish Lateglacial deposits. Percentages of *Rumex* are highest in those lakes (Loch Borralan and Lochan an Smuraich) where drift deposits surround the lake, and lowest in Loch Sionascaig. By contrast, percentages of *Lycopodium selago*, a common pioneer of new screes (*e.g.* quarry waste heaps) today, are highest at the sites (Loch Sionascaig and also Loch Droma (Kirk and Godwin, 1963)) nearest to steep mountains.

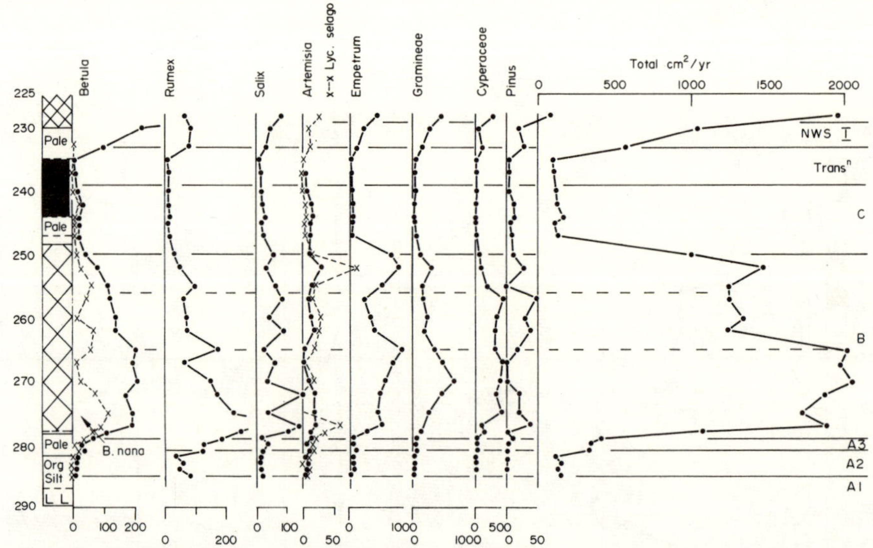

5d. *Lochan an Smuraich*: annual deposition rate (influx) of grains of selected taxa, using the accumulation rates shown in Fig. 3b.

The predominance of chionophilous plants and low values for *Artemisia* pollen suggest that A 1 pollen spectra reflect an oceanic type of Lateglacial environment, with comparatively little periglacial erosion of soils because of long winter snow cover.

2. A 2, 13,000 - 12,000 B.P. (*Rumex-Empetrum*, with *Juniperus* at some sites).
Since no species lists for *Empetrum* heath in Scotland today lack *Calluna*, the nearest comparable contemporary pollen spectra, from lakes in south Greenland, represent a vegetation in which species of *Betula*, *Salix* and *Empetrum* differ from those present or common in Scotland today (Fredskild, 1973). Total annual deposition in this zone at Lochan an Smuraich, *c*.330 grains/cm^2/year, agrees well with values of 400 and 300 today from the two Greenland lakes (Table 2c). Deposition rates for *Empetrum* and Cyperaceae also agree well, but persistence of much *Rumex* in this zone in the Scottish lakes, and presence of *Betula glandulosa* in Greenland, produce differences in the percentage frequencies. Though approximate only, this comparison of pollen data suggests that at the opening of the Bölling chronozone climatic amelioration in northern Scotland led to replacement

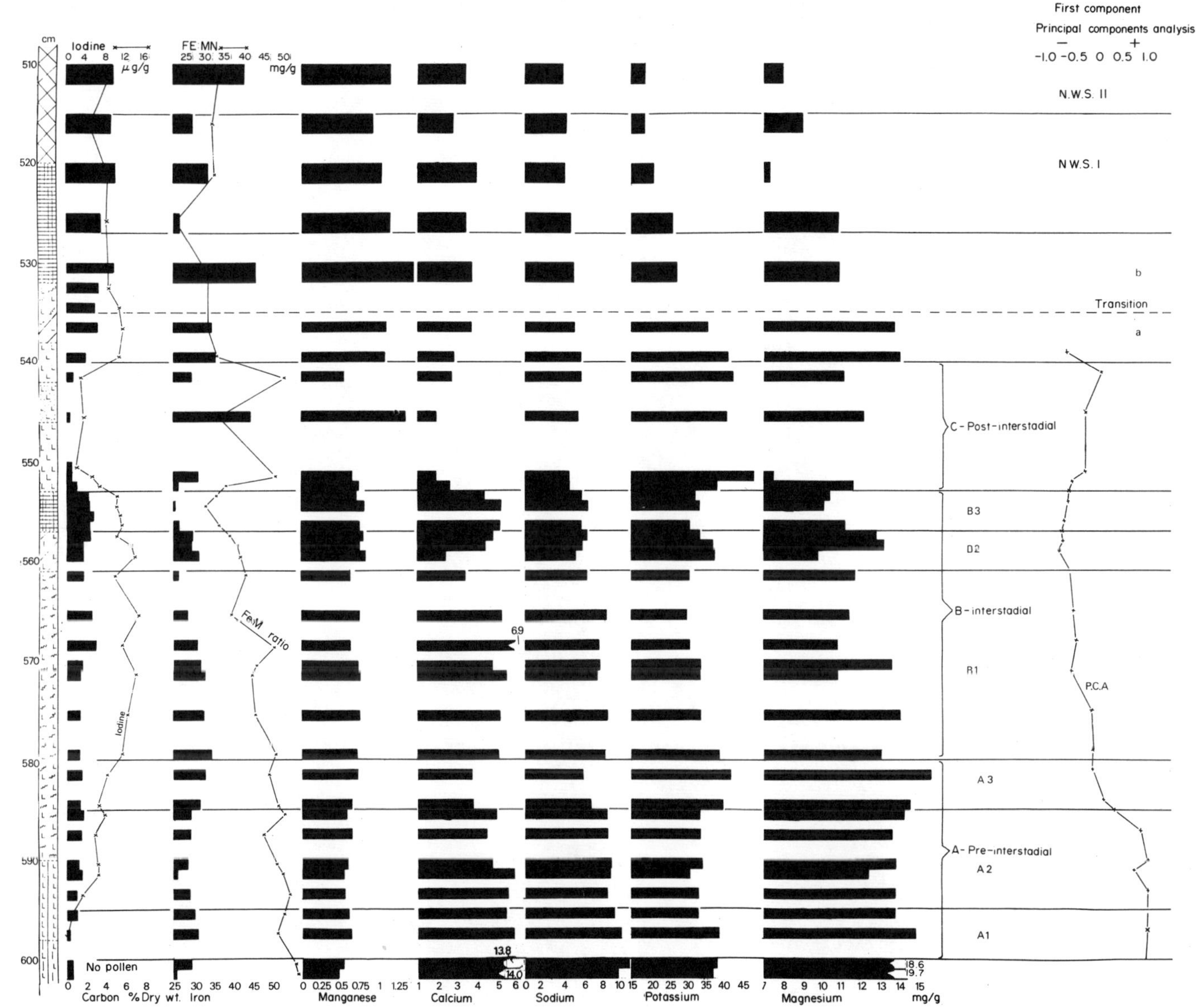

Fig. 6. *Loch Sionascaig*: Lateglacial chemical diagram, divided horizontally at pollen zone boundaries. Reproduced from *Phil.Trans.R.Soc.* B 264, 1972, by permission of the Royal Society.

TABLE 2A Comparison of Lateglacial percentage pollen spectra, Rumex Zone A, with modern samples

	SAMPLES ZONE A PERCENTAGE TOTAL POLLEN				NORWEGIAN MOSS POLSTERS PERCENTAGE TOTAL NAP				
					Nigardsbre			Austerdal	
	Sionascaig	Cam	Smuraich	Tarff	1	2	3	1	2
Pinus	4	3	-	4	80	27	13	83	110
Betula undiff.	8	2	3	6	126	26	8	50	90
Betula type nana	-	2	2	-	-	-	-	-	-
Juniperus	1	3	3	1	10	-	1	1	3
Picea					7	5	1	4	10
Alnus					27	2	1	13	7
Salix undiff.	-	-	3	-	3	1	-	1	9
Salix type herbacea	14	19	7	11	-	-	-	-	-
Empetrum	2	-	6	12	1	-	-	5	3
Calluna					-	1	1	2	13
Gramineae	21	27	9	17	33	52	30	14	31
Cyperaceae	14	10	8	11	5	-	2	5	1
Artemisia	5	1	1	2					
Rumex type acetosa	26	31	51	24	8	26	37	69	7
Oxyria	-	-	-	-	21	15	29	-	19
Cruciferae	-	2	1	-					
Compositae	2	-	1	2	-	-	-	-	7
Caryophyllaceae	-	-	2	-	-	1	-	2	-
Rosaceae	2	-	-	1	-	-	-	-	5
Ranunculaceae					1	-	1	-	-
Rubiaceae	1	-	1	1					
Umbelliferae					-	1	-	-	-
Leguminosae	1	-	1	-	-	1	1	-	5
Thalictrum	-	-	1	5					
Chamaenerion					1	1	-	-	5
Artica					5	1	-	-	1
Polygonum					1	-	-	-	-
Saxifraga					-	1	-	-	3
Hippophae	-	-	-	1					
Lycopod. selago	12	9	3	7	-	-	-	1	-
Lycopod.spp.					-	-	-	1	4
Dryopteris	-	7	-	-	-	1	2	8	13
Polypodium	5	-	1	-					
Cryptogramma					-	-	3	28	9
Polypodiaceae undiff.	24	17	-	6	20	18	25	50	49
Bryophytes undiff.	19	404	208	*	71	8	165	21	30

* not counted

TABLE 2B Comparison of pollen spectra, Rumex-Salix herbacea Sub-zone, A 1, with modern samples

	SAMPLES ZONE A 1				SURFACE SEDIMENTS OF NORWEGIAN LAKES		
	a Percentage total pollen		b		a Percentage NAP		b
	Cam	Smuraich	Cam	Smuraich	Leirvatn	Gjuvatn	Leirvatn
	(2 samples from each zone)						
Pinus	5,6	2,4	143,85	106,524	} High values-regional pollen		
Betula undiff.	-,3	6,2	- ,51	317,262			
Betula type nana	1,1	2,4	24,17	106,524			
Juniperus	-	-	-	-	20	2	2355
Salix undiff.	1,-	4,4	24,-	211,524	2	-	81
Salix type herbacea	22,18	14,13	598,270	739,1703	11.5	26	2842
Empetrum	1,1	1,4	24,34	53,524	4.5	-	893
Ericaceae undiff.					0.5	5	81
Gramineae	31,43	10,23	858,642	528,3013	38.5	50	9906
Cyperaceae	10,6	4,2	286,85	211,261	5	-	2274
Artemisia	3,1	1,1	120,17	53,131	1.5	5	244
Rumex + Oxyria	19,13	43,37	502,203	2270,4847	2	2	568
Cruciferae					0.5	-	81
Compositae	2,1	2,2	71,17	106,262	7	-	1380
Caryophyllaceae	-,1	-,3	-,17	-,393	0.5	-	81
Ranunculaceae					0.5	2	81
Rubiaceae	2,1	-,1	71,17	-,131			
Leguminosae	-	1,-	-	53,-			
Thalictrum					1.5	-	244
Polygonum					0.5	-	81
Saxifraga					-	2	-
Lycopodium selago	3,3	7,11	96,51	370,1441	1	-	162
Lycopodium spp.					5.5	-	1786
Selaginella					4	-	893
Dryopteris	-	-,1	-	-,131			
Polypodium	9,11	4,3	263,169	211,393	0.5	-	81
Polypodiaceae undiff.	22,32	6,3	622,473	317,393	43	26	12424
Bryophytes undiff.	1050,820	333,355	33750,12337	16896,46505	100	61	19730

a. Percentages
b. Concentrations (grains/cm^3)

of a discontinuous vegetation cover, composed of pioneer communities, by a more closed cover of dwarf shrub heath or tundra. *Juniperus* does not expand to high values at this horizon at any site in northern Scotland; maximum percentages were found in Loch Sionascaig on the western seaboard (Fig. 4a), while deposition rates at Cam Loch are about twice those now found in Greenland lakes where the ratio of cover of *Empetrum:Juniperus* in the vegetation is estimated at 50:1 (Fredskild, 1973).

3. A 3, 12,000 - 11,800 B.P. (*Rumex-Artemisia* with reduced *Empetrum-Juniperus*).
No contemporary pollen spectra resembling those of this sub-zone could be found. Accumulation rates cannot be determined with sufficient precision to decide whether or not an absolute increase in *Artemisia* pollen is registered;

TABLE 2C Comparison of Lateglacial pollen spectra, Empetrum Zones, A 2 and B, with modern samples

a. Percentages (Percentage total pollen)

	A 2		B		SURFACE SEDIMENTS OF GREENLAND LAKES†	
	Cam	Smuraich	Cam	Smuraich	Isoetes Sö	Klöftsö
Pinus	3	-	3	-	-	-
Betula undiff.*	3	2	5	4	35	40
Betula type nana	1	9	1	5		
Juniperus	12	-	4	2	+	1
Salix undiff.	1	1	1	4	} 3	} 3
Salix type herbacea	7	4	4	1		
Empetrum	9	20	12	41	25	15
Ericales undiff.					+	5
Gramineae	25	12	17	13	15	20
Cyperaceae	15	9	44	14	10	10
Artemisia	3	3	1	1		
Rumex + Oxyria	16	35	4	8	1	+
Cruciferae	1	-	-	-		
Compositae	1	1	1	3		
Caryophyllaceae	1	-	-	1	+	-
Ranunculaceae	-	1	-	-	-	+
Rubiaceae	-	-	1	-		
Umbelliferae	-	1	-	2	-	+
Leguminosae	-	2	-	-		
Thalictrum					+	-
Hippophae	1	-	-	-	-	-
Chenopodiaceae	-	-	1	-		
Lycopodium selago	2	3	2	1	1	+
Selaginella	-	-	2	-	-	-
Dryopteris	-	1	1	-	+	+
Polypodium	3	1	2	1		
Polypodiaceae undiff.	12	3	2	-		
Bryophytes undiff.	736	158	1750	153	Not recorded	

b. Annual deposition rates (grains/cm^2/year)

	A 2		B			
	Cam	Smuraich	Cam	Smuraich	Isoetes Sö	Klöftsö
Pinus	4	-	7	-	-	-
Betula undiff.* } Betula type nana	6	36	14	112	145	125
Juniperus	16	-	10	24	3	6
Salix undiff. } Salix type herbaceae	10	16	11	61	15	10
Empetrum	12	66	27	502	125	50
Ericales undiff.					5	15
Gramineae	34	39	39	160	50	75
Cyperaceae	20	30	101	171	50	30
Artemisia	4	10	1	12		
Rumex + Oxyria	22	116	8	98	5	+
Cruciferae	1	-	-	-		
Compositae	1	-	3	37		
Caryophyllaceae	1	-	-	12	+	-
Ranunculaceae	-	3	-	-	-	+
Rubiaceae	-	-	1	12		
Umbelliferae	-	3	-	37	-	+
Leguminosae	-	7	-	-		
Thalictrum					+	-
Hippophae	1	-	-	-		
Chenopodiaceae	-	-	3	-		
Lycopodium selago	3	10	3	12	3	+
Selaginella	-	-	3	-		
Dryopteris	-	3	2	-	-	+
Polypodium	4	3	4	12		
Polypodiaceae undiff.	20	10	4	-		
Total cm^2/year	148	330	247	1230	400	300

* In the Greenland samples *Betula* undiff. includes *Betula glandulosa*
† Fredskild (1973). + Signifies presence.

possibly only a fall in local *Empetrum* and juniper pollen took place.

4. B, 11,800 - 11,000 B.P. (*Empetrum* - Cyperaceae).
Table 2c shows the extent to which pollen deposition during the Alleröd chronozone at Cam Loch and Lochan an Smuraich resembles that in the surface samples from south Greenland. Total pollen deposition per unit area per year at the pollen-poor site, Cam Loch, falls near the Greenland figure, but at the pollen-rich site, Lochan an Smuraich, it was between 3 and 4 times the Greenland figure. About ten times as much *Empetrum* pollen, and five times as much sedge pollen, reached the sediments of Lochan an Smuraich each year, compared with present deposition rates in Greenland; all taxa recorded except total *Betula* show higher figures at Lochan an Smuraich. In view of the variations between recent pollen deposition rates in different parts of the same basin reported by Davis *et al.* (1973), such differences may well fall within the normal range of deposition rates from ericaceous dwarf shrub tundra. However in the Scottish profiles it is clear that maximal Lateglacial annual pollen deposition rates were found in Alleröd time, and it seems possible that optimal Lateglacial environmental conditions for this vegetation were experienced during this period. An alternative explanation is that the high pollen deposition rates represent intensified erosion of pollen-containing Alleröd soils from the catchment - even the very limited catchment of this enclosed kettlehole. Throughout Alleröd time, maximal Lateglacial deposition rates for tree birch pollen (Figs. 5c and 5d) indicate that tree birches must have approached nearest to the region during Alleröd time (*cf.* their presence near Loch Meodal on Skye at about this time deduced by Birks (1973, Fig. 27)).

5. C, 11,000 - *c.*10,400 B.P. (*Artemisia* - herbs).
As for sub-zone A 3, no contemporary pollen spectra of this kind have been found. Percentages and deposition rates of *Artemisia* show much local variation within the region, percentages varying from less than 10 at Lochan an Smuraich (where no absolute increase was involved) to more than 20 at Lochs Craggie, Tarff and Duntelchaig. Walker (1975a) suggests a correlation between low frequencies of *Artemisia* and much local snow cover at his sites in the Grampians. Certainly among the sites described in this paper, Lochan an Smuraich lay nearest (of the L sites) to Loch Lomond Readvance ice, but on the other hand the next nearest site is Loch Craggie (if the end moraines near Loch Ailsh are attributed to Loch Lomond Readvance ice (Sissons, 1974c)) and here *Artemisia* percentages are high. The characteristic association of high *Artemisia* percentages with pollen of families which include many herbs of open ground (Caryophyllaceae, Compositae, Cruciferae), indicates a break-up at these sites of the continuous vegetation cover, for which the most obvious explanation would be an increase in frost disturbance, during freeze-thaw cycles, of ground not protected by a winter snow cover.

6. *c.*10,400 - 10,000 B.P. (Transitional (herb) zone).
At most sites a Cyperaceae-*Rumex* zone is present between the top of the *Artemisia* zone and the *Empetrum* zone hitherto called NWS I (Pennington *et al.*, 1972). Radiocarbon dates from Cam Loch and Loch Sionascaig suggest that the horizon at 10,000 B.P. falls within the *Empetrum* zone (Fig. 4). This pollen sequence has been interpreted (Pennington *et al.*) as a repetition of the plant succession recorded at the base of the Lateglacial profiles, consequent on restoration of the glacial environment during Younger Dryas time. Radiocarbon dates from Loch Clair (8,910 ± 130 B.P.) and Loch Sionascaig (*c.*9,000 B.P.) for the base of the Flandrian *Betula* zone, show that the spread of forest in northern Scotland was delayed for *c.*1,000 years after the date of 10,000 B.P. which marks the beginning of the Flandrian.

Diatom Analysis

Dr. Elizabeth Y. Haworth writes: Diatom assemblage zones can be distinguished within the Lateglacial profiles of Loch Sionascaig (Haworth, *in* Pennington *et al.*, 1972) and lochs Borralan, Tarff and Cam (Haworth, 1974). The boundaries of these diatom assemblage zones correspond well with those of pollen zones A, B and C (Table 3). Changes in assemblages can be related to the environment within the lake and to local soil conditions (edaphic factors) and are interpreted as indicative of the overall control of the Lateglacial environment by those climatic variables which influence water and soil chemistry through their effect on the nature and stability of the land surface.

Loch Sionascaig Lateglacial diatom profile (see Table 3).
There are few diatom taxa in the basal biogenic sediments of Loch Sionascaig, but numbers of taxa increase in successive samples of this lowest diatom zone, and the assemblage is one that typically occurs on the benthos of lakes of alkaline water. The boundary DZ 1/2 (Table 3) is placed where there is the greatest increase in species diversity and where, in Loch Sionascaig, planktonic taxa first appear as a dominant part of the diatom assemblage. This feature is unique to this site, a large lake by British standards (Table 1) and so far the only site where a good conformable Lateglacial sequence has been obtained from the middle of a large lake. The plankton forms, *Cyclotella* spp., are those cited by Hustedt (1937-39) as typically dominant forms in oligotrophic (nutrient-poor), neutral pH waters of northern-subalpine, temperate lakes, but the benthic assemblage in the Sionascaig assemblage remains indicative of alkaline waters, and also includes taxa more typically found in nutrient-rich

TABLE 3 Diatom biostratigraphy

Diatom assemblages	Diatom assemblage zones (DZ)		Sediment stratigraphy	Pollen assemblage zones		Time-scale Cam Loch C14 dates
Fragilaria dominance with appearance or increase in plankton forms. Reappearance of many DZ2 taxa and arrival of new taxa indicative of continuing alkalinity.	4		grey-brown organic mud	N. W. S. I		Flandrian *c*.10,000 B.P.
Fragilaria elliptica often dominant especially in small lakes. *Cyclotella comensis* (plankton) dominant in Cam Loch. Aerophile taxa present. Many DZ2 taxa absent in this zone.	3		grey clay	C		Loch Lomond Stadial
Fragilaria, mainly *F. construens*, dominant in a mainly benthic assemblage typical of alkaline waters. Only in Sionascaig is plankton well represented by *Cyclotella* spp. (nutrient poor plankton forms). Number of taxa increases and includes those characteristic of nutrient rich waters, especially in Cam Loch. Suggests nutrient gradient from surface sediment up into open water.	2		grey-brown organic silt	B		*c*.11,000 B.P. Alleröd time *c*.11,800
Fragilaria elliptica prominent in assemblages and a few characteristically subaerial taxa.	1	1c	increase in grey clay particles	A	A3	Older Dryas time *c*.12,000
Fragilaria taxa dominant. Low species diversity. Assemblages mainly benthic and typical of alkaline waters.		1b	light-brown clay-silt slightly organic		A2 A1	Bölling time *c*.13,000 B.P.
This zone only apparent at Cam Loch. *Fragilaria* dominance. Pioneer community with low species diversity.		1a	soft grey clay	Pre A1		

(eutrophic) waters such as the north German lakes (Hustedt, 1950) or in Danish lakes (Jörgensen, 1948).

The DZ 2/3 boundary coincides with the lithological change to grey clay, and is marked by the disappearance of many of the taxa of the previous zones. There are taxa which appear restricted to this part of the profile, *e.g. Navicula fracta*, *Pinnularia cuchlandtii* (Pennington *et al.*, 1972, Fig. 7) and taxa that live subaerially on soils, wet rocks and among mosses are also more prominent in this zone than elsewhere, which supports the pollen and chemical evidence for renewed soil disturbance due to climatic recession.

Other Lateglacial diatom profiles. Three other Lateglacial

TABLE 4 Loch Tarff diatom profile

Diatom Zones	core depth (cm) / taxa	F. elliptica	F. pinnata	F. construens	F. brevistriata	Cal. bacillum	Cym. thumensis	Mel. teres	Cym. semisymmetrica	Gyro. attenuatum	Am. ovalis v. libyca	F. construens v. binodis	Am. ovalis v. pediculus	Dent. tenuis	Ni. fonticola	N. explanata	Sy. parasitica	N. vitabunda	Coc. diminuta	Am. eximia	Ac. laterostrata	N. järnefeltii	Coc. placentula v. lineata	N. pseudoscutiformis	Coc. placentula v. euglypta	Epi. turgida	G. intricatum v. pumila	Ac. levanderi	F. virescens	P. nodosa	T. flocculosa	Ac. suchlandtii	Ac. maxima	Ac. flexella	D. elliptica	Cym. aspera	Ac. linearis	Ac. calcar	P. mesolepta	An. exilis	Cyc. comta	Cym. ehrenbergii	Cym. ventricosa	Sta. phoenicenteron	Cyc. antiqua	E. valida	An. zellensis	P. sudetica	Cera . arcus	N. contenta v. parallela	Cym. sinuata	E. lunaris	Cym. cuspidata	G. constrictum	G. acuminatum	G. angustatum	[illegible]. semicruciata	Ac. clevei	Su. elegans	[illegible]. minima v. subatomoides	Ac. östrupii	Fr. vulgaris
	cm	1	2	3	4	5	6	7	8	9	10	11	12	13	14	15	16	17	18	19	20	21	22	23	24	25	26	27	28	29	30	31	32	33	34	35	36	37	38	39	40	41	42	43	44	45	46	47	48	49	50	51	52	53	54	55	56	57	58	59	60	61
4	227	235	155	122	52	7	2	12		1	10	4	9		1		14		17	4	2	18	1	+								12	+	+	2		1	1			2	2					+	+										10	+	274	1	+
	228	225	175	102	57	9	15	10	+	+	+	2	15		1	+	2		16	10	2	26	+	5		+		1	2			9		+	1	+		1			1	+		+			+	+										22	+	28	9	+
	229	338	207	87	101	3	14	29			2	2	11		4				92	4		13	1				+	1				7			+						1	+	2				+	+			+							11	+	53	2	
	230	664	108	3	135	1	3	+	5	5	2								15	5		3	1		+						3	1			1	+		1			1	+	1	8	+			+				1	1			2		6				
3	231	888	4	3	8	4			3		17									6		1	4					2			3			1	4				4	2				10	+	1	+	1				3	4									
	232	866	7	5	8	8			7		9									15		3	3					4			5				5				+	1		1	3	9	1			2		1	2	1	9	+	4	3	+					
	233	848	33	3	1						11			+				2	1	1		2									23			1				2	1	2			5	8				2		2		10	9	7	3	1						
	234	560	145	118	24	4		2	10		7	1	2							4		9	+	1				1	2	5	7	18	+	1	2		3		+	11	1		9	+		+	1	6	+	2	1	12	4	2	5	1						
2b	235	246	188	124	143	15	4	3	23	+	8	27	8	4				17	2	14	5	12	3	1		+	+	13	8	+	3	29		4	1		2	8	1	4	14	+	11	+		2	1	3	+	1	1	3	+							3		
	236	188	299	156	92	4	18	8	2	2	13	10	6			+	3	20	7	18	7	20	4	2	+		2	6	2	+	2	24			1			14			15		6	1	+	+	+	+												3		
	237	142	362	144	121	21	14	12	4	5	20	14	5			+	4	29	2		10	16	3	1	+	+	2				+	4						27	+	2	9	+	3	1																		
2a	238	228	402	199	83	5		1	10	+	8	2	7				1	14		4	5	+					2		3		1		+	2		+	6	8																								
	239																																																													
	240	232	395	180	44	10	1	2	13		14	1	9					16	1	5	10	7	5		2			2	6	1	2	9	+	3	2								1					+														
	241																																																												
	242	219	387	130	105	16	19	12	2	1	12	7	1			2	2	9	23	2	3	19	1	2	2	+	1	2	1												1																					
	243																																																												
	244	277	440	68	71	21	18	7	3	+	2	6	5		3	1		2	48	2	1	3	4	2																				+																		
	245																																																												
1b	246	361	182	112	149	5	46	8	9	2	5	26	5		22	2	10	12	6																		4							+														1				
	247																																																												
	248	399	218	183	99	5	26	18	2	1	2	13	19	1	2	+	4																									+	+	1			1	+												1		
	249																																																												
	250	408	341	167	53		4	+			1			1																																																
	251	575	183	110	73	21	7	+	1	9	16	2	1																																																	
1a	255	+																																																												
	260																																																													
	265	+	+	+	+			+													+																				+																					

+ = presence, nos. = per thousand

profiles from lake basins (lochs Cam, Borralan and Tarff (Fig. 1)) all support the theory that these diatom assemblage zones are of regional rather than local significance (Table 3). Unfortunately the core taken in 14 m of water in Cam Loch did not include the same planktonic phase as was found in Sionascaig in zone DZ 2 (core from *c.* 20 m of water), but the location of this core in a bay of Cam Loch biases the assemblage towards the benthic or littoral community, as in the two profiles from smaller, shallower lakes where the assemblages were similarly dominated by *Fragilaria* spp. Cam DZ 2 assemblage contained many taxa found in nutrient-rich waters (Haworth, 1974) and the interpretation that this was indeed a more productive phase is corroborated by the interpretation made on the basis of the organic constituents of the same period (see Cranwell, below). Overall it is suggested that the lake environment was similar to that found in present day alpine lakes (Pechlaner, 1971; Brettum, 1972) where a concentration gradient may become established above the major source of nutrients, the benthos, and where penetration of light to great depths allows the mainly non-diatomaceous plankton population to respond to this gradient (Haworth, 1974). All four profiles suggest a great uniformity in the type of lake following the retreat of the ice, as all have diatoms indicative of alkaline waters - and thus demonstrate a fundamental difference between lakes of the British Lateglacial and Norwegian lakes in presently glaciated areas, which tend to become acidotrophic once the flow of meltwater from active glaciers ceases.

Some diatom taxa occur constantly throughout the profiles but others show definite patterns of distribution (see the Sionascaig diagram in Pennington *et al.*, 1972, Fig. 7). The record for Loch Tarff (Table 4) shows that rather a small number of significant taxa were found at this site, but it still provides a good example of the general diatom stratigraphy, and includes the domination of *Fragilaria* spp. throughout the period, especially that of *Fragilaria elliptica* in DZ 3. Many taxa exhibiting patterns of possible ecological significance occur only at the level of presence or absence, presumably because they are at the limit of their ecological tolerance. Evidence of the reality of this pattern is thus considerably strengthened by its recurrence in several different diatom profiles, not only in northern Scotland but elsewhere in the British Isles. The advantage of comparing profiles from several sites has been in the recognition of several sub-zones in the diatom zonation, for which evidence at any one site is so slight that it may be overlooked. The most important of these is one which can be correlated with pollen and chemical indications of a slight climatic recession at the top of diatom zone 1 and pollen zone A (pollen sub-zone A 3). In this a few sub-aerial taxa and other elements of a diatom zone 3 assemblage (Table 3) occur. Other sub-zones may well become recognisable as expertise in analysis and interpretation reaches levels already attained in pollen analysis.

GEOCHEMISTRY

Inorganic

Four Lateglacial profiles, from Loch Sionascaig (Fig. 6), Cam Loch (Pennington and Sackin, 1975, Fig. 5b), Loch Tarff (Pennington *et al.*, 1972, Fig. 16) and Lochan an Smuraich (unpublished) have been analysed for a range of 8 - 9 chemical elements, and although it was not possible to distinguish "chemizones" it has been established that horizons of major change in sediment composition coincide with pollen zone boundaries.

Sediments older than 13,000 C14 years (pollen zone A 1 and below). In each profile the lowest sample to contain countable pollen is also that in which are found the first changes in sediment composition which indicate soil leaching (diminution in calcium, followed by sodium) and the onset of accumulation of organic matter in soils no longer subject to frost disturbance (increases in carbon, nitrogen and iodine). Below this horizon the barren sediments contain only traces of carbon, no iodine, and high proportions of bases - calcium, sodium, potassium and magnesium - indicating the presence of unleached skeletal soils in the catchments. Potassium and magnesium show a parallel tendency to diminish above this horizon owing to leaching of parent material, but the enrichment of the sediments of zone A 1 in these two elements indicates formation of potassium- and magnesium-rich clay minerals and transport of these into the lakes (Fig. 6). The sediments therefore record the processes of soil maturation - leaching, humus accumulation and chemical weathering - which set in as soon as the soils were no longer disturbed by periglacial processes. The time at which this change took place corresponded at each site with the change from "polar desert" to a partially vegetated land surface: it must have been well before 13,000 B.P. at Cam Loch.

Interstadial sediments 13,000 - 11,000 B.P. (pollen zones A 2, A 3 and B). The zone boundary A 1/2 coincides with a rise in organic matter (Figs. 4c and 6) to interstadial levels, and with further falls in calcium, sodium, potassium and magnesium. These changes are consistent with the progressive stabilisation of the land surface and leaching of these stabilised soils which is suggested by the pollen changes. Both the low values for carbon (only rarely exceeding 5 per cent of the dry weight) and the comparatively high ratios of iodine to carbon (Pennington and Lishman, 1971) distinguish the composition of these interstadial sediments from that of sediments of the same age from similar lakes in the Lake District lowlands where interstadial birchwoods developed. This suggests lower biological productivity and more acid organic matter in the interstadial soils of northern Scotland.

Division of the Interstadial (pollen zone A 3, = Older Dryas chronozone) is apparent on the chemical diagrams. Of the sites analysed, Lochan an Smuraich, the enclosed kettlehole basin, is the only one at which carbon values fall within the lithostratigraphic unit labelled "Pale" layer (Figs. 4b, 4c and 5); this layer was not distinguished lithostratigraphically in Loch Sionascaig (Fig. 6). The change in organic content within the sediments of pollen zone A 3 is therefore at most a small one. The chemical diagram from Loch Sionascaig (Fig. 6) shows however a maximum in this pollen zone of potassium and magnesium, associated with clay mineral concentration, and this evidence for increased input from weathered soils is supported by the diatom evidence for species of sub-aerial habitats in this zone at some sites (Haworth, 1974 and above). The evidence as a whole is therefore in favour of a climatic recession of small amplitude during the Older Dryas chronozone, with no general renewal of the processes of periglacial erosion.

In the later Interstadial (= Alleröd chronozone), maximum values for carbon and its associated elements in soil organic matter, nitrogen and iodine, are found. Principal Components analysis of the Cam Loch data emphasised the distinctive character of the sediments of this zone (Pennington and Sackin, 1975). It is postulated that under a vegetation of ericaceous dwarf shrub heath or tundra, continuous leaching since *c*.13,000 B.P. of soils whose bedrock is intrinsically poor in bases had, by Alleröd time, led to soil acidity and the development of an organic Al horizon. At the same time, an increasing tendency towards soil erosion throughout Alleröd time (Pennington, 1973; Craig, 1973) is postulated as the mechanism which led to redeposition in lake basins of the material of this organic soil horizon, formed at an earlier date on the catchments and therefore of older C14 age. The minima in total iron, and to a lesser extent in manganese, which characterise Alleröd sediments, could indicate either poverty of the organic soils in these elements, or loss from the ecosystem due to transport in solution of these elements under reducing conditions in soils and lake waters (Mackereth, 1966).

Sediments of the Loch Lomond Stadial 11,000 - *c*.10,400 B.P. (pollen zone C). At all sites, post-interstadial sediments corresponding with the Younger Dryas chronozone include a lower and lithologically distinct deposit corresponding with the *Artemisia* pollen zone, C, which is interpreted as the sediment of a period of climatic severity, when universal frost-disturbance of soils deposited mineral material from weathered soils at L sites, and the local presence of active ice at R sites led to deposition of unweathered mineral matter derived from fresh glacial erosion. This has been demonstrated at Lake District sites using iodine as a measure of matured interstadial soils (Pennington and Lishman, 1971), but our chemical methods, developed for analysis of fine-grained sediments (Mackereth, 1966) have not so far proved applicable to the coarser-grained sediments found in northern Scotland both at R sites and in the laminated post-interstadial sediments of Loch Stack.

Analysis of the fine-grained post-interstadial deposits at the L sites lochs Sionascaig, Tarff and Cam, showed that in composition they resemble the barren sediments at the base of the profiles rather than that of the Interstadial,* but nevertheless differences with respect to certain mineral elements between post-interstadial sediments and the barren sediments at the base of the profile point to a different, probably more local, origin for the post-interstadial sediment. The very low carbon values found in the sediments of pollen zone C, even within the enclosed kettlehole basin of Lochan an Smuraich (Fig. 4c) show that destruction of the organic interstadial soil horizon must have been complete, down to the lowest altitudes, and aquatic productivity low.

Transitional deposits (*c*.10,400 - 10,000 B.P.). Figure 6 shows the rapid change in sediment composition which begins at the upper boundary of pollen zone C, total carbon increasing steeply to values higher than those reached at any horizon in the Interstadial. Simultaneously the proportion within the mineral fraction of the erosion indicators potassium and magnesium (and to a lesser extent sodium) decreases equally rapidly. The changes in sediment composition resemble those found at the base of Interstadial sediments, but take place much more rapidly and progress further, so that within a period of not more than 600 years the sediment deposited has come to resemble in composition the Flandrian mean.

Organic

Humus type. Atherton *et al.* (1967) reported on the ESR spectra of humic acids as an index to soil humus type, and showed that the ESR spectra of Lake District lake sediments agree with biological evidence from pollen analysis in separating samples corresponding in time with circumneutral mull soils on the catchment from those corresponding with acid mors. Figure 6 shows that the ESR spectra of two samples from the later part of the Younger Dryas chronozone

*Figure 7 of Pennington and Sackin (1975) shows how this can be demonstrated by statistical analysis of the Cam Loch data.

at Loch Sionascaig (respectively from the transitional herb (Cyperaceae-*Rumex*) zone and from the *Empetrum* zone (NWS I)) show that the transition from circumneutral organic matter to acid organic matter of mor type took place within the period *c*.10,400 - 10,000 B.P.

Other organic indicators. Dr. P. Cranwell writes: In lakes formed by glacial activity, the sedimentary organic matter is derived partly from terrestrial biota, mobilised from the drainage basin by erosion and leaching, and partly from aquatic biota. Although in these open-water sediments there is little structurally recognisable material present, the sediments contain a wide range of organic compounds which are characteristic of the original biota. Separation of these compounds according to the functional group and type of structure present gives a series of distinct classes which constitute a chemical record of environmental change, complementary to that provided by microfossils in sediments. Thus changes in the balance between allochthonous and autochthonous input with depth in the sediment profile are reflected by changes in composition of the sterol and straight-chain saturated fatty acid fractions, while the branched/cyclic acids constitute an indicator, associated mainly with the microbial population, which can be correlated with the trophic status of the lake when the sediment was deposited. Other classes of compound and even single components which occur only in a small group of organisms have been used as indicators of the sources of sedimentary organic matter (Cranwell, 1973, 1974, and in press).

In the Lateglacial sediments of Cam Loch, while the total organic matter in the interstadial sediment is low, analysis as outlined above indicates that the component derived from aquatic biota is large, and there is a large microbial component. These features are observed in the recent sediments of modern productive lakes and suggest that Cam Loch had a high trophic status during the Interstadial. Further evidence may be obtained from the iron and manganese distribution in the profile, indicating that in Interstadial time mobilisation of these elements by reduction occurred, due to hypolimnetic oxygen depletion - a feature of modern productive lakes. Furthermore the diatom flora is also characteristic of modern eutrophic environments (see above). The nutrient-rich conditions implied by these observations probably resulted from the exposure of leachable rock surfaces by glacial action. Chemical records show that nutrient-rich conditions also existed in the earliest part of the Flandrian period but were then followed by progressive impoverishment of the drainage basin.

SYNTHESIS

On the evidence of sediments found in those lakes which were not subsequently affected by ice of the Loch Lomond Readvance, much of northern mainland Scotland must have been free of ice for some time before 13,000 B.P. Pollen and diatoms provide evidence for the existence of terrestrial and aquatic biota before 13,000, and changes in sediment composition before this date point to the end of permafrost in the lowlands. The earliest plant communities were of snowbed and pioneer type, identified with those now found in the glaciated mountainous region of south-west Norway.

About 13,000 B.P. a widespread climatic amelioration (synchronous with that recorded in marine faunas in deposits from North Jutland (Krog and Tauber, 1973), by vegetation changes in north-west England (Pennington and Bonny, 1970) and by changes in the insect fauna in North Wales (Coope and Brophy, 1972; Coope, 1975)) is deduced from a change in composition of lake sediments to more organic material, accompanied by a vegetation change (shown both by percentage and absolute pollen data) to dwarf shrub tundra, dominated by *Empetrum* and sedges, with some juniper. At the same horizon changes in sediment composition indicate an acceleration of processes of soil maturation pointing to further stabilisation of soils. Assemblages of benthic diatoms characteristic of base-rich and nutrient-rich waters agree with the chemical composition of sediments to indicate eutrophic conditions and hypolimnetic oxygen deficiency. An interstadial environment, now correlated with the Bölling chronozone of the Late Weichselian as defined by Mangerud *et al.* (1974), is indicated.

Evidence for a subsequent recession, dividing the Interstadial by the equivalent of the Older Dryas chronozone of the Late Weichselian, is weak but consistent, suggesting a short period of reduced pollen production by woody plants coincident with a temporary increase in the rate of soil erosion, which transferred into lake basins weathered material from mineral soils together with some diatoms of soil habitats - but there was no renewal of deep periglacial erosion of mineral soils.

The second part of the Interstadial, corresponding with the Alleröd chronozone (11,800 - 11,000 B.P.), corresponds first with maximal Lateglacial pollen production from the ericaceous dwarf shrub vegetation (comparable with contemporary pollen deposition in lakes of the Cape Farewell region of south Greenland), secondly with the nearest approach of birch woodland to this northern area of mainland Scotland during the Interstadial, and thirdly with chemical indications both of the development of acid organic soils on the catchments and redeposition of this organic matter in lakes, combined with continued high productivity by benthic algae. However, in the only loch (Sionascaig) in which planktonic diatoms were found in Lateglacial sediments, the species assemblage, indicative of nutrient-poor waters, provided evidence for declining base-status of lake waters in this period.

At *c*.11,000 B.P. there was a sudden vegetation change to communities for which no modern equivalent has been found, and this coincided with an equally sudden lithological change to minerogenic sediment resembling in composition the barren sediments at the base of the profiles. At all sites except those from the north of Sutherland, pollen of *Artemisia* and other taxa of disturbed soils reached its maximum, both in Lateglacial deposition rates and in percentage frequency, within the minerogenic sediment which is identified as the product of renewed periglacial erosion of mineral soils. It must be supposed that lowland soils were not protected by a winter snow cover from the effects of freeze-thaw cycles. At sites within the limits of the Loch Lomond Readvance the deposits of this period (radiocarbon dated by backward projection of the depth-time-scale for Loch Clair (Pennington *et al.*, 1972, Fig. 18)) are of coarse sand or fine gravel from which no pollen could be recovered; just outside the ice limits, varved sediments, derived from seasonal outwash from active ice into lakes which froze each winter, accumulated in lowland lakes, *e.g.* lochs Maree, Ness and Stack. Both low temperatures, and a more continental type of cold climate than that now found at higher latitudes and altitudes in north-west Europe, must be supposed.

Deposition of this mineral sediment ceased by *c*.10,400 B.P. In sediments dated by radiocarbon assay to the period 10,400 - 10,000 B.P., pollen data indicate a rapid transition to closed plant communities resembling those of the interstadial, and the composition of sediment changes rapidly to that of the Flandrian muds. Diatom evidence shows that in aquatic as well as in terrestrial biota there was, after the end of the Loch Lomond Stadial, a succession comparable with that found at the base of the Interstadial (indicative of renewal of the glacial environment during the Loch Lomond Stadial) but completed within a much shorter time - suggestive of more rapid climatic amelioration to interglacial temperatures. Radiocarbon dating of two Flandrian pollen diagrams (lochs Sionascaig and Clair) showed that the local spread of birch woodland did not take place until 1,000 years after the opening of Flandrian time at *c*.10,000 B.P.

ACKNOWLEDGEMENTS

My thanks are due to: The Royal Society and the Editors of Boreas for permission to reproduce figures, Mrs. J. P. Lishman for inorganic chemical analyses, Miss A. P. Bonny for the pollen counts for lochs Sionascaig and Tarff, Mrs. K. J. Clark who drew the new figures, and to Mr. B. Walker, Mr. W. Askew, Mr. P. Cubby and Mr. J. Freeman who obtained the cores from Scottish lochs. Permission from the landowners to sample their lochs is gratefully acknowledged.

NOTE ADDED IN PROOF

In a publication now in the press (*Phil. Trans. R. Soc.* Series B) I have described the lithology, geochemistry and palaeobotany of the now closely radiocarbon-dated Late-Devensian section from Low Wray Bay, Windermere, Cumbria, and Dr. G. R. Coope has described the animal remains which he recovered from a parallel core. We have jointly suggested this as a reference site for the Late-Devensian Interstadial. The lower boundary of interstadial sediments falls at different horizons according as to whether lithostratigraphic, palaeobotanical or palaeoentomological criteria are adopted. The sediments of the period 13,000 - 11,000 B.P., which has in my present paper been referred to as the Interstadial, constitute in Windermere a distinct woodland biozone which represents the Palaeobotanical Interstadial in the Windermere section. The Windermere evidence therefore supports that from northern Scotland in demonstrating a vegetation change, interpreted as the consequence of climatic amelioration, at *c*.13,000 B.P. In the older Windermere sediments the absence of Arctic insect species and the presence of temperate ones, at least in the *Rumex*-Gramineae zone, leads Dr. Coope to include these deposits within the interstadial, though since within them the plant evidence indicates no expansion of woody plants (trees and shrubs) the sediments of this age cannot be included in an interstadial defined on palaeobotanical criteria. In the Windermere section the base of an Upper Laminated Clay falls at *c*.11,000 radiocarbon years B.P., and represents the beginning, in north-west England, of Loch Lomond Stadial sedimentation, consequent on recrudescence of glaciation in the Lake District mountains.

Radiocarbon Dating of the Lateglacial and Early Flandrian Vegetational Succession in the Scottish Highlands and the Isle of Skye

YRJÖ VASARI (University of Oulu, Finland)

ABSTRACT

Radiocarbon evidence is presented in order to date the Lateglacial and early Flandrian vegetational succession in the Scottish Highlands and the Isle of Skye. In addition to 16 new radiocarbon dates, 10 previously published ones are considered. In the light of these a three-fold geographical division of the area is suggested for the Lateglacial period.

In north east Scotland the vegetational (and climatic) succession is easily divisible into conventional pollen zones which can be correlated with the continental chronozone scheme. Climatic amelioration began in pre-Bölling time, soon after 13,000 B.P. The dates for the zone boundaries are:- I/II, c.11,850 B.P.; II/III, c.10,650 B.P.; and III/Flandrian, c.10,145 B.P.

In the central and western Highlands the Lateglacial biostratigraphy cannot be divided into conventional pollen zones. The climatic amelioration began around 13,000 B.P. and proceeded progressively until c.11,400 B.P. when the general stadial cooling began. The average of the Lateglacial/Flandrian boundary dates is 10,120 B.P.

In the Isle of Skye the vegetational succession is regarded as having begun during the Loch Lomond Stadial only, not much earlier than 10,250 B.P. The average age for the Lateglacial/Flandrian boundary is 10,157 B.P. The early Flandrian pioneer vegetation phase was evidently longer here than in the east.

INTRODUCTION

A recent study of the pattern of floristic change in Scottish lochs in Lateglacial and Flandrian times (Vasari and Vasari, 1968) demonstrated that the zones distinguishable in the vegetational development of northern Scotland could be correlated with the Jessen-Godwin zonation system. It was assumed that the zones reflect climatically-distinct periods that are broadly synchronous (*cf.* Van der Hammen *et al.*, 1971; Mitchell *et al.*, 1973; Godwin, 1975, p.45). In zoning the Scottish pollen profiles it was recognised that, because of regional variations of environmental factors, different rates of migration *etc.*, regional differences must have characterised the vegetation of the past as well as that of the present (see Birks *et al.*, 1975; Birks and Saarnisto, 1975; Vasari, 1975). Thus some modifications were necessary with respect to the basic criteria used when distinguishing the zones (*cf.* Vasari and Vasari, 1968, pp. 9-10).

The zonation scheme used in the previous work, although intended to represent synchronous phases, was based strictly only on biostratigraphical assemblage zones, since it lacked the support of radiocarbon dates. Material for radiocarbon analyses was therefore collected from sites previously investigated by Vasari and Vasari (1968) in order to make reliable correlations between the pollen zones. At the same time additional biostratigraphical data were collected in order to elucidate further the vegetational history of northern Scotland during the Lateglacial and early Flandrian.

Since the 1960's much attention has been devoted to the basic principles involved in Quaternary stratigraphy and it has become increasingly accepted that, until a proper chronological scheme can be established, zonation systems that are based upon a combination of biostratigraphical, climatostratigraphical and chronostratigraphical criteria should be avoided. Instead, stratigraphical subdivisions should be

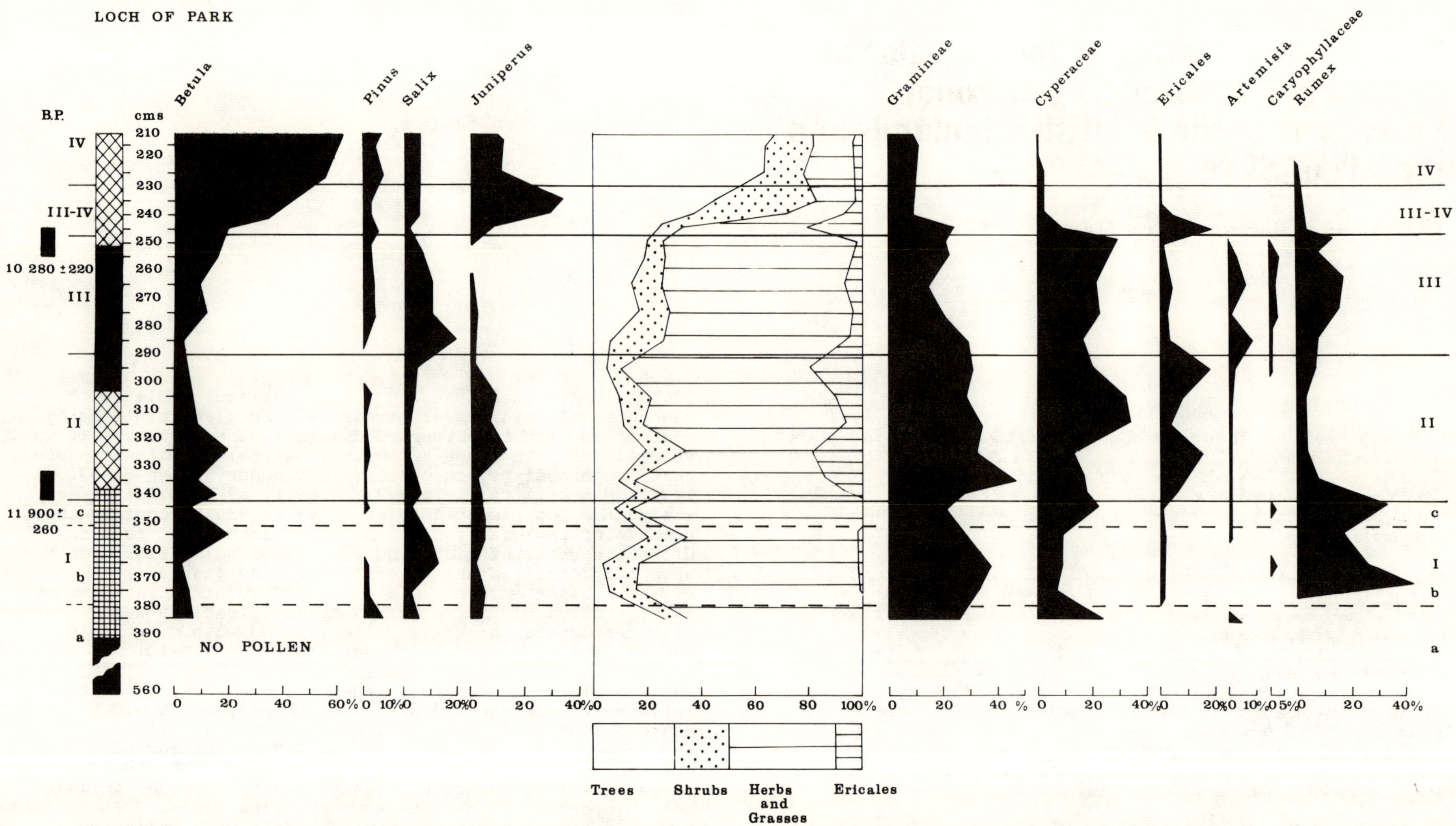

Fig. 1. Loch of Park: partial pollen diagram. (Reproduced from *Acta Bot. Fenn.* 80, 1968).

based on one type of criteria only. Thus, in studies concerned with past vegetation, pollen diagrams should be divided into local pollen assemblage zones which may later be combined into regional pollen assemblage zones possibly tied, in due course, to chronostratigraphy (*cf.* West, 1970; Pennington *et al.*, 1972; Birks, 1973; Mangerud *et al.*, 1974).

To some extent an overall view of the pattern of such pollen assemblage zones can be obtained by grouping them together according to different principles. A three-fold Lateglacial scheme ("pre-interstadial", "interstadial", and "post-interstadial"), based on climatic succession, has been established for west Britain. This corresponds with pollen zones which can be further divided into pollen sub-zones (see

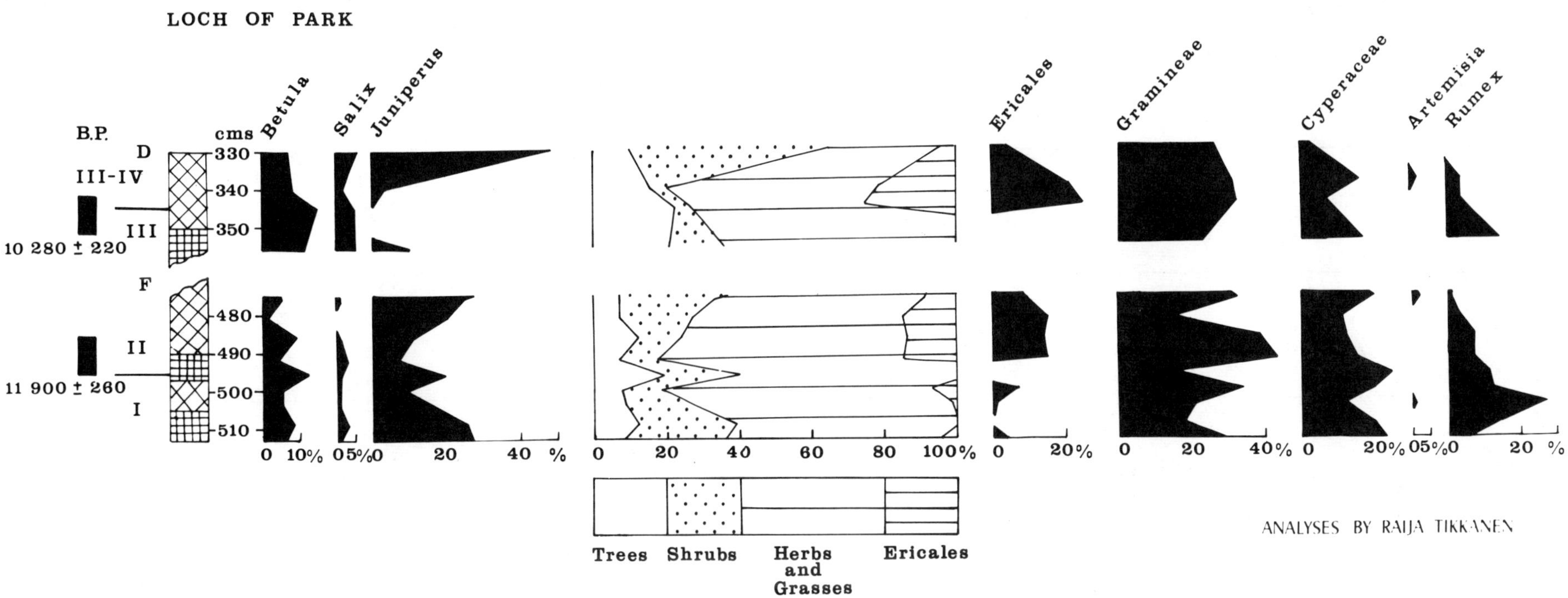

Fig. 2. Loch of Park: radiocarbon dates and selected pollen frequency curves.

Pennington, 1970, 1975a; Pennington and Bonny, 1970; Pennington *et al.*, 1972; Pennington and Sackin, 1975). This scheme was found to be in good agreement with that proposed on the basis of coleopteran evidence (Coope, 1970; Coope *et al.*, 1971; Coope and Brophy, 1972). Recently, however, Pennington has proposed a more detailed scheme correlated with the continental chronostratigraphy (Pennington, 1975b, this volume).

MATERIAL

The material discussed in this paper relates mainly to sites previously investigated by Vasari and Vasari (1968), *viz.* Loch of Park and Loch Kinord (Dee Valley, Aberdeenshire), Muir Park Reservoir (near Drymen, Loch Lomond area), and sites in the Isle of Skye. Additional material has also been collected from a site in Abernethy Forest originally studied by H. H. Birks (1970). Radiocarbon dates from these and from other sites in the Scottish Highlands will be presented and a final section will be devoted to discussion of dates from the Isle of Skye. In this paper "Scottish Highlands" includes the Muir Park Reservoir site.

Samples for radiocarbon assay were collected using a slightly modified Livingstone piston sampler, with two chambers providing cores of 35 mm and 45 mm diameter. Critical horizons in the sediment cores for radiocarbon dating were determined by rapid pollen inspection, involving counts of 100 grains from samples at 10 cm or 5 cm intervals. Samples of sediments usually 10 cm in thickness were then cut from the cores, preferably across a critical pollen zone boundary. Normally material from two cores was required for assay, but occasionally material from one core was sufficient. The relationship of the dated samples to pollen stratigraphy is shown in Figs 1 - 10. In those figures that are reproductions of original diagrams the dates are shown in relation to pollen stratigraphy only; in diagrams

TABLE 1 Location and description of the Loch of Park radiocarbon dates

Depth of Sediment (cm)	Horizon Dated (after Vasari and Vasari, 1968)	Pollen-stratigraphy	Litho-stratigraphy	Laboratory Sample Number	Date (radiocarbon years B.P.)
340-351 (core D) 350-360 (core H)	Zone III/III-IV Boundary	Boundary between *Rumex* and *Empetrum* maxima	Boundary clay-gyttja/ gyttja	HEL-416	10,280 ± 220
485-495 (core F) 483-493 (core E)	Zone II Beginning	Rise of *Empetrum*, Fall of *Rumex*	Gyttja/ clay-gyttja/ gyttja	HEL-417	11,900 ± 260

illustrating the cores obtained more recently by the present author the dates are plotted against depth and pollen stratigraphy.

SCOTTISH HIGHLANDS

Loch of Park, Aberdeenshire (NO/772988; *c.* 70 m O.D.)
Pollen analytically, the profile analysed from Loch of Park was regarded as representing a full Lateglacial sequence (Vasari and Vasari, 1968, Fig. 4; Vasari, 1970, Fig. 3; *cf.* also Mitchell *et al.*, 1973, Table 8). The bottommost part consists of a 170 cm thick layer of bluish clay which is almost entirely devoid of pollen. This is followed upwards by 50 cm of clay-gyttja, 35 cm of gyttja, and 52.5 cm of clay. From 252.5 cm upwards there is only gyttja which gradually becomes coarser, with an admixture of sand in the uppermost parts. The three-fold basal stratigraphy of clay, (clay-)gyttja, clay was interpreted as representing the Lateglacial sequence which appeared to be corroborated by pollen evidence.

A new boring, made in 1972 from as near the original spot as possible, revealed basically an identical stratigraphy. Because of the poor organic content of the bottom sediments and the difficulty in reliably identifying stratigraphic horizons in the field, samples for radiocarbon analysis were taken only from the lower boundary of the Lateglacial gyttja layer and from the lowermost part of the Flandrian clay-gyttja. The information concerning the dates from Loch of Park is given in Table 1 and Figs 1 and 2.

Pollen-stratigraphically, the bottom clay and clay-gyttja were originally ascribed to the three sub-zones of Jessen-Godwin Zone I. Sub-zones Ia (mainly sterile clay) and Ib (clay-gyttja) reflect first a progressive climatic amelioration which is then followed by a colder phase (Ic). It was not certain whether this three-fold pollen stratigraphy indicated a phenomenon of wide significance or whether it was of local importance only (Vasari and Vasari, 1968, p.17; Vasari, 1970, pp. 63-66). The possibility of a climatic basis for this sequence is now supported by new lithostratigraphical evidence, since pollen sub-zone Ib appears in the 1972 cores as a layer with richer organic content between two layers of clay-gyttja.

It is significant that the lower of the radiocarbon samples, from the Ic/II pollen zone boundary, gave an age which corresponds almost exactly with the accepted date for the I/II pollen zone boundary for wide areas of north-western Europe. The conventional age for this boundary in Britain is close to 11,950 years B.P. (Godwin and Willis, 1959b; Godwin, 1975, p.18) while recent dates from western Britain suggest *c.* 12,000 B.P. (Pennington and Sackin, 1975; Pennington, 1975a). The ages obtained in southern Fennoscandia range between 12,070 and 11,730 B.P., being on average 11,840 B.P. (Mangerud *et al.*, 1974) and in the Netherlands the beginning of the Allerød is estimated to be about 11,800 B.P. (Van der Hammen *et al.*, 1967).

Similarly, the age obtained for the III/III-IV pollen zone boundary (10,280 ± 220 B.P.) is in very close agreement with the prevalent concept of the age of the Lateglacial/Flandrian

TABLE 2 Location and description of the Loch Kinord radiocarbon dates

Depth of Sediment (cm)	Horizon Dated (after Vasari and Vasari, 1968)	Pollen-stratigraphy	Litho-stratigraphy	Laboratory Sample Number	Date (radiocarbon years B.P.)
465-475 (core E)	Zone III-IV/IV Boundary	Boundary between *Juniperus* and *Betula* maxima	Gyttja	HEL-421	9,820 ± 250
496-506 (core E)	End of Zone III	Boundary between *Artemisia* and *Empetrum* maxima	Gyttja	HEL-420	10,010 ± 220
550-560 (core C)	Beginning of Zone III	Boundary between *Juniperus* and *Artemisia* maxima	Boundary gyttja/ clay-gyttja	HEL-419	10,640 ± 260
580-590 (core C)	Zone IIa/IIb Boundary	Boundary between *Empetrum* and *Juniperus* maxima	Gyttja	HEL-418	11,520 ± 220
630-640 (1969)	Zone I	*Betula-Rumex* Assemblage	Clay-gyttja	HEL-174	11,480 ± 440

boundary in Britain, *viz.* 10,250 B.P. (Godwin and Willis, 1959b; Godwin, 1960, 1975).

The basal clay, attributed to pollen zone Ia, exhibits a curious combination of pollen sterility with the presence of macroscopic remains indicative of fairly favourable climatic conditions, and this was interpreted as a consequence of very rapid sedimentation under periglacial conditions. Support for such an interpretation can be found in a recent discussion by Pennington *et al.* (1972, p.274) of bottom deposits that are also devoid of microfossils. They make comparisons with the Norwegian lake Vågåvatn which, though presently surrounded by pine forests, has bottom sediments containing few micro-organisms due to turbulence and rapid through-put of water.

Loch Kinord, Aberdeenshire (NO/435997; *c.* 175 m O.D.)
Samples for radiocarbon analysis were taken from site B which contains the thickest Lateglacial sediments (Vasari and Vasari, 1968; Vasari, 1970). The pollen flora from the basal 8 cm of clay-gyttja is characterised by *Betula*, *Salix*, Gramineae, *Empetrum* and *Rumex* (= pollen zone I). This is overlain by a layer of gyttja, 114 cm in thickness, which is assigned to pollen zone II, but which is divisible into two distinct parts (zones IIa and IIb), a lower *Empetrum-Salix*-Gramineae pollen assemblage, and an upper *Betula-Juniperus*-Gramineae assemblage. This thick gyttja layer is succeeded by 38 cm of clay (pollen zone III) and 5 cm of clay-gyttja before the beginning of purely organogenic deposition.

One of the dates (HEL-174) was obtained on material collected in 1969 with a 35 mm core. The rest of the samples were obtained in 1972 with a 45 mm core. At this site samples from one core only were sufficient for radiocarbon dating. The dates are listed in Table 2 (see also Figs 3 and 4).

The two lowermost samples, from distinctly separate stratigraphical horizons, have yielded practically the same

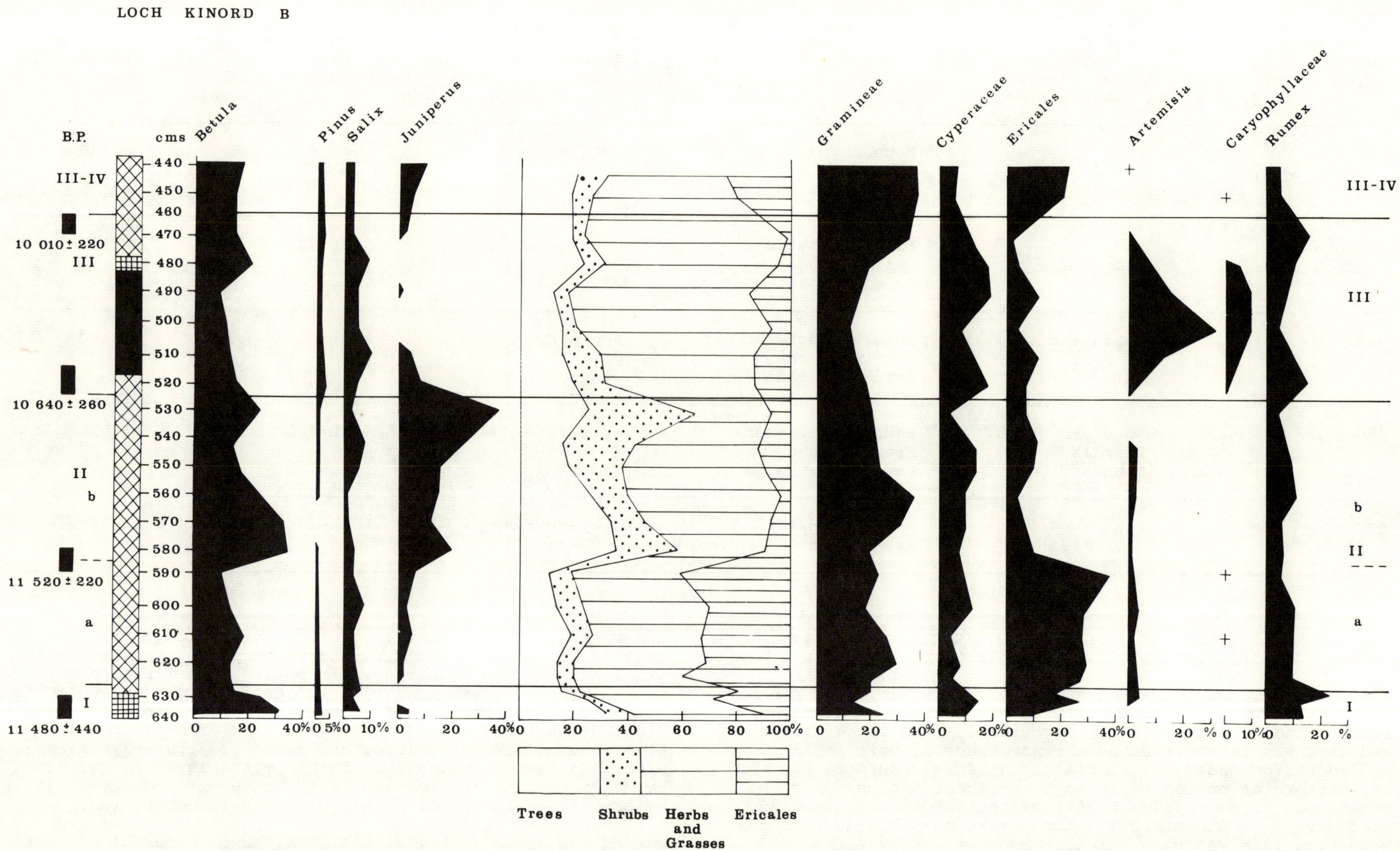

Fig. 3. Loch Kinord: partial pollen diagram. (Reproduced from *Acta Bot. Fenn.* 80, 1968).

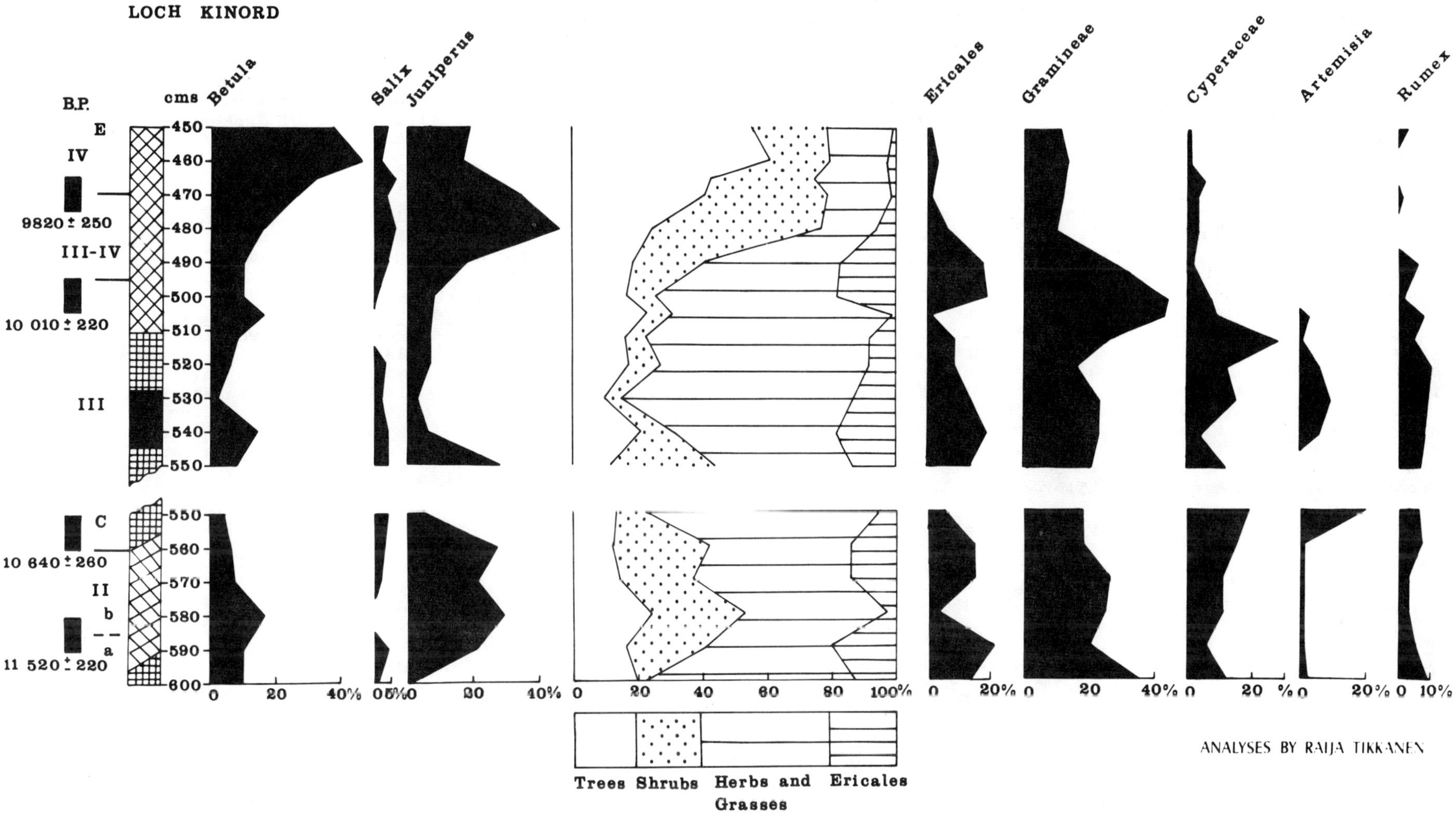

Fig. 4. Loch Kinord: radiocarbon dates and selected pollen frequency curves.

radiocarbon age. It would appear that the lowermost sample, taken in 1969, is not as reliable as HEL-418 which is consistent with the other dates. The probable unreliability of HEL-174 is most likely due to the fairly small size of the sample which has resulted in a date with a much higher standard deviation than the other dates from Loch Kinord.

Sample HEL-418 dates the IIa/IIb boundary described above. This division of pollen zone II at Loch Kinord has no connection with the subdivision of the Alleröd in Denmark and southern Sweden (*cf.* Iversen, 1954; Krog, 1954; Berglund, 1966). The age obtained, 11,520 ± 220 B.P., agrees well with the supposed Alleröd age of the gyttja deposits since it falls well within the accepted limits of the Alleröd chronozone, *viz.* 11,800 - 11,000 B.P. (*cf.* Van der Hammen *et al.*, 1967; Mangerud *et al.*, 1974; Pennington, 1975b).

Coleopteran evidence from northern Wales (Coope and Brophy, 1972) and pollen evidence from north-western parts of England and Scotland (Pennington, 1975b) have shown that maximum Lateglacial temperatures were reached in western Britain before the beginning of Alleröd time and that there was a more or less continuous fall in temperature throughout the Lateglacial. In southern Fennoscandia the temperature rise culminated in the later part of the Alleröd oscillation, after a slight fall in the middle of it (Berglund, 1966; Iversen, 1973, p.31). In the case of Loch Kinord a temporary climatic deterioration in the middle of Zone II cannot be detected as this zone is divisible into two parts only. The succession from a *Rumex-Empetrum* to a *Juniperus-Betula* assemblage is very similar to the corresponding assemblages in the Lateglacial/Flandrian transition. The shift in dominance between birch and juniper in favour of juniper towards the end of this period indicates the beginning of climatic cooling. The two pine needles found in the Loch Kinord pollen zone II deposits (Vasari and Vasari, 1968) characteristically belong to the later part of this period, for in south-eastern Scandinavia pine also appears towards the end of the Alleröd (Berglund, 1966).

If the I/II pollen zone boundary in the Loch Kinord profiles corresponds with the chronozone boundary Older Dryas/Alleröd (11,950 - 11,800 B.P.), the duration of the *Empetrum*-dominated phase in this part of the vegetational succession can be estimated to have lasted about 300-400 years.

Sample HEL-419 dates the II/III pollen zone boundary to 10,640 ± 260 B.P. This is clearly younger than the generally accepted lower limit of the Younger Dryas period, *viz.* 11,000 B.P. (Van der Hammen *et al.*, 1967; Mangerud *et al.*, 1974; Pennington, 1975b). However, it agrees closely with the age of the lower limit of Zone III defined by Godwin (10,750 B.P.) on the basis of dates from Scotland and northern England (Godwin and Willis, 1959b; Godwin, 1975). The corresponding date in Pennington's (1975b) Cam Loch series from northwestern Scotland (10,698 ± 490 B.P.) and the age obtained by Sissons and Walker (1974) from the central Grampian Highlands for a sample in a similar stratigraphical position (10,764 ± 120 B.P.), are in close agreement with HEL-419.

Sample HEL-420 dates the beginning of the rise of the *Empetrum* curve at the Lateglacial/Flandrian boundary. The age 10,010 ± 220 B.P. agrees with expectations. The uppermost sample, HEL-421, dates the end of the early Flandrian juniper maximum and the beginning of the birch maximum (used to define the III-IV/IV pollen zone boundary in Vasari and Vasari, 1968) to 9,820 ± 250 B.P. This Flandrian date is in reasonably good agreement with dates listed by Smith and Pilcher (1973) from sites in Ireland and southern Scotland. Samples HEL-420 and HEL-421 limit the III-IV pollen zone (Lateglacial/Flandrian transition phase), characterised by successive maxima of *Empetrum* and *Juniperus*, to a period of about 200 years.

Banffshire

Two other sites with Lateglacial deposits and relevant radiocarbon dates will be briefly discussed in order to augment concepts concerning the vegetational history of north-eastern Scotland.

Garral Hill, Keith (NJ/444551; *c.* 168 m O.D.). Donner (1957) has published a Lateglacial pollen diagram from Garral Hill on the north-eastern edge of the Grampian Highlands and collected samples for which several radiocarbon dates have been published by Godwin and Willis (1959b). The pollen diagram concerns only the supposed Lateglacial Interstadial deposits of either purely or mainly organogenic composition between two thick layers of silt or silty-clay with stones.

The bottommost part of the pollen profile is characterised by high values of Gramineae and *Rumex* pollen and extremely low counts of *Betula* (Donner, 1957, Fig. 24). Lithostratigraphically this phase (235 - 220 cm) corresponds to dark brown, silty clay. The upper part of the profile, corresponding to compressed silty *Hypnum* peat (mostly formed of *Drepanocladus fluitans*) and the lower part of grey silt with stones, shows a slight rise in the *Betula* curve and a strong Ericales (Ericaceae and *Empetrum*) maximum. The curves of *Rumex* and some other pollen types, for example Caryophyllaceae, have fallen significantly. The author tentatively places the I/II pollen zone boundary at the limit between these two different vegetational phases (220 cm). The radiocarbon date (215 - 220 cm, Q-101) of 9,930 ± 225 B.C. (*c.*11,880 B.P.) corresponds almost exactly with the lower limit of pollen zone II in the Loch of Park series (11,900 B.P.) and thus confirms this hypothesis. This being so, it seems highly probable that the thin (5 cm) grey, silty layer between the *Hypnum* peat (corresponding to pollen zone II) and the underlying silty clay-mud represents the Older Dryas phase (Jessen-Godwin Zone Ic).

Loch Builg (NJ/189028, *c.* 525 m O.D.). Clapperton *et al.* (1975) have published a preliminary report of Lateglacial deposits with some pollen data and a radiocarbon date of 11,770 ± 87 B.P. (SRR-305). This date gives the age for the lower part of a 12 cm thick gyttja layer of supposed

TABLE 3 Macroscopic plant remains from Abernethy Forest

Taxon	Type of remains	Pollen Zones (as defined in Vasari and Vasari, 1968)					
		I	II	III			III-IV
		Samples (depth in cm)					
		535-528	518-513	503-498	498-487	487-477	467-460
Chara	oos	*c.* 30	3	1	-	2	-
Ranunculus							
(Batrachian) spp.	fr	1	-	2	17	18	-
Potamogeton							
cf. berchtoldii	fr	-	-	-	-	-	2
filiformis	fr	1	-	-	-	-	-
Equisetum	frgm	pc	-	-	pc	pc	pc
Carex	frgm	-	pcc	pc	sp	sp	pc
Potentilla palustris	fr	6	1	2	-	8	-
Rumex cf. tenuifolius	fr	13	1	2	2	14	-
Betula nana	fr	5	2	-	-	-	-
Salix herbacea	lvs	-	-	1	14	10	-
Erica tetralix	lvs	1	-	-	-	-	-
Ericaceae sp.	fr	-	-	-	1	-	-
Empetrum sp.	fr	3	1	-	-	1	1
- " -	lvs	1	3	-	-	-	-
Betula cf. pubescens	fr	1	1	-	-	1	10
- " -	sc	-	-	-	-	-	3
Wood	frgm	pcc	-	-	-	-	-
Bryidae	frgm	pcc	pcc	pc	sp	pc	pcc

fr = fruit frgm = fragments lvs = leaves
oos = oospores sc = scales

sp = scattered pc = scarce pcc = very scarce

Lateglacial Interstadial origin. The pollen consists mainly (80 per cent) of *Empetrum*, Gramineae, Cyperaceae, *Salix*, *Juniperus* and various open land herbs. This date corresponds closely with those obtained for the I/II pollen zone boundary in Loch of Park and the corresponding date from the Garral Hill series. In addition, below the dated horizon at Loch Builg there is a stratigraphy consisting first of a 6 cm thick layer of clay-gyttja, then 3 cm of gyttja, 7 cm of clay gyttja, and finally 14 cm of

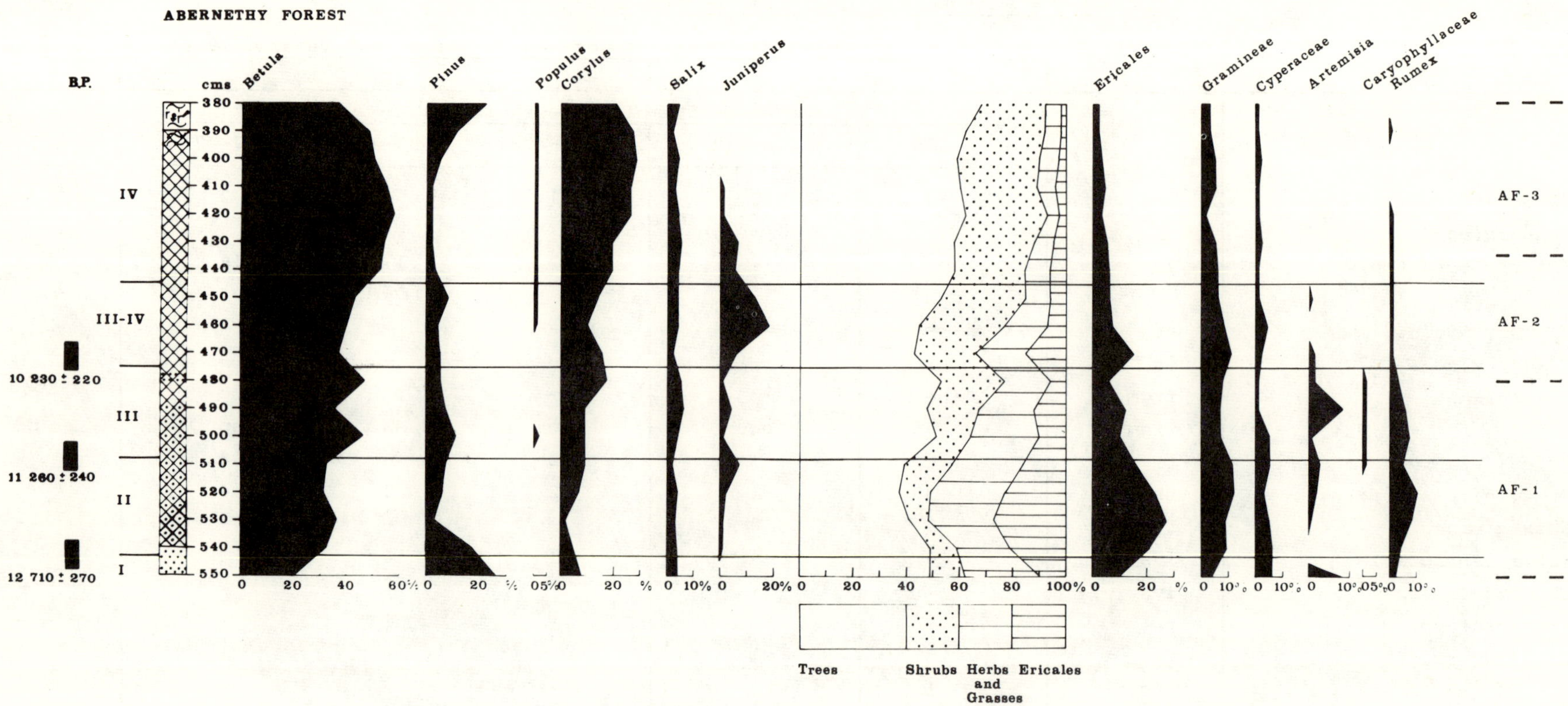

Fig. 5. Abernethy Forest: partial pollen diagram. (Reproduced from J. Ecol. 58, 1970).

clay above the basal gravel. That this reflects a pre-Alleröd succession of a warm interval (equivalent to the Bölling, Ib) between two colder periods (Ia and Ic) seems a reasonable explanation (*cf.* Clapperton *et al.*, 1975).

Abernethy Forest, Inverness-shire (NH/967175; *c.* 221 m O.D.)

Samples were collected in 1972 from the Abernethy Forest site originally studied by H. H. Birks (1970) from as close as was thought possible to her sampling point. This site was included in the present study first because of its close proximity to sites studied by Vasari and Vasari (1968), lying as it does in Strath Spey on the northern side of the Grampian Highlands, and secondly because of its seemingly complete Lateglacial vegetational sequence. The latter factor affords an excellent opportunity for comparisons between the use of a local pollen assemblage zonation and the Jessen-Godwin Lateglacial zonation system.

The lower part of H. H. Birks' Abernethy Forest diagram (1970, Fig. 3) lends itself easily to a zonal division according to the principles suggested by Vasari and Vasari (1968) for use in north-eastern Scotland. This diagram is here redrawn in part and divided into conventional Lateglacial zones. Macrosubfossil samples were analysed by Mrs. A. Vasari in order to elucidate further the character of the Lateglacial flora within this site. The samples were taken from core D, the stratigraphy of which is illustrated in Fig. 6. Table 3 gives the results of these analyses.

Material from two sediment cores was required for radiocarbon assay. The cores were stratigraphically identical so that

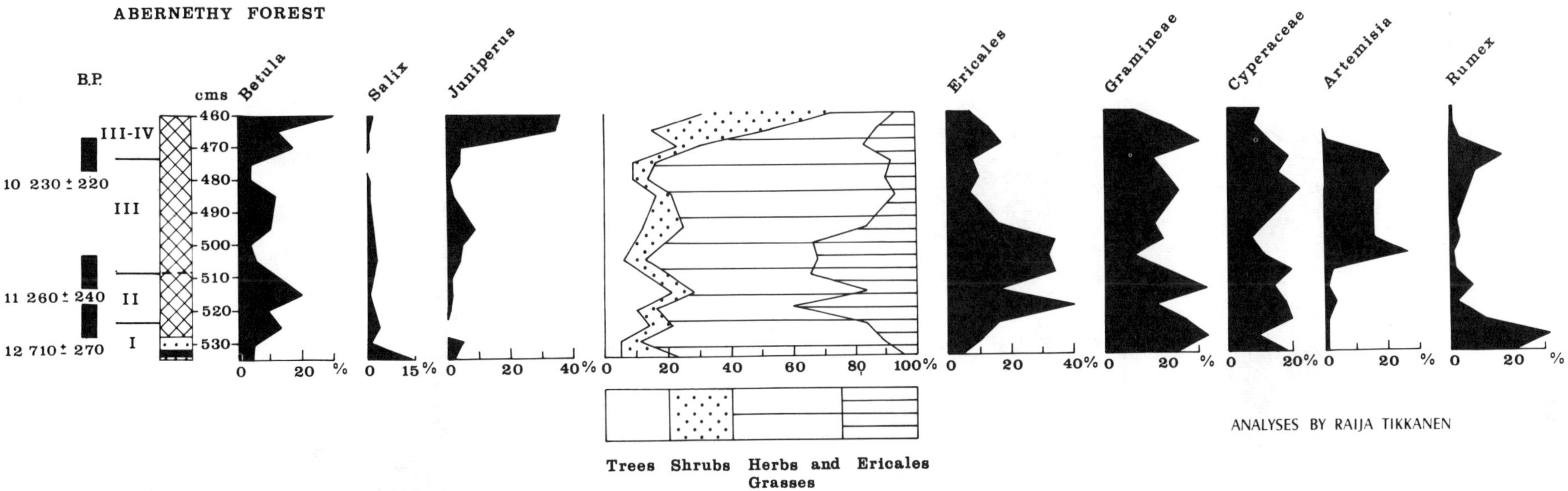

Fig. 6. Abernethy Forest: radiocarbon dates and selected pollen frequency curves. In stratigraphy, at 508 cm level, is a boundary between dark, banded gyttja (below) and light-coloured (silty?) gyttja (above).

TABLE 4 Location and description of the Abernethy Forest radiocarbon dates.

Depth of Sediment (cm)	Horizon Dated (as defined in Vasari and Vasari, 1968)	Pollen-stratigraphy	Litho-stratigraphy	Laboratory Sample Number	Date (radiocarbon years B.P.)
467-477 (cores D & B)	Zone III/III-IV Boundary	Ericales maximum, rise of *Juniperus*	Gyttja	HEL-422	10,230 ± 220
503-513 (core D) 498-508 (core B)	Zone II/III Boundary	Rise of *Artemisia*	Boundary dark, banded gyttja/ light-coloured gyttja	HEL-423	11,260 ± 240
518-528 (core D) 512-522 (core B)	Zone I/II Boundary	Boundary between *Rumex* and Ericales maxima	Boundary sand/ sandy gyttja	HEL-424	12,710 ± 270

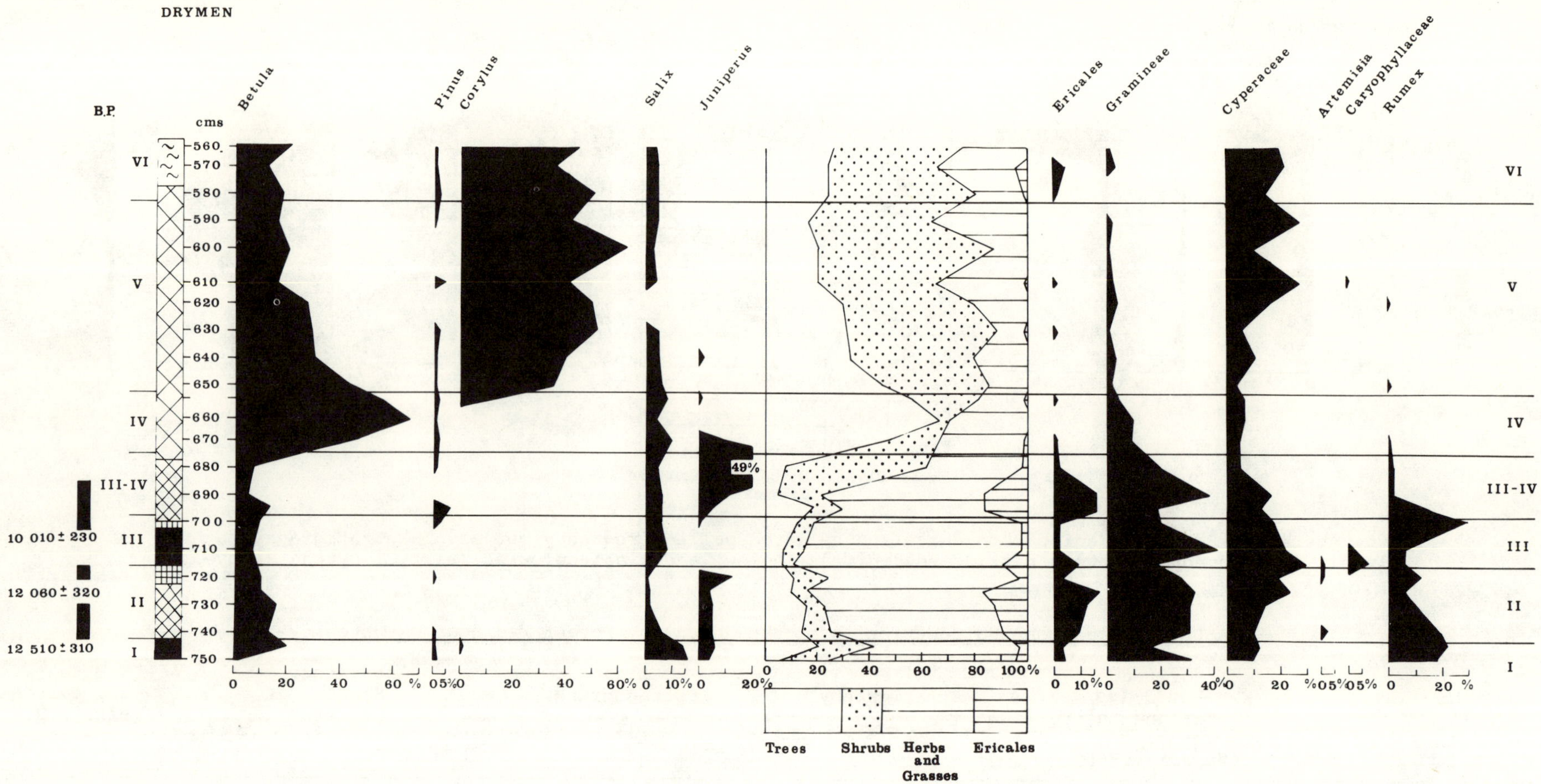

Fig. 7. Muir Park Reservoir (Drymen): partial pollen diagram. (Reproduced from *Acta Bot. Fenn.* 80, 1968).

pollen analyses were required from one core (D) only. The resultant radiocarbon dates are given in Table 4 and Figs 5 and 6.

The dates above convincingly justify the concept that the lowermost Gramineae-*Rumex*-*Artemisia* pollen assemblage zone (H. H. Birks, 1970, zone AF 1) covers the whole Lateglacial vegetational succession. However the dates are almost consistently higher than those obtained from corresponding horizons in the Loch of Park and Loch Kinord sites. The macrosubfossil list (Table 3) includes remains of *Chara* and *Potamogeton filiformis*, indicating rather alkaline conditions during the time of deposition. This suggests the possibility of a "hard-water effect" in the dates. (*N.B.* also the occurrence of *Najas flexilis* in the material of H. H. Birks, 1970.) In addition, leaves of *Dryas octopetala* and *Salix herbacea*, both known basiphilous plants, were found in the Lateglacial sediments during a recent visit to this site (VIIth. International Meeting of European Quaternary Botanists, 1973).

It is likely that the bottom sediments at this site

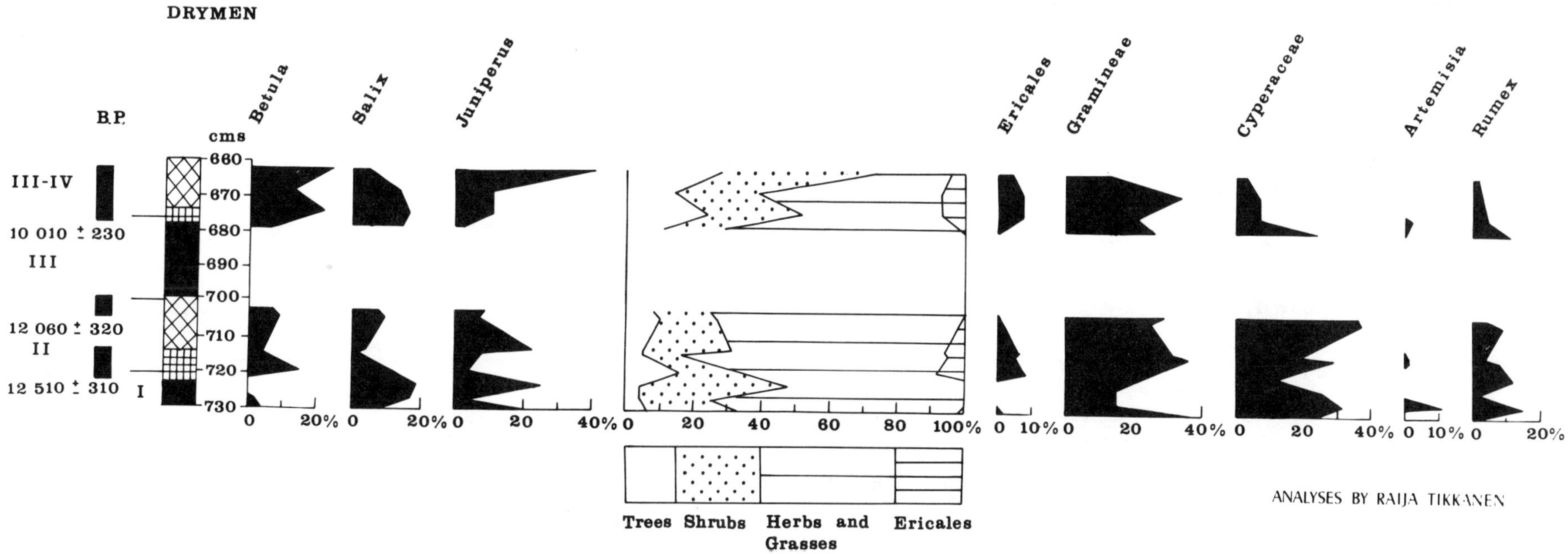

Fig. 8. Muir Park Reservoir (Drymen): radiocarbon dates and selected pollen frequency curves.

accumulated in a "slow-moving fluviatile environment" (H. H. Birks, 1970, p.836). Under such conditions an admixture of allochthonous material is likely to affect the reliability of radiocarbon dates. However, the uppermost sample shows quite a good agreement with the corresponding date from Loch of Park (10,280 ± 220 B.P.) and thus may be more reliable. By that time the base status of the site seems to have shifted substantially towards less alkaline conditions.

With an error of the order of only 250-500 years, the age of the II/III pollen zone boundary would agree well with current concepts (*cf.* Godwin, 1975, p.18; Pennington, 1975b). However, the lowermost date would still fall well before the lower limit of the Alleröd chronozone.

As an alternative hypothesis it could be supposed that the two lower dates are correct within the limits of the standard deviations. This would mean that the lowermost sample dates the beginning of organic deposition to somewhere within the Bölling chronozone, and that no equivalents to the Bölling and Alleröd interstadials can be recognised. It has indeed been shown that in many cases it is extremely difficult to find a correlative for the sub-zone Ic in British pollen diagrams (Pennington, 1970, 1975a, 1975b; Pennington and Bonny, 1970; Pennington *et al.*, 1972; Pennington and Sackin, 1975; Lowe and Walker, this volume).

On the basis of the material available from the Abernethy Forest site, it appears that basiphilous plant communities with some tree birches developed during the Lateglacial Interstadial. The climatic deterioration of the Loch Lomond Stadial meant an almost complete elimination of tree birches from the area. At the beginning of the Flandrian, when birch trees returned, the prevalent vegetation was clearly less basiphilous, both in and around the basin (*cf.* H. H. Birks, 1970).

Muir Park Reservoir, near Drymen, Stirlingshire (NS/490923, *c.* 220 m O.D.)

The Drymen profile has generally been interpreted as indicating a "complete" Lateglacial sequence, containing the Jessen-Godwin pollen zones I, II and III (Donner, 1957; Vasari and Vasari, 1968; Mitchell *et al.*, 1973).

TABLE 5 Location and description of the Muir Park Reservoir (Drymen) radiocarbon dates.

Depth of Sediment (cm)	Horizon Dated (after Vasari and Vasari, 1968)	Pollen-stratigraphy	Litho-stratigraphy	Laboratory Sample Number	Date (radiocarbon years B.P.)
662-678	Zone III/III-IV Boundary	Fall of *Rumex*, rise of *Juniperus*	Boundary clay-gyttja/gyttja	HEL-162	10,010 ± 230
699-704	End of Zone II	Fall of *Empetrum*, rise of Cyperaceae	Boundary gyttja/clay	HEL-161	12,060 ± 320
715-723	Zone I/II Boundary	Fall of *Salix*, rise of *Empetrum*	Boundary clay/clay-gyttja	HEL-160	12,510 ± 310

Radiocarbon samples were obtained from this site in 1969 in order to date the I/II pollen zone boundary (after Vasari and Vasari, 1968), the end of pollen zone II, and the beginning of the III-IV transitional pollen zone (see Figs 7 and 8). In this case pieces of only one core (35 mm diameter) proved to be sufficient. Details of the radiocarbon dates from Drymen are given in Table 5.

The dates in this series agree closely with those from Abernethy Forest. The age obtained for the beginning of organic sedimentation, 12,510 ± 310 B.P., is practically the same as that from Abernethy Forest, 12,710 ± 270 B.P. The supposed II/III pollen zone boundary is in this case also "too old" by about 1,000-1,300 years. The Lateglacial/Flandrian boundary however, as in Abernethy Forest, agrees well with current concepts.

Again three possible explanations can be considered:
1. "Hard-water effect": In this site there are also indications of high base status in and around the loch basin (*cf.* Vasari and Vasari, 1968). However, sample HEL-162 gave an expected date.
2. Disturbances in the sedimentation: There are very distinct biostratigraphical contrasts between the upper part of pollen zone II and the lower part of zone III in the Drymen series. It seems possible that either erosion of previously formed sediments caused a hiatus in the profile (*N.B.* also the shortness of the Lateglacial part of the profile) or, as a result of a deteriorating climate, older material from the catchment area was eroded and washed into the basin. This latter process appears to have affected the Cam Loch series of Lateglacial dates where "... soil erosion now recognised to have taken place in Allerӧd time ... transferred to Cam Loch organic soils from an area of basin peat higher in the catchment" (Pennington, 1975b, p.165).
3. Acceptable dates: The dates can be regarded as reliable within the limits of the standard deviations, thus presenting a pattern rather similar to that of Abernethy Forest.

ISLE OF SKYE

According to Vasari and Vasari (1968) vegetational succession began in sites in the Isle of Skye at about the Lateglacial/Flandrian boundary. During the Loch Lomond Stadial accumulation of organic material began in sufficient quantities to make botanical analyses possible. This, however, was not the case everywhere, for clays extremely poor in organic content were still formed where more extreme conditions prevailed (*e.g.* the contrast between Loch Fada and Loch Cuithir). Deposits of the Loch Lomond Stadial in Skye differ from those studied on the Scottish mainland, for instance in their lack of subfossils of tree birches, their smaller percentages of pollen of *Rumex* type, and their poor organic content in general. The early Flandrian succession of Skye was basically similar to that of other areas in northern Scotland, but differed in detail. Thus the *Empetrum* (and Ericaceae) maximum was not nearly as marked as elsewhere, and the *Juniperus* maximum overlapped with the *Betula* maximum, maintaining significant representations in the diagram into the dominant *Corylus* phase.

As a result of Birks' (1973) palaeobotanical work the vegetational history of Skye during part of the Late Devensian and early Flandrian is known in greater detail

TABLE 6 Radiocarbon dates of Birks (1973) in relation to vegetational succession according to the scheme of Vasari and Vasari (1968)

Site	Laboratory Sample Number	Date (radiocarbon years B.P.)	Horizon Dated and Pollen-stratigraphy	Litho-stratigraphy
Loch Meodal	Q-960	9,482 ± 150	End of Zone IV; rational limit of *Corylus*	Fine detritus mud
Loch Fada	Q-961	7,500 ± 120	End of Zone IV; rational limit of *Corylus*	Boundary silty fine detritus mud/Fine detritus mud
Loch Cill Chriosd	Q-956	9,655 ± 150	Zone IV/V Boundary; rational limit of *Corylus*	Boundary Silty fine detritus mud with fine sand/Slightly fibrous fine detritus mud
Lochan Coir' a' Ghobhainn	Q-958	8,650 ± 150	End of Zone IV; Boundary between *Juniperus* maximum and rational limit of *Corylus*	Slightly fibrous fine detritus mud
	Q-957	9,420 ± 150	End of Zone III-IV; Boundary between Ericaceae and *Juniperus* maxima	Slightly fibrous fine detritus mud
	Q-956	9,691 ± 150	Zone III-IV; Beginning of main Ericaceae maximum	Silty diatomaceous fine detritus mud
	Q-955	10,254 ± 220	Zone III/III-IV; Boundary, between *Oxyria* and Ericaceae maxima	Boundary Silt and fine sand/Silty diatomaceous fine detritus mud

than perhaps that of any other area of comparable size in Scotland. Though he employed a higher level of accuracy in pollen determinations and pollen analyses than in those diagrams previously published by Vasari and Vasari (1968), correlations can be made between available pollen diagrams for Skye, especially within the Trotternish peninsula (*cf.* Birks, 1973, p.351).

Birks has obtained seven radiocarbon dates for important horizons in pollen profiles from Skye, but concludes that three of these are anomalous in that they do not fit with pollen stratigraphy. Thus he relied mainly upon estimates based on sedimentation rates when dating the vegetational history of the Isle of Skye. His conclusions are reviewed in the light of new radiocarbon dates, and in Table 6 his radiocarbon-dated pollen diagrams have been zoned according to the principles outlined in Vasari and Vasari (1968). This facilitates the correlation of the pollen stratigraphy and the appraisal of the radiocarbon dates.

Attempts to sample Loch Fada in 1972 had to be abandoned because the sediments, containing a high proportion of fine minerogenic material, proved too compact for penetration with a large capacity piston sampler. Loch Cuithir (NG/474597), however, was an excellent site for sampling, and although two cores of 45 mm diameter were required for sufficient material for radiocarbon assay, the cores proved, in pollen-analytical checks, to match each other to a centimetre. The dated horizons were (1) the beginning of the vegetational succession at this site (approximately at the III/III-IV

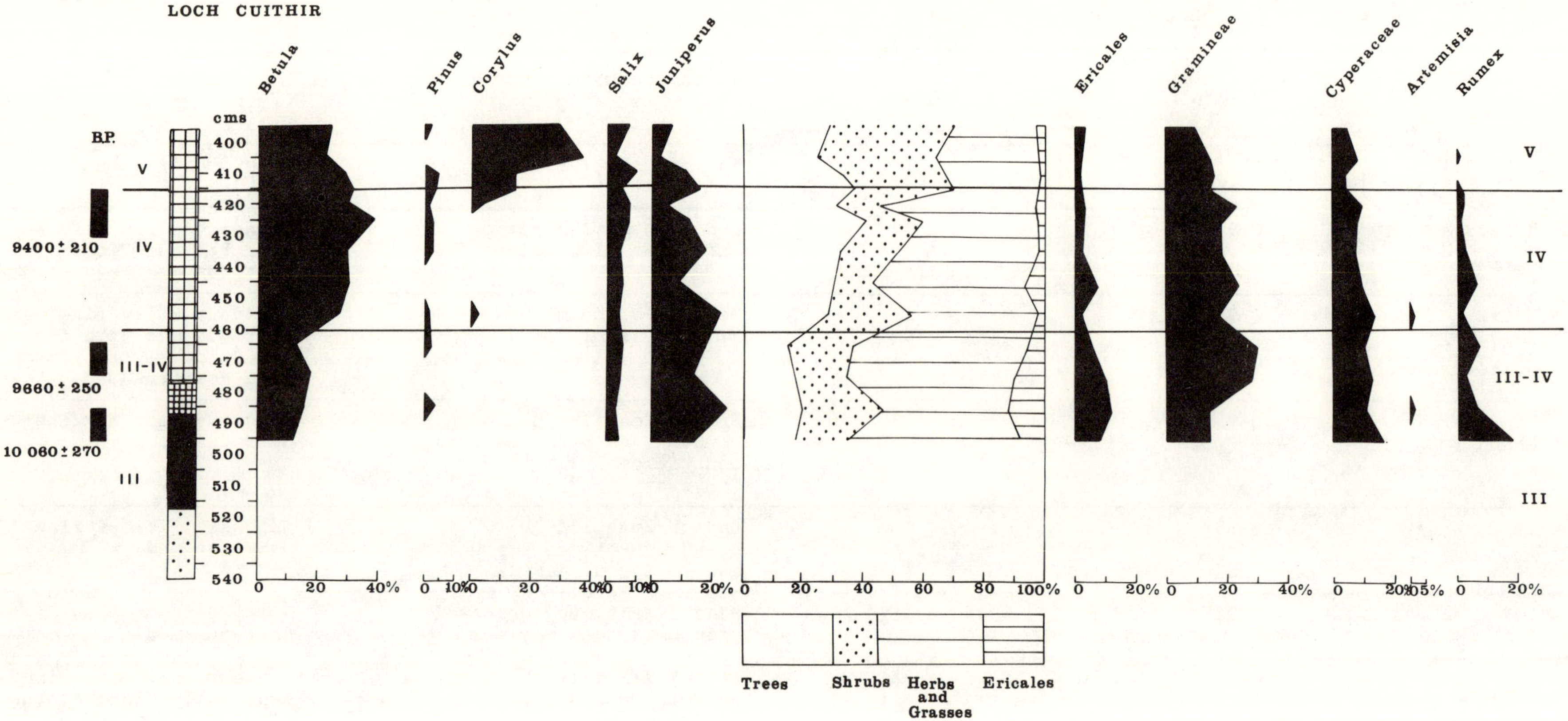

Fig. 9. Loch Cuithir: partial pollen diagram. (Reproduced from *Acta Bot. Fenn.* 80, 1968).

pollen zone boundary of Vasari and Vasari, 1968), (2) the end of zone III-IV, and (3) the end of the *Juniperus* maximum, slightly above the rational limit of Coryloid pollen (see Figs 9 and 10). Details of the radiocarbon dates from Loch Cuithir are given in Table 7.

Discussion of the Isle of Skye Radiocarbon Dates

There are only two profiles with more than one radiocarbon date from the Isle of Skye, Lochan Coir a' Ghobhainn (Birks, 1973) and Loch Cuithir. Within these, the dates agree quite well with each other. Indeed, of the total of 10 radiocarbon dates from Skye there appear to be only two which are anomalous when examined closely in relation to pollen stratigraphy. Both of these (Q-961 and Q-958) refer to the rational limit of *Corylus*.

There is a time difference of 194 radiocarbon years between the lowermost samples of Lochan Coir a' Ghobhainn and Loch Cuithir. The horizon dated in both cases is roughly the same, *viz.* the Lateglacial/Flandrian boundary as defined on the basis of successive maxima of *Rumex (-Oxyria)*, *Empetrum* (-Ericaceae) and *Juniperus*. In both cases the beginning of organic deposition has been dated. The Lochan Coir a' Ghobhainn sample is the older one and there is reason to believe that this reflects the true situation since it dates the very end of the *Rumex acetosa* - *Oxyria* maximum which remains just below the dated sample in the Loch Cuithir diagram.

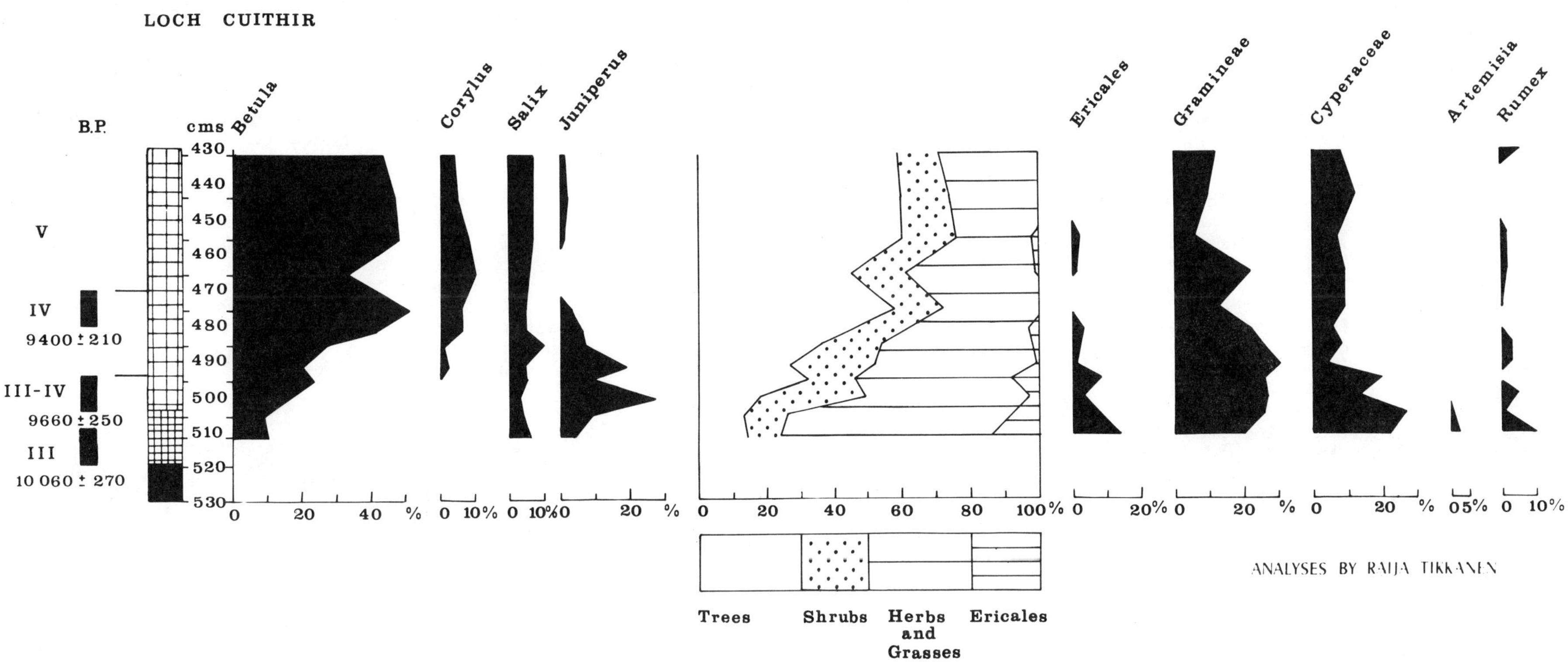

Fig. 10. Loch Cuithir: radiocarbon dates and selected pollen frequency curves.

TABLE 7 Location and description of the Loch Cuithir radiocarbon dates

Depth of Sediment (cm)	Horizon Dated (after Vasari and Vasari, 1968)	Pollen-stratigraphy	Litho-stratigraphy	Laboratory Sample Number	Date (radiocarbon years B.P.)
470-480	Zone IV/V Boundary	Rational limit of *Corylus*	Diatom gyttja	HEL-502	9,400 ± 210
494-504	End of Zone III-IV	Rational limit of *Betula*, second *Juniperus* maximum	Boundary clay-gyttja/ diatom gyttja	HEL-503	9,660 ± 250
509-519	Zone III/III-IV Boundary	Beginning of vegetational succession	Boundary clay/ clay-gyttja	HEL-504	10,060 ± 270

There are, in all, five dates from Skye for the rational limit of *Corylus*, used to define the pollen zone boundary IV/V. Of these, three agree reasonably well with each other thus dating the spread of *Corylus* to between 9,655 and 9,400 years B.P. These values are very close to corresponding dates from southern Sweden but are significantly older than other available dates for this limit in Britain (*cf.* Hibbert *et al.*, 1971). This appears to fit with Deacon's concept (1974) of possible refugia of *Corylus* off the west coast of the British Isles and in south-west Scandinavia during the Devensian glaciation (*cf.* also Godwin, 1975). The dates for the *Corylus* maximum in Loch Fada (Q-961; 7,500 ± 120 B.P.) and Lochan Coir a' Ghobhainn (Q-958; 8,650 ± 150 B.P.) are considerably younger (*cf.* also Birks, 1973, Fig. 27), though these too remain within the limits given by Smith and Pilcher (1973) for the rational limit of hazel.

The remaining dates all refer to phases of vegetational development of supposedly early Flandrian age, *viz.* beginning of Ericales maximum (Q-956), beginning of *Juniperus* maximum (Q-957) and *Juniperus* maximum (HEL-503). The time range between them is remarkably small (between 9,691 and 9,420 B.P.). Together they seem to indicate that a forestless period lasted in Skye through the early Flandrian until about 9,500 B.P., when the landscape became covered by advancing birch-hazel forests of a "Boreal" character. This again bears resemblance to developments in southern Scandinavia (*cf.* Mangerud *et al.*, 1974, and references therein).

It is therefore concluded that the early vegetational development described by Vasari and Vasari (1968) and Birks (1973) is better understood if it is regarded as having begun during the Loch Lomond Stadial (Jessen-Godwin Zone III) and the early Flandrian only, rather than having extended as far back as 14,000 - 13,000 years B.P. It is of interest to compare the results of Jóhansen (1975) from a somewhat similar though more extreme area, the Shetland Mainland. There, too, the vegetational succession has been found to have begun close to the end of the Lateglacial (10,400 ± 100 B.P.), with a similar uninterrupted succession of *Rumex*, Ericales and *Juniperus* maxima coalescing with the *Betula-Corylus* maxima. Since this is also clearly visible in early Flandrian spectra described by Pennington *et al.* (1972) from north-west Scotland, this must represent an extensive regional pattern. Such a pattern involves special difficulties when dividing the early Flandrian vegetational successions into pollen zones that may then be correlated with the conventional Jessen-Godwin zonation system.

An interesting problem is presented by evidence of a clear lithostratigraphical interval with increased minerogenic influx in limnic sediments in several of the sites described by Birks (1973), *viz.* Loch Meodal, Loch Mealt, Loch Cill Chriosd and Lochan Coir a' Ghobhainn. This interval is found within the correlative of Jessen-Godwin Zone IV and it evidently does not deviate much in age from the Piottino Oscillation which has been detected in pollen diagrams in a wide area from the North Sea coasts to the Alps (*e.g.* Behre, 1967). The possibility that a cold spell equivalent to the Piottino Oscillation can be detected in pollen profiles from the Isle of Skye, perhaps even in a larger area in western Britain, is worthy of further study. Sissons (1974c, p.320) also calls attention to the fact that "following the main cold period of the stadial, there was a slight amelioration of climate or, more probably, a slight amelioration interrupted by cold spells".

DISCUSSION

Regional Variation in Lateglacial Developments

In the light of the material presented above the area under examination appears to be divisible into three parts.

North-eastern Scotland. In sites from Aberdeenshire (Loch of Park, Loch Kinord) and Banffshire (Garral Hill, Loch Builg) sedimentation began during the Lateglacial shortly after 13,000 B.P. A colder phase, dividing the Lateglacial Interstadial into two parts, is clearly discernible both litho- and biostratigraphically. Three dates for the pollen zone boundary Ic/II, giving an average age of 11,850 B.P., fall between the chronozone limit 11,950 B.P. of Godwin (1975) and that of 11,800 B.P. (Van der Hammen *et al.*, 1967; Mangerud *et al.*, 1974; Pennington, 1975b).

The Zone IIa/IIb boundary and end of Zone II at Loch Kinord have been dated to 11,520 B.P. and 10,640 B.P. respectively. The Lateglacial/Flandrian boundary is dated to between 10,280 and 10,010 B.P.

On the whole the radiocarbon dates establish a close correlation between the Late Devensian pollen zones in the sense of Mitchell *et al.* (1973) with the Late Weichselian chronozones of the adjacent European continent (Mangerud *et al.*, 1974, for Scandinavia; Van der Hammen *et al.*, 1967, for the Netherlands) in the manner outlined by Pennington (1975b, this volume). The chronozone correlations thus allow a comparison of possible vegetational differences between different areas.

In the light of the available evidence the duration of stadial conditions in north-eastern Scotland seems to be somewhat shorter than the Younger Dryas period on the continent. The marked cooling is likely to have began later (a feature characteristic for east Britain ?, *cf.* Godwin and Willis, 1959b) and ended earlier than on the continent. This would also at least partly explain the differences between the chronozone boundaries of Godwin (1975) and Mangerud *et al.* (1974).

Central and western Highlands. Radiocarbon dates from sites in the central and western Highlands indicate that organic deposition began at about 13,000 B.P. (Kirk and Godwin, 1963; Sissons and Walker, 1974; Pennington, 1975b, Lowe and Walker, this volume). Dates presented in this paper from Abernethy Forest and Drymen are in agreement with this hypothesis.

In the Abernethy Forest and Drymen series the radiocarbon dates do not support the zonation scheme defined for north-east Scotland (*cf.* Vasari and Vasari, 1968). The vegetational (and climatic) development seems to follow a pattern similar to that described for various parts of west Britain (Pennington, 1970, 1975a, 1975b; Pennington and Bonny, 1970; Pennington *et al.*, 1972; Coope and Brophy, 1972). The essential features of this pattern are that the climatic amelioration began very early, at about 13,000 B.P., and proceeded seemingly without interruption until about the middle of Alleröd time, *i.e.* until around 11,400 B.P. The climatic deterioration thus set in considerably earlier than the conventional lower limit for the Younger Dryas chronozone, and a well defined lower limit for stadial conditions is difficult to detect. In other words the limits for the cold period are wider here than in north-east Scotland. Pennington (1975b, this volume) has recently suggested a full correlation of the vegetational succession in west Britain with that of the continent. The fact remains, however, that the vegetational (and climatic) succession in the west is easier to fit into a flexible "pre-interstadial - interstadial - post-interstadial" pattern (Pennington *et al.*, 1972) than into conventional pollen zones.

The radiocarbon dates for the Lateglacial/Flandrian boundary are, however, very similar to those from the north-east. The relatively narrow time-range, 10,230 - 10,010 B.P., suggests a wide regional synchroneity.

Isle of Skye. This area presents quite a special case, as has been discussed earlier. The pollen record does not extend further back than into the Loch Lomond Stadial. The oldest available radiocarbon dates are 10,254 B.P. (Q-955; Birks, 1973) and 10,060 B.P. (HEL-504). Together they fix the Lateglacial/Flandrian boundary at roughly the generally accepted time.

It is problematic why deposits older than this have not yet been found. It may be a feature of wider significance in the Islands of Scotland (*cf.* Jóhansen, 1975), though Moar (1969b) has reported pollen profiles from the Orkney Mainland that evidently reach further back into the Lateglacial period.

The Lateglacial/Flandrian boundary

A regular feature of the vegetational succession across the Lateglacial/Flandrian boundary in Scotland is the close succession of *Rumex*, *Empetrum*, *Juniperus* and *Betula* maxima. In Vasari and Vasari (1968) the *Rumex* maximum is included in the Lateglacial, whereas the others have been regarded as belonging to the Flandrian pioneer vegetation succession. They have been grouped together into an initial phase of the Flandrian period, called transitional zone III-IV. The numerous radiocarbon dates now available help to obtain a better idea of the chronological position of this phase.

Six radiocarbon dates place the Lateglacial/Flandrian boundary to between 10,280 and 10,010 B.P. (Loch of Park, Loch Kinord, Abernethy Forest, Drymen, Loch Cuithir, Lochan Coir a' Ghobhainn). The average of these dates is 10,141 B.P. This fits extremely well with the generally accepted age of the beginning of the Flandrian, *viz.* 10,000 B.P. Notably, however, the maximum ages are very close to the age suggested by Godwin (1975) and Berglund (1966) for this same boundary, *viz.* 10,250 B.P.

Most of the dates that can be used to define the upper boundary of pollen zone III-IV relate to sediments from the Isle of Skye, where early Flandrian conditions were in some ways exceptional. According to them, the *Empetrum-Juniperus* phase lasted in the west until about 9,500 B.P. The date of 9,820 B.P. from the north-east area (Loch Kinord) for this same horizon suggests a somewhat shorter duration of the initial phase in the north-east of Scotland.

There are a number of possible errors inherent in the radiocarbon dating method as well as in the nature of the samples, but the results obtained clearly support the concept of synchronous vegetational changes which, caused ultimately by a major climatic shift, took place in north-western Europe between 10,250 and 10,000 radiocarbon years B.P. The characteristic vegetational pattern is duplicated in pollen diagrams from a large number of sites along a wide arc from the Fennoscandian Arctic coast (Hyvärinen, 1975) over southern Scandinavia (*e.g.* Berglund, 1966) to the British Isles.

The *Rumex* peak generally dated to the end of the Loch Lomond Stadial probably reflects an abundance of *Rumex tenuifolius* (Vasari and Vasari, 1968), though in some areas (*e.g.* the

Isle of Skye) it can also be due to *Oxyria* (Birks, 1973). As suggested by Vasari and Vasari (1968), it may be supposed that the *Rumex* peak forms the first sign of the beginning of leaching of developing soils. The predominant vegetation was probably similar to the grass heaths and grass meadows as described by Berglund (1966). Tree birches were also present, with the possible exception of Skye (*cf.* Vasari and Vasari, 1968).

The *Empetrum* maximum reflects a stage when dwarf-shrub heaths characterised a generally open landscape. There were regional differences, as described earlier by Vasari and Vasari (1968) and Walker (1975a), especially in the role of trees, mainly birch. The existence of *Empetrum* heaths was closely connected with special edaphic conditions, such as a fairly advanced state of leaching of the soil (*cf.* Berglund, 1966; Vasari and Vasari, 1968; Pennington *et al.*, 1972).

An extensive juniper cover must have dominated the landscape before the major expansion of tree birches, which must have been present in small numbers throughout the Lateglacial period, at least in north-eastern Scotland.

In conclusion, this Lateglacial/Flandrian transitional phase was probably only 300-500 years in duration (*cf.* Pennington and Bonny, 1970). It was characterised by a vegetation which physiognomically resembled park-tundra because of the sparseness of trees. The stabilisation of soils and the presence of a strong thermophilous element fully justifies its incorporation into the Flandrian, as also suggested by the chronostratigraphic data.

ACKNOWLEDGEMENTS

I wish to express my sincere thanks to my wife, Annikki, for her help in the field and laboratory, and for her general encouragement. Many people in Scotland gave help and encouragement, but special thanks for their generosity are due to Major and Mrs. F. B. Foster (the owners of Loch of Park), Dr. and Mrs. S. E. Durno, Mr. and Mrs. A. R. Gunson, Mrs. Agnes Walker and family, Mr. and Mrs. J. J. Lowe, Dr. and Mrs. J. H. Dickson, and the Forestry officials at Drymen. I owe a particular debt to Mr. J. J. Lowe and Dr. J. M. Gray for correcting the language of this paper.

I would like to thank also Prof. Sirkka Kupila-Ahvenniemi for making available the facilities of the Department of Botany at Oulu, Mr. H. Jungner and the entire staff of the Radiocarbon Dating Laboratory, University of Helsinki, Miss Raija Tikkanen for help with the pollen analysis and cartography, my son Pekka for his help in the field, and Drs. H. Hyvärinen and M. Saarnisto for fruitful discussions. The work was financially supported by the National Research Council for Sciences, the Finnish Cultural Foundation (Suomen Kulttuurirahasto) and the University of Helsinki.

The Scottish Lateglacial Environment: A Synthesis

J. M. GRAY (Queen Mary College, London)
and
J. J. LOWE (City of London Polytechnic)

ABSTRACT

Recently-defined chronostratigraphic boundaries for Britain and north-west Europe may be markedly divergent from times of significant climatic change. The Scottish Lateglacial can be divided into two main climatostratigraphic units, viz. the Lateglacial Interstadial and the Loch Lomond Stadial.

The Lateglacial was a time of great environmental change in Scotland. At the beginning of the period much of the country was ice covered but most areas were deglaciated by 12,500 B.P. as a result of the rapid thermal recovery that brought summer temperatures at least as warm as the present time by 13,000 B.P. As the ice wasted away it was replaced by the sea at coastal localities although relative sea-level was falling quickly during the early Lateglacial Interstadial due to rapid glacio-isostatic recovery of the land. Open herbaceous communities with a basiphilous component and interspersed with large areas of bare ground were present on land at this time but towards the end of the Interstadial a closed vegetation cover of dwarf-shrub heath, herbaceous grassland and scattered birch copses had become established over large parts of the country. At the end of the Interstadial the temperature decline that began at 12,500 - 12,000 B.P. but which was most marked at about 11,000 B.P. resulted in a break-up of the closed vegetation cover, the beginning of extensive soil erosion and eventually in the build up of the Loch Lomond Readvance glaciers. At this time the firnline rose markedly from west to east, mean July temperatures at sea-level were 5-9°C, and the atmospheric circulation was much more vigorous. Periglaciation was extremely severe, and other important geomorphological processes included very active fluvial erosion and deposition, and rapid marine erosion to form the platforms and cliffs of the Main Lateglacial Shoreline. Very rapid climatic recovery from the minimum temperatures of the Loch Lomond Stadial occurred before 10,000 B.P. and caused widespread stagnation of the glaciers and a vegetation succession that resulted in the replacement of open herbaceous grassland by widespread birch woodland within about 1,000 years.

INTRODUCTION

Although Scottish Lateglacial events have been considered in previous review articles these have been concerned with a much longer time period or with a more limited subject matter than the present review and do not discuss in detail the problematic task of defining precise boundaries for stratigraphic units. This paper aims to synthesise present knowledge on the environments of the Scottish Lateglacial as a whole and to discuss the stratigraphic subdivision of this period. Figure 1 seeks to summarise the events and their chronology. The terms Lateglacial, Lateglacial Interstadial and Loch Lomond Stadial are defined in the introduction to this volume and their use is explained in the section of this paper on terminology and chronology.

DEGLACIATION AND READVANCES

The Lateglacial was a time of great change in the ice cover of Scotland. At the beginning of the period much of Scotland was ice covered yet within about 1,000 years the ice had disappeared or virtually disappeared only to build up again during the Loch Lomond Stadial and waste away completely early in the Flandrian. Information about the pattern and timing of ice-wastage is very variable, and it is not surprising that a number of major controversies regarding deglaciation have arisen in recent years.

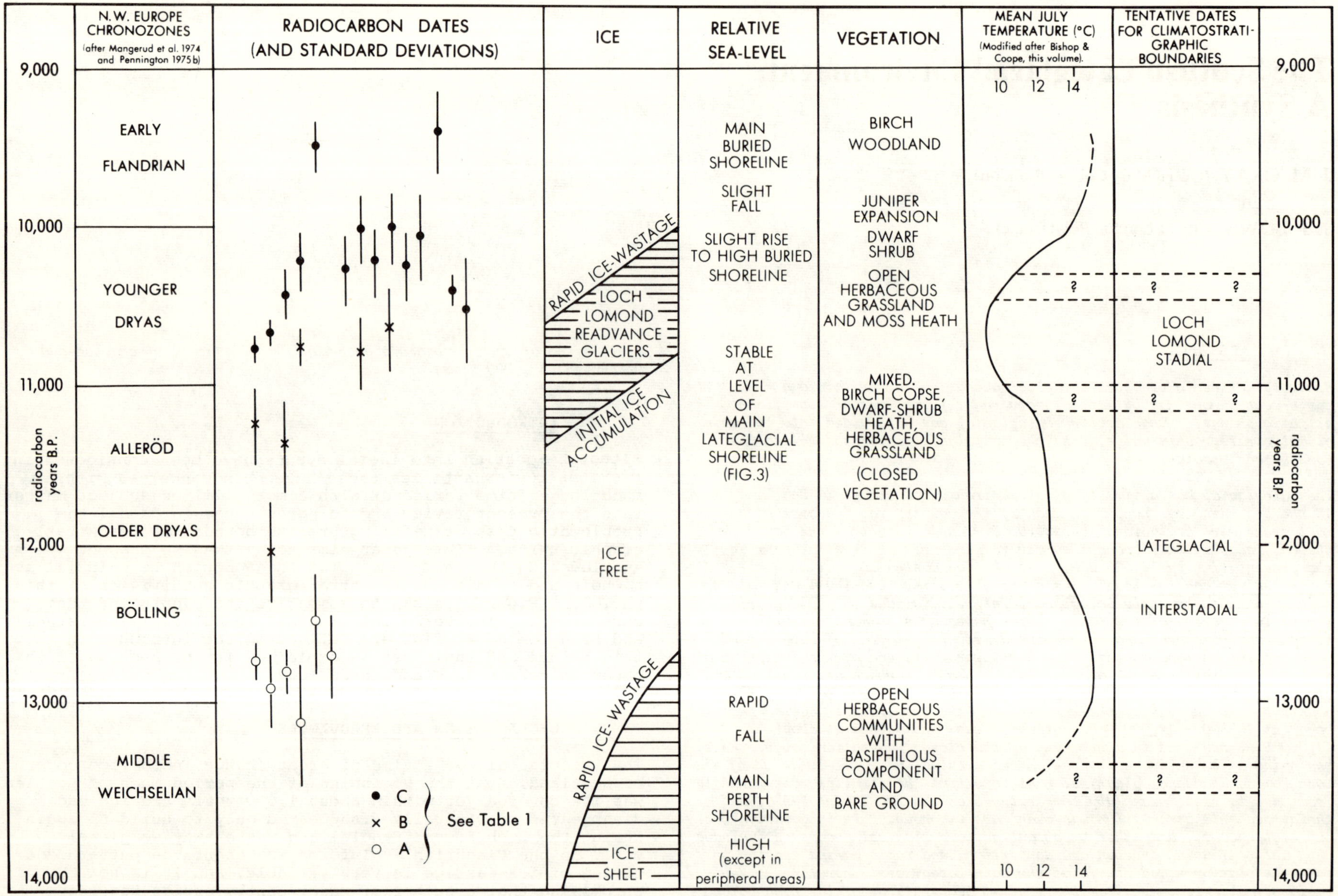

Fig. 1. Summary of Lateglacial and early Flandrian environmental changes in Scotland.

The Late Devensian ice sheet reached its maximal extent shortly after 18,000 radiocarbon years ago (Penny *et al.*, 1969) when most of Scotland was ice covered. Charlesworth (1955), Synge (1956) and Fitzpatrick (1972) proposed that this ice sheet did not extend over a large part of Buchan but, from the distribution of meltwater channels, fluvioglacial deposits and consideration of other evidence, Clapperton and Sugden (1975, this volume) have suggested that the whole area was covered by the Late Devensian ice sheet and have proposed tentative flow lines for the ice at that time.

In the last 50 years a number of readvances during deglaciation have been postulated but many of these have now been rejected because of inadequate or suspect evidence. Among these have been Charlesworth's (1926) Lammermuir - Stranraer Readvance (Sissons, 1961b), Synge's (1956, 1963) Aberdeen and Dinnet readvances (Clapperton and Sugden, 1972, this volume) and Sissons' (1967a) Aberdeen - Lammermuir Readvance (Sissons, 1974c, 1976b).

More controversial is the Perth Readvance first described by Simpson (1933) and subsequently mapped over much of Central Scotland by Sissons (1963b, 1964). Several pieces of evidence have cast doubt upon this readvance. Dating of a mammoth tusk reported to have been obtained from below several metres of till at Kilmaurs, Ayrshire to 13,700 +1300, -1700 B.P. (Sissons, 1967c) apparently demonstrated a readvance of ice subsequent to this time, but more recently a reindeer antler supposedly from the same horizon yielded a date of > 40,000 B.P. (Shotton *et al.*, 1970) thus making the dating of the overlying till much less certain. Further, a date of > 36,900 B.P. (Shotton and Williams, 1973) for marine shells from fluvioglacial deposits of the supposed Perth Readvance near Stranraer does not accord with the proposed readvance.

Paterson (1974) has questioned Simpson's original interpretations in the Perth area. Simpson believed that a readvance was demonstrated by a section at Almondbank where till is overlain by what he believed to be 640 varves above which is outwash, the surface of which is kettled in places. Paterson, however, has suggested that the rhythmites are not annual additions but may have accumulated by the frequent discharge of sediment down the front of an advancing delta, as described, for example, by Harrison (1975). The outwash was considered to be deposited over the fine sediments as the delta continued to advance seawards. Paterson also found that there is no change in the trend of fluvioglacial landforms across Sissons' limit of the readvance in Strathmore, a fact that throws doubt upon the significance of the mapped line.

In both the Tay and Forth valleys fluvioglacial deposits at the limit of the proposed readvance correlate with the most extensive visible Lateglacial shoreline in eastern Scotland, the Main Perth Shoreline, thus demonstrating that the limits in the two valleys are contemporaneous. The marked development of this shoreline and the subsequent abrupt drop in the marine limit (see below) suggest that the fluvioglacial landforms mark at least a significant halt stage. Similarly D. G. Sutherland (pers. comm.) has suggested a halt during the general retreat of the last ice sheet in order to explain the drop in the marine limit of over 20 m that occurs in four valleys in the Cowal peninsula in the southwest Highlands.

Gray and Sutherland (this volume) have reappraised the evidence for the Oban-Ford Moraine believed by Synge (1966) and Synge and Stephens (1966) to mark a readvance or significant, synchronous, recessional stage approximately equivalent to the Perth Stage (F. M. Synge, pers. comm.). Gray and Sutherland conclude that the ice-limits more probably represent only fairly brief, diachronous halts in the general recession of the ice. They may in fact span part or all the time of the ice still-stand in Cowal. Thus, there is justification for agreeing with Sissons' (1976b, p.87) conclusion that "there is at present no firm evidence for a Perth Readvance or for other major readvances supposed to have interrupted the decay of the last ice-sheet."

Sissons (1967a, p.137) proposed that the "widespread deglaciation that followed the Perth Readvance may well have resulted in glaciers disappearing completely from Scotland" prior to the Loch Lomond Readvance which occurred at the end of the Lateglacial (see below). Peacock (1970), however, considered that the large volume of ice that existed in west Inverness-shire during the latter readvance could not have accumulated in the short time available and instead believed that active ice existed throughout the Lateglacial in that area. Sugden (1970), from his morphological mapping in the Cairngorms, proposed that the Loch Lomond Readvance in the Cairngorms could be represented by one of three situations:-

(i) a major ice sheet over the mountains and the adjacent Spey valley that had existed through the Lateglacial;

(ii) a minor fluctuation during the decline of this ice sheet;

(iii) a few corrie glaciers.

Subsequently a considerable debate took place between Sissons and Sugden, the former advocating complete deglaciation prior to the Loch Lomond Readvance but the latter defending the possibility of the existence of ice throughout the Lateglacial in the Cairngorms (Sissons, 1972a, 1973a, 1973b, 1974a; Sissons and Grant, 1972; Sugden, 1973a, 1973b, 1974).

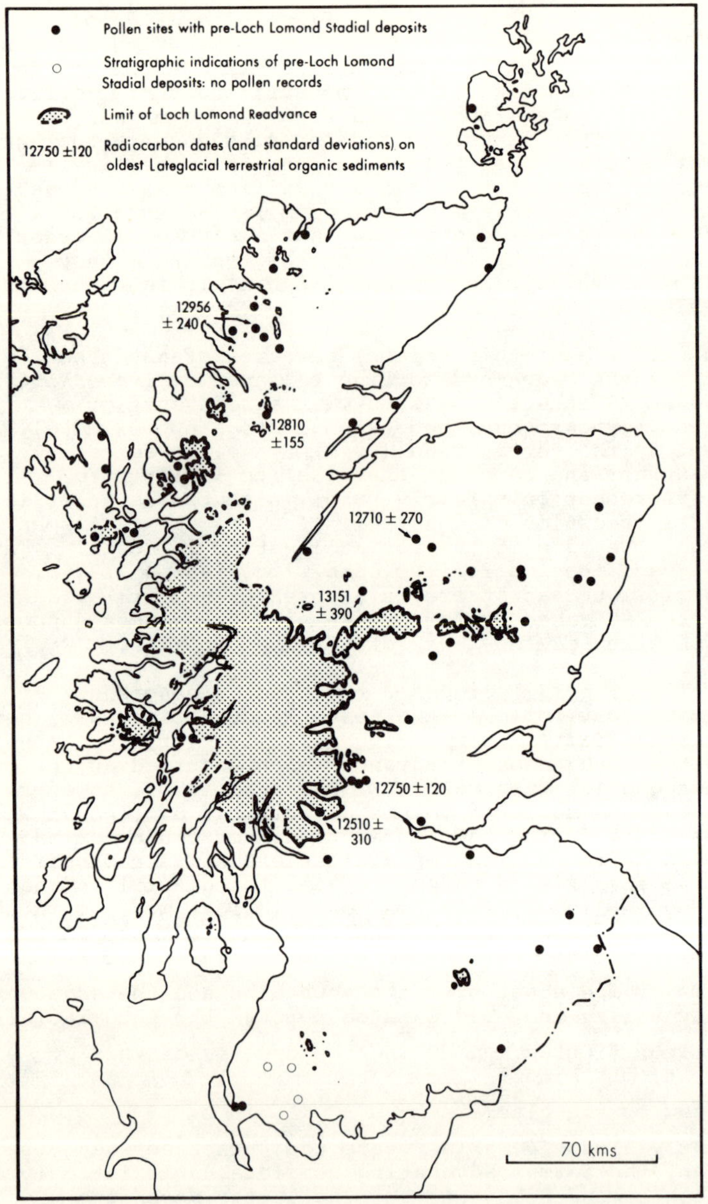

Fig. 2. Lateglacial pollen profiles, early Lateglacial radiocarbon dates on limnic-terrestrial sediments and limits of the Loch Lomond Readvance (modified after Sissons, 1976b).

More recently a radiocarbon date of 13,151 ± 390 from Loch Etteridge in the Spey valley (Sissons and Walker, 1974) excluded the first hypothesis while Sugden and Clapperton (1975) concluded that new evidence seems to point to the third possibility.

Moraines occur in about a third of the Cairngorm corries and are typically constructed of loose boulders sometimes arranged in arcuate ridges 2 - 4 m high and occasionally having clear down-valley limits. Sugden and Clapperton (p.35) argued that if the Loch Lomond Readvance is represented by the corrie moraines the readvance in the Cairngorms "is smaller by an order of magnitude than the reconstructed limits postulated elsewhere in the eastern Grampians" by Sissons (1972a, 1974b) and Sissons and Grant (1972). Thus they went on to question the evidence such as hummocky moraine and periglacial features used by Sissons and others (see below) to delimit the extent of the Loch Lomond Readvance. However, if the calculations of Sissons and Sutherland (1976) are correct the presence of small glaciers in the Cairngorms is quite compatible with more extensive glaciers in the south-east Grampians due to the rise in the firnline north-westwards from the Highland edge (see below).

Presently there are at least 58 Lateglacial pollen sites in Scotland (Fig. 2) that together demonstrate that large areas of the country were deglaciated during the Lateglacial Interstadial. The basal organic deposits in some kettle holes have been radiocarbon dated (see below), some of the dates indicating deglaciation at these sites by about 13,000 B.P. In addition the beetle evidence from Roberthill which shows that summer temperatures were as warm as at present by about 13,000 years ago (Bishop and Coope, this volume) and the apparent retreat of polar waters from the west of Britain around 13,500 B.P. (Ruddiman and McIntyre, 1973) have led Sissons (1976b, p.90) to reiterate his (1967a, p.137) proposal in more chronologically precise terms: "total deglaciation by 12,500 radiocarbon years ago seems a conservative suggestion".

However, the controversy over deglaciation during the Lateglacial Interstadial may not yet be resolved if the curve presented by Peacock *et al.* (this volume) is correct. A number of workers have shown that in the upper Forth and Tay valleys and in the south-west Highlands relative sea-level fell from levels of around 30 - 40 m to quite low levels while there was still a substantial ice mass over Scotland (Sissons and Smith, 1965; Cullingford, this volume; Gray and Sutherland, this volume; D. G. Sutherland, pers. comm.), and Peacock *et al.* show this drop underway at Lochgilphead at about 12,400 B.P. and continuing until about 12,000 B.P. by which time sea-level had fallen below 10 m O.D. If the curve is correct then either there was still a substantial mass of ice in Scotland after 12,500 B.P. or Lateglacial relative sea-level movements were much more complex than presently supposed.

The radiocarbon dates so far obtained are insufficient to allow reconstruction of the pattern of deglaciation in the same way as in North America (Bryson *et al.*, 1969) but studies of the marine limit have provided a limited amount of information. In south-east Scotland, for example, the wastage of ice westwards up the Forth and Tay estuaries is now fairly well established (Sissons and Smith, 1965; Cullingford, this volume). In the Clyde estuary a marine limit of 36 - 40 m in southern Cowal (D. G. Sutherland, pers. comm.) compares with only 25 m in the Glasgow area and suggests that the former area was deglaciated first and that ice wastage took place up estuary contrary to previous views (Sissons, 1974c, 1976b).

The widespread questioning of readvances does not extend to the Loch Lomond Readvance (or Loch Lomond Advance as Sissons (this volume) points out may be a more appropriate name if the ice built up afresh) though not all workers agree on the criteria used to map its extent. A readvance of ice following the Lateglacial Interstadial is proved by the fact that marine clay with shells dated to 11,800 - 11,300 B.P. is either incorporated in or lies below terminal moraines of the readvance at Loch Lomond, Menteith, Loch Creran and Loch Spelve (Sissons, 1967c; Peacock, 1971a; Gray and Brooks, 1972). The only period climatically severe enough and long enough for the development of ice masses of the required magnitude is the Loch Lomond Stadial.

The extent of the readvance can be partly delimited by careful stratigraphical studies of sediments in former or present day lakes. Thus the Lateglacial pollen sites already mentioned show that the readvance did not reach these sites (Fig. 2). On the other hand a large number of lake sites within the limit contain only Flandrian sediments or Flandrian and Lateglacial/Flandrian transition sediments (Sissons *et al.*, 1973). In addition morphological evidence has been used to delimit the extent of the readvance (Sissons, 1972a, 1974b, this volume, 1977a; Sissons and Grant, 1972; Thompson, 1972; Gray and Brooks, 1972). The evidence used has included the distribution of fresh hummocky, fluted, end and lateral moraines, drift and boulder limits, together with contrasts with periglacial features and supposed ice sheet landforms and sediments outside the readvance limits. Soil mapping on Mull (J. S. Bibby, pers. comm.) has shown that the Lussa Association closely corresponds with the limit of the readvance mapped independently from morphological evidence (Gray and Brooks, 1972). Sugden and Clapperton (1975) and Clapperton and Sugden (this volume), however, have proposed an alternative explanation for the occurrence of hummocky moraine at the heads of glens for they believe this may reflect the distribution of former patches of ice stagnating *in situ* and resulting in concentrations of hummocky deposits. However, hummocky moraines are not always confined to the heads of glens and similar situations. Furthermore, features such as boulder limits and clear lateral and end moraines are difficult to reconcile with the stagnation hypothesis. It is clear, however, that careful, detailed mapping is required since areas of hummocky moraine occur outside the Loch Lomond Readvance limits as defined by end moraines and landform associations (Sissons, this volume).

Figure 2 shows that the Loch Lomond Readvance consisted of one major ice cap over the western Highlands surrounded by some smaller ice caps and a large number of valley, piedmont and corrie glaciers that have been clearly described by Sissons (1976b, pp. 93-102). In the Southern Uplands a number of small glaciers were present in the Loch Doon area (R. Cornish, pers. comm.) and in the Tweedsmuir Hills. Also associated with the Loch Lomond Readvance and its subsequent deglaciation were a number of fluvioglacial and glacio-lacustrine features. Outwash terraces are found at, or a short distance within, the readvance limit at a number of places in western Scotland (McCann, 1966a; Peacock, 1970; Gray, 1975c; Gray and Sutherland, this volume) while kame terraces are particularly well developed near Loch Etive (Gray, 1975c) and Callander (Thompson, 1972). The most famous ice-dammed lakes associated with the readvance are those in the Glen Roy area (Sissons, 1976b, pp. 70-72) though there are less well known examples near Loch Tulla, south of Rannoch Moor (Thompson, 1972) and in Glen Doe (Sissons, 1977a). In the last area sudden drainage of the lakes brought about substantial washing of bedrock surfaces. In the Gaick area a number of delicate channel systems are associated with the Loch Lomond Readvance ice cap there (Sissons, 1974b).

The occurrence of outwash spreads and kame terraces immediately inside the limit of the Loch Lomond Readvance may be partly the result of climatic fluctuations during the initial stages of deglaciation (Gray, 1975c) and a similar explanation can be applied to the multiple end-moraines that occur in a narrow zone within the maximal limits of the former readvance glaciers (Sissons, this volume). The absence of end moraines farther up valley from these maximal limits suggests that following the initial fluctuations of the snout stagnation was rapid.

PERIGLACIATION

Although there is some evidence of Flandrian periglaciation

in Scotland, including present day activity, most of the evidence for periglaciation dates from the Lateglacial. There is much direct and indirect evidence of periglaciation during the Loch Lomond Stadial but the situation as regards the earlier period of ice sheet wastage is much less certain.

With only a few exceptions major periglacial features are believed to be restricted to the ground outside the limit of the Loch Lomond Readvance (Sissons and Grant, 1972; Sissons, 1972a, 1974b, this volume). Some of the exceptions come from north-west Scotland where a fossil rock glacier (Sissons, 1975a) and some solifluction lobes (Sissons, this volume) have been described from inside the readvance limit. However, the former is attributed to reactivation of dead ice following a major landslide and thus to special circumstances, while downslope movement of material inside the readvance limit has been very limited as shown by the presence of delicate features and undisturbed loose debris. Further exceptions are the three frost wedges that have been described from immediately inside the limit; one in the north-west (Sissons, this volume) and two on Mull (J. S. Bibby, pers. comm.). These features imply that very low temperatures persisted after the ice reached its greatest extent.

In contrast, however, outside the readvance limits extensive solifluction lobes and sheets, blockfields, screes, and numerous frost wedges occur. The evidence strongly suggests that the major periglacial activity in Scotland occurred contemporaneously with the Loch Lomond Readvance glaciers and/or at the time of deglaciation from the Late Devensian maximum.

Direct evidence of periglaciation during the Loch Lomond Stadial includes the layer of solifluction gravel at Bigholm Burn dated to between about 11,500 and 9,500 B.P. (Bishop and Coope, this volume), the soliflucted till overlying peat dated to 11,650 - 11,150 B.P. at Robroyston and Springburn (Dickson *et al.*, 1976), the minerogenic sediments comprising Loch Lomond Stadial deposits in numerous pollen sites stretching back into the Lateglacial Interstadial (*e.g.* Lowe and Walker, Pennington, and Vasari, this volume), and the three frost wedges just inside the readvance limit mentioned above. In addition Rose (1975) argued that two deep frost wedges in beach gravels at Old Kilpatrick on the Clyde developed after the gravels had become fossil and concluded that they were formed during the Loch Lomond Stadial.

Whether periglaciation occurred during the decay of the Late Devensian ice sheet is uncertain. Some workers have argued for its effectiveness (*e.g.* Ryder and McCann, 1971; Clapperton and Sugden, this volume) but there is as yet no evidence of it. Sissons (1974a) suggested that major periglacial features in Scotland were largely or entirely produced at the time of the Loch Lomond Readvance.

Other periglacial processes were also operative during the Loch Lomond Stadial. Nivation associated with snow patches must have been very effective and Sissons (1977b) suggests that a massive protalus rampart near Gairloch in Wester Ross was formed at that time. It is 1 km long, up to 55 m high and the volume of debris contained within it (about 600,000 m^3) implies that the bedrock faces that supplied it retreated by some 17 m during the Loch Lomond Stadial. Fluvial erosion and deposition were probably also very effective at the time as a result of rapid run-off due to factors such as a sparse vegetation cover and frozen ground (Sissons, 1976b). Similarly marine erosion resulting from slumping and flow of unconsolidated sediments and frost shattering of rock in the semi-diurnally wetted intertidal zone and cliff base has been suggested to explain the considerable marine planation that evidently occurred at the end of the Lateglacial (see below).

SEA-LEVEL CHANGES

Lateglacial sea-level changes in Scotland were mainly due to the interplay between the isostatically rebounding crust and the eustatically rising sea. It is probable that neither of these movements was simple and in addition they were probably complicated by, for example, hydroisostatic movements offshore. The result was a complex history of sea-level change that has yet to be fully deciphered.

At present it is not possible to give any accurate figure for the amount of glacioisostatic depression of Scotland during the Late Devensian maximum but it was probably in the order of 250 - 300 m. As the ice-sheet thinned over Scotland and the ice loading was reduced a certain amount of restrained rebound (Andrews, 1970) must have occurred. At the same time the ice margin was wasting back and as it did so the sea was able to flood in after it and form shorelines, initially in peripheral parts of Scotland but progressively towards the heads of sea lochs and firths as deglaciation proceeded.

Unfortunately very little is known about this early phase of sea-level change except in the Forth-Tay area, though even here none of the early shorelines has been accurately dated. The earliest shorelines to be formed were the six East Fife features (Cullingford and Smith, 1966) which have gradients ranging from 1.26 m/km for the oldest to 0.60 m/km for the youngest. That they were formed as the ice decayed is indicated by the merging of some into glacial outwash and by the fact that in general they extend farther west with decreasing age. As ice-wastage continued a number of other

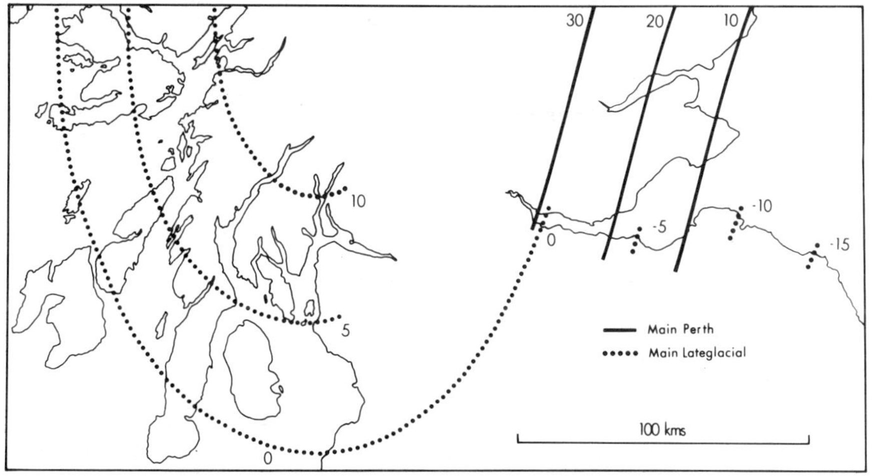

Fig. 3. Isobases (in metres) on the Main Perth Shoreline (after Smith *et al.*, 1969) and the Main Lateglacial Shoreline (after Sissons, 1976a; Gray, in press).

raised beaches were formed, the best developed of which are associated with the Main Perth Shoreline (Sissons and Smith, 1965; Cullingford, this volume). Isobases computed for this shoreline (Smith *et al.*, 1969) indicate a gradient of 0.43 m/km down towards E17°S (Fig. 3). Cullingford (this volume) has calculated that if this shoreline was formed about 13,500 B.P. isostatic uplift of 90 - 95 m at Forgandenny in Strathearn since the formation of the shoreline is indicated and the corresponding figure for Plean in the Forth valley is 98 - 103 m.

There is evidence that subsequently relative sea-level fell rapidly but as it did so a number of lower shorelines were formed in both the Forth and Tay valleys, though there are major differences between the two areas. In the Forth two lower shorelines were formed, the upper one being only 2 - 2½ m below the Main Perth and the lower one, which occurs on the Carron delta near Falkirk, being 8 - 9 m below it (Sissons and Smith, 1965). In the Tay, however, representatives of these two shorelines are very poorly developed but 4 clear shorelines occur at lower altitudes together with some less well formed ones (Cullingford, this volume).

Elsewhere in Scotland the Lateglacial shoreline sequence is much less well known. There are large areas where no detailed work has been carried out and even where reliable field data are available it has proved very difficult to reconstruct Lateglacial shorelines. Gray (1974b, p.132), for example, could only tentatively identify two short shorelines in the Firth of Lorn area and remarked that "The apparent absence of widespread, clear, lateglacial shorelines is probably due to factors such as the steepness of the coastline and the rapid rate of isostatic uplift ... lateglacial beaches were apparently only formed intermittently where favourable local conditions occurred such as a sediment supply or a sheltered locality".

Evidence for the rapid fall of sea-level following the Main Perth Shoreline comes from various parts of the country. At Stirling and Falkirk, for example, there is evidence that relative sea-level fell by 17 - 18 m while the ice margin retreated only a short distance (Sissons and Smith, 1965) while for similar situations figures of 9 m in Strathearn and 17 m in the Tay valley have been obtained (Cullingford, this volume). Abrupt falls in the marine limit also occur in southern Cowal (D. G. Sutherland, pers. comm.) while channelled and dissected outwash terraces between Oban and Lochgilphead confirm the rapid fall in sea-level in that area (Gray and Sutherland, this volume). Peacock *et al.* (this volume) show the fall underway at Lochgilphead at about 12,400 B.P. and continuing until about 12,000 B.P. but these dates conflict with evidence of the timing of deglaciation (see above). A freshwater peat on the foreshore at Redkirk Point on the Solway Firth that has been dated to 12,290 ± 250 B.P. indicates that sea-level there had fallen to below Ordnance Datum before that time (Bishop and Coope, this volume).

The general fall in relative sea-level that occurred subsequent to the formation of the Main Perth Shoreline was succeeded by a period of low, apparently stable, sea-level during which a remarkable amount of marine planation occurred. At Grangemouth, for example, an area of planated Lateglacial marine sediments, till and bedrock 28 km^2 in extent lies buried below the Flandrian carse clay (Sissons, 1969). The planation has been traced farther east along the Forth valley (Sissons, 1976a) where it changes to a submerged platform up to 1 km wide cut across till and sedimentary rocks. The related shoreline, termed the Main Lateglacial Shoreline, slopes from slightly above O.D. at Grangemouth to about -20 m O.D. at the mouth of the Tweed. Sissons (1974d) argued that this feature correlates with the well-developed rock platform and associated features in the south-west Highlands which had previously been referred to an interglacial and correlated with a till-covered platform in east Ireland (Stephens, 1957; McCann, 1966b; Synge and Stephens, 1966; Sissons, 1967a; Gray, 1974a). Sissons' (1974d) suggested correlation has recently been confirmed by work in the south-west Highlands which has shown that the well-developed platforms there do not correlate with the east Ireland features. The shoreline has a pattern of deformation (Fig. 3) that accords with that to be expected on the basis of an ellipsoid of glacioisostatic uplift aligned approximately north-south over the western Highlands though this simple pattern is interrupted by a number of irregularities (Gray, in press). Sissons (1974d) has explained the extensive erosion in terms of the effect of frost shattering and other periglacial processes on the semi-diurnally wetted platforms and cliff bases during the conditions of intense cold and increased storminess at the end of the Lateglacial Interstadial and succeeding Loch Lomond Stadial (see below). The low sea-level during this period accords with the low levels to which outwash formed at or near the limit of the Loch Lomond Readvance descends (McCann, 1966a; Sissons, 1967a, 1976b; Gray, 1975c). At the end of the Loch Lomond Stadial evidence in the western Carse of Stirling and near Grangemouth and Falkirk indicates a transgression to the level of the High Buried Shoreline followed by a regression while ice stood at the Menteith Moraine (Sissons, 1966, 1969).

ENVIRONMENTAL IMPLICATIONS FROM BIOSTRATIGRAPHY

Although some studies have been carried out on macrofossil plant remains, soil profiles, and faunal assemblages, present knowledge of Lateglacial biostratigraphy and its relationship to climatic factors rests heavily on palynological studies. There are, however, a number of problems associated with the interpretation of Lateglacial pollen diagrams that affect the recognition of major subdivisions of Lateglacial successions, and the accurate reconstruction of the vegetation and an assessment of its palaeoenvironmental implcations at each Lateglacial stage (Faegri and Iversen, 1964; Birks, 1973). Nevertheless, a large number of detailed Lateglacial pollen profiles are now available from Scotland (Fig. 2) and together with some macrofossil data and recent refinements in palynological interpretations (see below) it is possible to reconstruct, in general terms, the nature of vegetational successions for the Lateglacial.

It has now been established that over much of Highland Britain there are marked discrepancies between lithostratigraphic, biostratigraphic, and climatostratigraphic boundaries that were originally thought to be synchronous. Thus in recent years there has been a move away from the use of the Jessen-Godwin pollen zonation scheme for the Lateglacial period (Zones I, II and III), with particular reservations being expressed about the interpretation of 'pollen Zone I' and the 'Zone I/II boundary' (see Lowe and Walker, this volume). Zone II has been equated with the Alleröd interstadial of north-west Europe, and at some sites in northern Britain (see *e.g.* Oldfield, 1960; Bartley, 1962; Vasari and Vasari, 1968) Zone I has been further subdivided into Zones Ia, Ib and Ic, interpreted as indicating a climatic oscillation equivalent to the Oldest Dryas - Bölling - Older Dryas sequence of north-west Europe. However, Coope (1975) has inferred from detailed coleopteran studies that a large scale climatic oscillation occurred between about 13,000 and 11,000 B.P. in Britain, that cannot be subdivided into separate Bölling and Alleröd interstadials. This oscillation, termed the Lateglacial Interstadial (Sissons and Walker, 1974; Coope, 1975), had a thermal maximum that differed both in timing and intensity from that inferred from palynological evidence. It has been suggested that this anomaly is best explained in terms of contrasting rates of response of different components of the biota to environmental changes during the Lateglacial (Coope, 1975).

Coleopteran studies have failed to detect distinct oscillations within the Lateglacial Interstadial that could be equated with the Bölling and Alleröd oscillations (Coope, 1970, 1975; Coope and Brophy, 1972; Bishop and Coope, this volume). The dominance of the coleoptera *Haliplus obliquus*, *Haemonia appendiculata* and *Eubrychius velutus* at Roberthill indicates that by about 13,000 B.P. average July temperatures were at least as warm as the present day, and evidence from Redkirk Point and Bigholm Burn, showing clearly the gradual replacement of relatively southern species by species with decidedly northern ranges, suggests a marked deterioration of climate at, or slightly earlier than 12,000 B.P., followed by a continued but slow decline of summer temperatures between 12,000 and 11,000 B.P. The insect assemblages from south-west Scotland provide further support for the recently proposed curve of average July temperature fluctuations for lowland Britain (*cf.* Bishop and Coope, this volume, Fig. 11, and Coope, 1975, Fig. 3), suggesting a wider application for this scheme of Lateglacial climatic variation.

Birks (1973, p. 383) has concluded from a number of sites in Skye that there was no evidence of any pollen stratigraphical or environmental changes that could be equated with the Bölling interstadial or Older Dryàs stadial, but that "... there was a progressive unidirectional vegetation succession from low-alpine or mid-alpine communities to sub-alpine juniper scrub, presumably in response to a climatic amelioration starting at about 12,800 radiocarbon years B.P.". Some doubt has been expressed about the age of basal sediments from these sites (Vasari, this volume), but recently-obtained radiocarbon dates have confirmed the identification of Lateglacial Interstadial sediments on Skye (H. J. B. Birks, pers. comm.). On the basis of radiocarbon-dated pollen profiles from the Grampian Highlands, Lowe and Walker (this volume) agree with Birks' concept that vegetational succession and soil development progressed without interruption from about 13,000 B.P. until the beginning of the Loch Lomond Stadial.

However, several sites have been discovered in Scotland that appear to contain a record of a minor climatic recession within the Lateglacial Interstadial. Basal sediments and palynological data from Loch of Park in the Dee Valley (Vasari and Vasari, 1968; Vasari, this volume) have been interpreted as indicating a climatic sequence equivalent to the Bölling - Older Dryas - Alleröd sequence of north-west Europe. On lithostratigraphic criteria Vasari (this volume) suggests that an Older Dryas recession is also represented at Garral Hill near Keith (Donner, 1957) and at Loch Builg in the eastern Cairngorms (Clapperton *et al.*, 1975). A radiocarbon-dated absolute pollen diagram from Cam Loch, Sutherland, has been subdivided into Bölling, Older Dryas, Alleröd and Younger Dryas chronozones (Pennington, 1975b, this volume), with biostratigraphic and lithostratigraphic evidence for a minor recession in Older Dryas time (12,000 to 11,800 B.P.). A comparison of pollen curve fluctuations suggests that the Older Dryas climatic recession may be represented in records from at least three other sites in northern Scotland (Pennington *et al.*, 1972; Pennington, 1975b).

With the exception of only one site, Cam Loch, the recognition of an Older Dryas climatic recession is based mainly on small percentage declines in the representation of woody plants, notably *Empetrum* and *Juniperus*. A major problem therefore exists, in that such small percentage changes may only reflect statistical artefacts (Davis and Deevey, 1964) and may not provide certain evidence for a fall in temperature (Pennington *et al.*, 1972). In addition, minor lithological changes may only reflect slight variations in the rate of supply of minerogenic or organic sediments, which need not necessarily be related to climatic factors or to erosion of soils from surrounding catchments. Recent evidence from sites in Perthshire, based on detailed analysis of lithostratigraphy and deteriorated pollen grains, suggests that collapse of sediments from the edge of basins, and erosion of unconsolidated basin-edge deposits may well have operated throughout the Lateglacial and even continued well into the Flandrian (Lowe, unpublished). Such processes would have affected both lithostratigraphy and, as a result of the redeposition of pollen grains, biostratigraphy.

Interpretation of the complex stratigraphies of such sites as Loch Builg and Garral Hill, is not straightforward (Gunson, 1975), and absolute pollen data is required to establish whether or not the sedimentary and pollen assemblage changes at these and other sites in northern and north-eastern Scotland necessarily reflect climatic changes. Fluctuations in percentage and absolute pollen data from Cam Loch, indicating an Older Dryas recession, have been shown to be statistically significant (Pennington and Sackin, 1975). However, until more absolute pollen diagrams are available it is difficult to estimate the significance of the Cam Loch data in terms of their regional application. Recognition of an Older Dryas climatic recession at sites in north-west England (Oldfield, 1960; Walker, 1966; Pennington, 1975b) and in Yorkshire (Bartley, 1962) as well as in northern and north-east Scotland is difficult to reconcile with the fact that detailed radiocarbon-dated pollen profiles from the Grampian Mountains, lowland Aberdeenshire, and the Isle of Skye (Birks, 1973, pers. comm.; Lowe and Walker, this volume; Vasari, this volume) have failed to detect any climatic oscillation during the Lateglacial Interstadial. In addition, an Older Dryas climatic recession has not been detected by coleopteran studies from north Wales, northern England, and south-west Scotland (Coope, 1970, 1975; Coope and Brophy, 1972; Bishop and Coope, this volume). The evidence for a climatic recession within the Lateglacial Interstadial of Scotland thus remains equivocal.

Most Scottish Lateglacial profiles accord in indicating a period of low pollen production at the base of the profiles. This initial vegetational phase was characterised by open-habitat taxa including Gramineae, Cyperaceae, Chenopodiaceae, Caryophyllaceae, *Salix* (*cf. herbacea*), and *Lycopodium*. In some cases relatively high percentages of arboreal pollen are found suggesting over-representation of grains derived by long-distance transportation or secondary redeposition as a result of the low production of local pollen (Gunson, 1975; Walker, 1975a; Lowe and Walker, this volume). Commonly represented are pollen and spore types of taxa that are characteristic today of calcareous or base-rich conditions, and assemblages suggest the development of species-rich grassland throughout Scotland. This is consistent with the concept of plant colonisation of newly-deglaciated bare mineral soils at about or shortly after 13,000 B.P.

Faunal remains not only indicate that by about 13,000 B.P. conditions were as warm as today, but also that soon after there developed a rich aquatic and semi-aquatic flora. This would seem to contrast with the low organic content of associated sediments and with the contemporary terrestrial vegetation. However, it may be that the increased warmth of this time resulted in increased decomposition of organic matter, and the open-habitat tree-less vegetation, originally interpreted as a cold 'Zone I' phase in palynological reconstructions, may actually represent pioneer vegetation under a temperate climatic regime. Pollen diagrams may be misleading not only as a basis for climatic inference but also as records of the variation of terrestrial flora during the early Lateglacial Interstadial. The discovery of the weevil *Cryptorhynchidius lapathi* in early Interstadial deposits from south-west Scotland is important since this insect drills into *Salix* branches that must be at least 5 cm in diameter (G. R. Coope, pers. comm.). Thus it may be that, in south-west Scotland at least, a rich terrestrial flora flourished locally early in the Lateglacial Interstadial, with *Salix* bushes quite common in the dominantly herbaceous and dwarf-shrub vegetation (*cf.* Moar, 1969a).

Lowe and Walker (this volume) suggest that a closed vegetation cover, consisting mainly of grassland, developed over the lower and more sheltered parts of the Grampains, with juniper, dwarf birch, willow and occasional copses of tree birch on the valley floors (*e.g.* the Teith valley). There is no macrofossil evidence for the presence of tree birch, and their inferred presence rests on comparatively high relative percentages of what appear to be arboreal birch pollen. Moss heaths and poor grassland communities were probably more characteristic of upper slopes and more exposed localities. At some sites, however, soil instability and open-habitat communities appear to have persisted throughout the Lateglacial Interstadial. This is indicated by the continuous accumulation of minerogenic sediments, by the consistently high representation of deteriorated pollen grains (mainly broken and crumpled grains), and by the continued presence of herbaceous taxa indicating disturbed soils.

According to Vasari and Vasari (1968), in lowland Aberdeenshire an open vegetation dominated by *Rumex* was gradually replaced by a more closed vegetation characterised by large areas of *Empetrum* heath. This in turn was followed by a phase of dominant juniper heath towards the close of the Interstadial, whilst tree birches existed as scattered copses only, without forming any real forests. In eastern Aberdeenshire, for example at Loch Kinord, some pines may also have occurred. Macrofossil evidence supports these conclusions. As in the Grampians farther to the south and west there are marked contrasts in the importance of certain taxa at different sites within north-east Scotland, *e.g.* in the frequencies of *Empetrum*, Ericaceae, and *Betula*, and in the aquatic flora. These may be explained by differences in both soil and climate (Gunson, 1975).

Pennington's (this volume) absolute pollen analysis of Lateglacial sediments from Cam Loch demonstrates that deposition rates for total pollen at that site do not exceed totals published elsewhere for present-day dwarf-shrub tundra. In addition deposition rates for tree birch suggest that at no time during the Lateglacial Interstadial did birch woodland develop near Cam Loch. Late Lateglacial Interstadial vegetation in northern Scotland appears to have been mainly characterised by an ericaceous dwarf-shrub heath (*Empetrum* heath) and grassland, with a very variable representation of juniper (Pennington *et al.*, 1972).

In an attempt to overcome some of the problems inherent in inferring past vegetation from fossil pollen assemblages (see above and Davis, 1963) Birks (1973) compared Lateglacial pollen assemblages from sites in Skye with modern pollen assemblages derived from a study of the relationships between modern pollen rain and vegetational types in Scotland. A variety of communities was represented in the late Lateglacial Interstadial vegetation, reflecting a marked edaphic differentiation within Skye during the Lateglacial Interstadial. Birch woodland is inferred for one site, Loch Meodal, where high percentages of *Betula* (undiff.) pollen and macrofossils of *Betula pubescens* are recorded. It appears, however, that there are few modern analogues to some of the shrub-dominated (*Juniperus* and *Betula nana*) communities recorded. Similarly the high percentages of *Empetrum* without *Calluna* in the Lateglacial Interstadial pollen spectra cannot be matched with any published surface pollen spectra from Highland Britain today (Pennington, this volume; *cf.* Birks, 1973). Modern analogues are perhaps not to be expected if, as coleopteran evidence suggests, an unstable equilibrium between biota and climate existed during the Lateglacial Interstadial (Coope, 1975; Bishop and Coope, this volume).

Evidence from other Scottish Lateglacial pollen diagrams (Donner, 1957; Kirk and Godwin, 1963; Newey, 1970; Moar, 1969a; Pennington and Lishman, 1971) agrees with the general conclusion of Birks (1973) and of Pennington, Vasari, and Lowe and Walker (this volume) in suggesting a complex vegetation pattern during the Lateglacial Interstadial, composed mainly of species-rich grassland and dwarf-shrub heath, with occasional copses of tree birch, and with variable quantities of *Empetrum*, *Juniperus* and *Salix*. It is thus apparent that the general tree-line (birch) did not reach as far north as Scotland, but that birch woodland flourished only locally in areas of particularly favourable aspect. The northern limit of the general distribution of birch during the Lateglacial is thought to have been in the vicinity of the Lake District (Pennington, 1970).

While the bulk of the palynological evidence from Scotland suggests that expansion of juniper heath and the development of copses of birch woodland occurred in Alleröd time (11,800 - 11,000 B.P.), faunal evidence from Redkirk Point and Bigholm Burn indicates that the thermal maximum of the Lateglacial Interstadial occurred at or slightly before early Bölling time, and that a considerable climatic deterioration actually took place around 12,500 - 12,000 B.P., after which temperatures remained relatively cool. The consistent increase in beetles with dominantly northern geographical distributions, such as *Elaphrus lapponicus*, *Amara torrida*, *Agonum consimile*, and *Olophrum boreale*, suggests that by about 12,000 B.P. the average July temperature had already decreased to about 12°C. Bishop and Coope (this volume, p.86) suggest that conditions at that time were comparable to those near the tree-line today in the Scandinavian Mountains, and that this was an environmentally critical period when "Any small scale fluctuation in temperature ... may well cross and recross the threshold limiting tree growth and thus cause apparently large scale changes in pollen assemblages, without any comparable large scale changes in insect assemblages", a suggestion that is entirely in keeping with the palynological evidence for the limited growth of birch woodland at this stage of the Lateglacial Interstadial.

Interstadial pollen assemblages, macrofossils, and litho-stratigraphic data, supported by important chemical studies (Pennington and Lishman, 1971; Pennington *et al.*, 1972; Pennington, this volume), suggest that the intricate variation in edaphic, topographic, climatic and biotic factors that is characteristic of Scotland today must also have operated in the past to produce a complex mosaic of plant communities and extremely variable soil depths and degrees of stability. The exact climatic parameters for the Lateglacial Interstadial are difficult to determine from palaeobotanical data (see Birks, 1973), but it may be that certain local site factors, such as degree of insolation, frost incidence and wind exposure, were even more critical during at least the later part of the Lateglacial Inter-stadial than they are today, resulting not only in marked vegetational differentiation over very short distances, but also in the prevention of a closed vegetation cover and the continuation of disturbed soil conditions throughout this period.

In most Scottish Lateglacial pollen profiles the climatic deterioration of the Loch Lomond Stadial is reflected in lower pollen concentrations and poorer pollen and spore preservation, by a reduction in the representation of shrubs and trees in pollen assemblages, by a return to predominantly minerogenic sedimentation (often of coarse sediments), and by the virtual exclusion of aquatic pollen in pollen spectra at some sites (Walker, 1975a) or by marked changes in macro-fossil records of aquatic plants at others (Vasari and Vasari, 1968). In a number of profiles the highest Lateglacial percentages of *Artemisia* pollen are recorded in the Loch Lomond Stadial, and there are associated high percentages of spores of the Clubmosses *Lycopodium* and *Selaginella* and of pollen of families which include herbs characteristic of open ground and disturbed soils, such as Compositae, Caryophyllaceae, Cruciferae and Chenopodiaceae. The presence of these taxa in significant quantities demonstrates that over virtually the whole of Scotland there was a dramatic break-up of the vegetation cover and destruction of soils that had gradually developed during the Lateglacial Interstadial.

The most commonly accepted explanation of this marked vegetational revertence phase is that it is related to an increase in frost disturbance of ground not protected by a winter snow cover (Pennington, this volume). Though there appears to be some uniformity in the pollen assemblages recorded for the Loch Lomond Stadial throughout Scotland, regional and local differences are also apparent, which may have been related in part to thickness and duration of snow cover. Walker (1975a) has suggested that variations in the relative representation of *Artemisia* between different sites in the central and eastern Grampians may partly reflect conditions of snow cover since this plant is known to be somewhat chionophobous (Andersen, 1961). Birks (1973) has shown that at a number of sites in Skye the Loch Lomond Stadial vegetation was predominantly of snow-bed type, and that low-alpine and mid-alpine chionophilous vegetation persisted into the Flandrian in sheltered localities on the island.

Pollen assemblage evidence for climatic deterioration and soil disturbance during the Loch Lomond Stadial is supported by chemical analysis of sediments (Pennington *et al.*, 1972; Pennington, this volume), and by analysis of diatom assemblages (Haworth *in* Pennington, this volume), aquatic macrophytes (Vasari and Vasari, 1968), and deteriorated pollen grains and spores (Lowe and Walker, this volume). There is little faunal evidence from Scotland for environ-mental conditions during the period from 11,000 to 10,000 B.P. However, the species-poor fauna recorded from Bigholm Burn which is dominated by the low temperature stenothermic species *Olophrum boreale*, indicates rigorous climatic conditions (Bishop and Coope, this volume). Taking the evidence as a whole there thus seems to be ample support for Pennington's contention (this volume, p.139) that "... destruction of the organic interstadial soil horizon must have been complete, down to the lowest altitudes ..."

The floristic composition of the Loch Lomond Stadial vegetation in some respects resembled that of the colonising plant communities recorded in early Lateglacial Interstadial sediments. The large areas of fresh mineral soils in the vicinity of the basins are reflected in the significant percentages of plants from the Stadial pollen record that are characteristic of base-rich conditions. These include calciphilous and basiphilous species that are common in Scotland today in 'flushed' situations (McVean and Ratcliffe, 1962), on shell sands and coastal limestones (Poore and McVean, 1957), base-rich rocks, especially serpentine (Proctor and Woodell, 1971; Spence, 1957), and on screes and mountain cliff-ledges (see especially discussion of calciphilous and basiphilous indicator species in Birks, 1973). Loch Lomond Stadial plant communities have been likened to such present-day communities as the shrub-dominated associations of sub-alpine and low-alpine zones in Scandinavia (Birks, 1973) and plant assemblages on the northern slopes of the Sierra de Gredos in central Spain (Pennington *et al.*, 1972). However, both Birks (1973) and Pennington (this volume) point out that some of the Stadial spectra have no modern analogues, and thus their interpretations are not straightforward. Pennington draws particular attention to the very high percentages of *Artemisia* in some Loch Lomond Stadial deposits, and concludes that some Stadial pollen spectra from the northern Scottish mainland represent types of vegetation not found in northern Europe today.

It is generally agreed that the transition to warmer and more stable conditions at the close of the Loch Lomond Stadial occurred rapidly. This is reflected in a usually abrupt change from minerogenic to organic sediment accumulation, and by a sharp fall in percentages of deteriorated pollen in a number of sites (Lowe and Walker, this volume). Chemical data also reveal a sharp change in sediment composition at the Loch Lomond Stadial/Flandrian boundary, with carbon content increasing markedly to values higher than those reached at any horizon in the Lateglacial Interstadial (Pennington, this volume). Coleopteran evidence from early Flandrian deposits at Brighouse Bay in south-west Scotland supports evidence from the English Midlands (Ashworth, 1973) for a climatic transition at the beginning of the Flandrian which occurred with "extraordinary rapidity", possibly within a period of less than five hundred years, and which resulted in a change from a climate of arctic severity to a climate similar to that of south-west Scotland today (Bishop and Coope, this volume). At the dated horizon of 9,640 ± 180 B.P. at Brighouse Bay all northern insect species had disappeared and in their place a rich temperate fauna had become established. *Empetrum* and *Juniperus* heaths expanded quickly, and birch woodland was widespread in Scotland in less than (or about) 1,000 years after the beginning of the Flandrian, defined at 10,000 B.P. (Pennington *et al.*, 1972; Birks, 1973; Walker, 1975b; Lowe and Walker, Pennington, Vasari, this volume).

The conclusions in this section are based primarily on palynological studies, coleopteran assemblages, and associated sediments. Very little direct evidence is available for soil types and edaphic variation in the Lateglacial, and very little is known about animal populations in general. Although the coleopteran studies from south-west Scotland reported here represent a significant advance in our state of knowledge, clearly many more detailed studies of this type are required. Other faunal remains have been reported from Lateglacial deposits throughout Scotland, but by and large these have not been dated accurately, and often their exact stratigraphic significance is uncertain (Sissons, 1976b; Bishop and Coope, this volume).

Information on *in situ* Lateglacial marine faunas has recently increased substantially, partly as a result of North Sea oil developments. Two main faunal assemblages of Lateglacial age have been identified; these have been termed "arctic" and "subarctic" by Sissons (1967a) and "Errol Beds" and "Clyde Beds" by Peacock (1975). Deposits of the "arctic seas", which for the most part comprise clay, silt and fine sand with occasional ice-rafted stones, were formed first as the sea flooded in to replace the wasting ice sheet. They are mainly confined to the coastal and offshore areas of eastern Scotland south of the Moray Firth including the Forth and Tay estuaries. Several arctic species, including the bivalves *Portlandia arctica* and *Palloillum groenlandicum*, the ostracod *Rabilimus mirabilis* and the seal *Pagomys foetidus*, are present in the generally impoverished faunas of the deposits.

Following the amelioration of climate and the retreat of polar waters by 13,000 B.P. (Bishop and Coope, this volume; Ruddiman and McIntyre, 1973) the deposition of arctic clays was replaced by the accumulation of thick "subarctic" marine sediments that overlie in places the "arctic" beds of the east coast but in the west lie directly on bedrock, till or fluvioglacial deposits. They are particularly well developed in the Clyde estuary but occur as far north as Loch Linnhe. In general they are "characterised by a richer high boreal fauna, the present-day equivalent being broadly southern Iceland, the Faroes, and Norway north of the Lofotens" (Peacock, 1975, p.46).

Investigations of the faunal remains in sediments from a borehole east of Colonsay (Binns *et al.*, 1974) have revealed three "palaeoclimatic zones" lying above till and poorly sorted, sandy muds. The lower zone has a foraminiferal assemblage dominated by the warm water species *Ammonia batavus* and a rich population of dinoflagellate cysts. The middle zone contains a cold water assemblage of foraminifera dominated by *Elphidium clavatum* and sparse dinoflagellate cysts, though this may be related to a change in the volume of suspended sediment rather than to low water temperature. The upper zone shows a decline in the cold water foraminifera and has been radiocarbon-dated to the early Flandrian. This suggests that the three zones may correspond with the Lateglacial Interstadial, Loch Lomond Stadial and early Flandrian respectively. Mollusca, ostracoda and foraminifera from a borehole in the Cromarty Firth (Peacock, 1974) indicate a similar succession.

CLIMATE

In previous sections of this paper evidence has been presented that demonstrates or implies climatic change during the Lateglacial. The present section will be restricted to considering evidence that indicates quantitative or semi-quantitative estimates of climatic parameters. At present only a limited amount of such information is available but it involves both geomorphological and biological evidence.

The frost wedges present in sands and gravels just inside the maximal extent of the Loch Lomond Readvance on Mull and in Wester Ross (see above) imply a mean annual temperature at sea-level no higher than -1°C and possibly lower than -6 to -8°C (Péwé, 1966) shortly after the maximum of the readvance. Using the method of Sissons (1974b), a firnline altitude of 250 m has been calculated for the Loch Lomond Readvance ice cap on Mull, an altitude that implies a sea-level mean July temperature of 5°C (Sissons, 1976b). Thus the mean January sea-level temperature for Mull was probably as low as -7°C and possibly below -17 to -19°C at that time. Clearly the climate was more continental than now though doubt has been expressed about Péwé's thermal limits (Sissons, 1974c, 1976b; Black, 1976).

Although the Loch Lomond Readvance firnline was at about 250 m on Mull the level must have risen steeply eastwards, for a figure of 790 m was obtained by Sissons (1974b) for the Gaick ice cap in the central Grampians. More recently Sissons (this volume, Fig. 11) has shown the steep rise inland from the north-west Scottish coastline and Sissons and Sutherland (1976) have computed a regression surface for the S.E. Grampian glacier firnlines (incorporating variations in insolation) that shows a rise from about 500 m close to the Highland edge to 850 m north of Lochnagar, a gradient of 14.5 m/km towards N30°30'W. This rise is explained in terms of the dominance of south and south-east winds in bringing snow-falls to that part of the Grampians. Using data for modern Norwegian glaciers as an analogue they have made a number of climatic inferences. Compared with present values, a July sea-level temperature about 9°C lower (at 6°C), a greater cloud cover, at least in summer, and a similar total precipitation though rather differently spatially distributed have been calculated. Since factors such as lower atmospheric moisture content and at least a partly frozen North Sea in winter must have operated to reduce precipitation in the south-east Grampians during the Loch Lomond Stadial, the fact that precipitation was about the same as at present has been explained by a much more vigorous atmospheric circulation with more frequent fronts, stronger winds and a greater cloud cover. In turn it has been speculated that the more vigorous circulation may have been due to the interaction between thermally contrasting air masses over the warm water of the North Atlantic Drift and the polar water that may have returned temporarily to the vicinity of the British Isles during the Loch Lomond Stadial (Sissons and Sutherland, 1976).

Bishop and Coope (this volume) have constructed a curve of mean July temperatures throughout the Lateglacial in south-west Scotland based on coleopteran evidence. The curve shows that about 13,000 radiocarbon years ago the mean July temperature was about 15°C and the climate was not much different from that in south-west Scotland today. This was followed by a thermal deterioration possibly in two steps (Fig. 1) which culminated in the Loch Lomond Stadial when the mean July temperature fell to 8-9°C. This compares quite well with the figures of 5 and 6°C for Mull and the south-east Grampians respectively calculated from firnline evidence (see above) and "at or a little below 10°C" calculated from coleopteran evidence in "lowland Britain" (Coope *et al.*, 1971, p.95). The coleopteran assemblage from the Brighouse Bay peat dated to 9,640 ± 180 B.P. shows that the climatic recovery following the Loch Lomond Stadial must have been very rapid for by that time the climate of south-west Scotland "was at least as warm or even slightly warmer" than today.

Peacock *et al.* (this volume, p.97), from marine shell and microfaunal evidence, have suggested that the climate of Scotland "was probably more severe than at present during the early part of the Lateglacial Interstadial" thus partly contradicting the conclusions of Bishop and Coope. However, there must be some doubt concerning, for example, the use of marine molluscs in determining mean July air temperatures.

It is difficult to establish thermal parameters from palaeobotanical data. Detailed autecological studies are rare and usually indicate taxa or communities to have a wide range of ecological and climatic tolerances. Further, plant distributions may be controlled at any one time not only by thermal conditions, but also, for instance, by other climatic parameters, edaphic factors, competition from other species, and migration rates following major environmental changes. In determining the ecological factors that controlled plant distributions in the past a multi-factorial approach is required when interpreting palaeobotanical data (Conolly, 1961; Walker, 1966; Brown, 1971).

However, it has been suggested by Iversen (1954) that the presence of *Betula pubescens* in the local flora implies a mean July temperature of at least 12°C. The macrofossil records of *Betula pubescens* from Skye (Birks, 1973) and Abernethy Forest (Vasari, this volume), and the development of tree birch copses in many areas of Scotland during the Lateglacial Interstadial, indicate that temperatures of about or in excess of 12°C were maintained long enough to permit the general immigration of this tree. Consistent records of *Littorella uniflora* (Shore-weed) pollen from the latter part of the Lateglacial Interstadial in Skye have also been interpreted as indicating mean July temperatures of 14°C or more for this time (Birks, 1973). However, the calculation of temperature depressions for times in the past, based on the occurrence of low-alpine vegetation at sea-level and assuming a temperature lapse-rate similar to that of today, must be regarded as questionable in view of the possibility of erroneous climatic inferences when using palaeobotanical data alone (Coope, 1970, 1975; Coope and Brophy, 1972; Birks, 1973).

TERMINOLOGY AND CHRONOLOGY

In the introduction to this volume it was pointed out that the term "Lateglacial", including its alternative forms, has been used in several ways. This section is intended to focus attention on the problems of definition and subdivision of the Lateglacial in the hope that subsequent discussion and research will lead to a terminology and chronology that will gain general acceptance.

First, it is possible to adopt chronostratigraphic boundaries for the Lateglacial. This approach is favoured for Norden where Mangerud *et al.* (1974) define the Late Weichselian as the period between 13,000 and 10,000 B.P. (Fig. 1), and Pennington (1975b, this volume) has applied this scheme to Britain. The date of 10,000 B.P. is the internationally accepted Weichselian/Flandrian chronostratigraphic boundary (Mitchell *et al.*, 1973; Mangerud *et al.*, 1974), while 13,000 B.P. is favoured by many workers for the beginning of the Scottish Lateglacial since this is about the time of the development of woody taxa and also coincides with the earliest post - Late Devensian ice sheet radiocarbon dates (Birks, 1973; Sissons, 1974c, 1976b; Pennington, 1975b, this volume).

Chronostratigraphic boundaries, however, need not correspond exactly with times of significant environmental change. The most important environmental variable is, of course, climate since it has such a profound effect on biota and on geomorphological processes. Thus Quaternary subdivision on the basis of climate has been a traditional objective. But, since climatic changes may be time-transgressive and are not observed directly but only in their indirect effect on animals, plants, glaciers, sediments, *etc.*, there has been a trend towards either chronostratigraphic subdivision or subdivision on the basis of biostratigraphy or lithostratigraphy. However, this may result in different definitions for the beginning or end of a particular stadial or interstadial. The lower boundary of the Lateglacial Interstadial in north-west England, for example, falls at different times according to the criteria used (Pennington, this volume).

In an area the size of Scotland, though biostratigraphic and lithostratigraphic boundaries may be time-transgressive, it is unlikely that the major Late Devensian climatic changes were greatly non-synchronous. Also the development of sophisticated techniques such as absolute pollen analysis and coleopteran studies is leading to a much more accurate knowledge of the intensity and timing of climatic changes. Mitchell *et al.* (1973, p.1) "... accepted climatic fluctuation as the guiding principle for defining the stages of the British Quaternary", though a chronozone scheme, based on radiocarbon dating, was preferred for the Lateglacial and Flandrian periods (see also Shotton and West, 1969). It is, however, important to determine to what extent the chronostratigraphic boundaries for the Lateglacial reflect times of significant climatic change. Mangerud *et al.* (1974, p.118) stated that "all Late Weichselian chronozone boundaries seem to be climatically conditioned within southern Scandinavia" and an attempt is made here to assess to what extent the same holds for Scotland.

In terms of climatostratigraphy the Lateglacial should refer to the period between the thermal improvement that led to final widespread decay of the Late Devensian ice sheet and the thermal improvement that resulted in the final disappearance of ice from Scotland following the Loch Lomond Readvance (see below). The lower climatostratigraphic boundary for the Lateglacial is very difficult to define at present. Six radiocarbon dates have been obtained for basal organic deposits in former or present day lakes (Table 1 (A) and Fig. 1). The dates, if valid, indicate that much of Scotland was deglaciated by 13,000 B.P. but certain factors must be taken into account in using these dates to calculate the time of deglaciation and the chronology of climatic change. First, the basal lake deposits occur in kettle holes where dead ice could have survived for some time after general deglaciation of surrounding areas (*cf.* Porter and Carson, 1971). Secondly, in no case was the basal organic matter in these lakes the first material to accumulate in the kettle holes, for minerogenic deposits underlie the lowest organic horizon. There may also have been a time-lag between deglaciation of the kettle holes and initiation of sediment accumulation. Thirdly, the radiocarbon dates have usually been obtained on sediment slices several centimetres thick and thus the dates may be significantly younger than the time of initial organic growth. This factor is particularly important in view of the probable slow initial rate of organic accumulation and the likely high compression

TABLE 1 Radiocarbon dates

	Location	Reference	Thickness (cm) of sample	Material	Date (C14 years B.P.)	No.
A. Oldest limnic-terrestrial sediments	Tynaspirit, Perthshire	Lowe and Walker*	2.0	gyttja	12,750 ± 120	HV-4989
	Cam Loch, Sutherland	Pennington (1975b)	10.0	organic silt	12,956 ± 240	SRR-253
	Loch Droma, Ross and Cromarty	Kirk and Godwin (1963)	4.0	grey-brown silt with macroscopic plant remains	12,810 ± 155	Q-457
	Loch Etteridge, Inverness-shire	Sissons and Walker (1974)	3.0-4.0	clay-gyttja	13,151 ± 390	SRR-304
	Drymen, Stirlingshire	Vasari*	8.0	clay-gyttja	12,510 ± 310	HEL-160
	Abernethy Forest, Inverness-shire	Vasari*	10.0	sandy-gyttja	12,710 ± 270	HEL-424
B. Biostratigraphic boundaries indicating beginning of Lateglacial vegetational revertence	Abernethy Forest, Inverness-shire	Vasari*	10.0	banded gyttja	11,260 ± 240	HEL-423
	Drymen, Stirlingshire	Vasari*	5.0	gyttja	12,060 ± 320	HEL-161
	Tynaspirit, Perthshire	Lowe and Walker*	2.0	gyttja	11,385 ± 290	HV-4987
	Loch Etteridge, Inverness-shire	Sissons and Walker (1974)	1.5	gyttja	10,764 ± 120	SRR-302
	Loch Kinord, Aberdeenshire	Vasari*	10.0	gyttja/clay-gyttja	10,640 ± 260	HEL-419
	Garral Hill, Banffshire	Godwin and Willis (1959b)	5.0	silty *Hypnum* peat	10,800 ± 230	Q-104
C. Biostratigraphic boundaries indicating end of Lateglacial vegetational revertence	Amulree, Perthshire	Lowe and Walker*	2.0	gyttja	10,770 ± 90	HV-5644
	Mollands, Perthshire	Lowe, unpublished	2.0	gyttja	10,670 ± 85	HV-5647
	Tynaspirit, Perthshire	Lowe and Walker*	2.0	gyttja	10,420 ± 160	HV-4985
	Cam Loch, Sutherland	Pennington (1975b)	10.0	organic mud	10,226 ± 190	SRR-247
	Blackness, Angus	Walker (1975a)	1.5	gyttja	9,490 ± 160	HV-5649
	Rannoch Moor, Argyllshire	Lowe and Walker (in preparation)	1.0	gyttja	10,520 ± 330	Birm-723
	Loch of Park, Aberdeenshire	Vasari*	10.0	gyttja/clay-gyttja	10,280 ± 220	HEL-416
	Loch Kinord, Aberdeenshire	Vasari*	10.0	gyttja	10,010 ± 220	HEL-420
	Abernethy Forest, Inverness-shire	Vasari*	10.0	gyttja	10,230 ± 220	HEL-422
	Drymen, Stirlingshire	Vasari*	16.0	gyttja/clay-gyttja	10,010 ± 230	HEL-162
	Lochan Coir' a' Ghobhainn, Skye	Birks (1973), reinterpreted by Vasari*	2.5	diatomaceous fine detritus mud	10,254 ± 220	Q-955
	Loch Cuithir, Skye	Vasari*	10.0	clay-gyttja	10,060 ± 270	HEL-504
	Loch Etteridge, Inverness-shire	Sissons and Walker (1974)	4.0	clay-gyttja	9,405 ± 260	SRR-301
	Murraster, Shetland	Jóhansen (1975)	13.0	detrital gyttja	10,400 ± 100	K-1865

* this volume

of the basal layers. It is suggested therefore that a 10 cm thick slice of basal organic material may have accumulated over several hundred years. Possible sources of error affecting these dates include "hard water error" (*e.g.* at Abernethy Forest (Vasari, this volume)) and, in the case of the thinner slices, the bulking of multiple shot samples, but, if valid, the dates are minimal for deglaciation.

In addition to the dates on basal organic material a date of 12,940 ± 250 B.P. has been obtained from a piece of wood in a peat layer at Roberthill near Lockerbie (Bishop, 1963; Bishop and Coope, this volume), while the oldest dates on marine shells in the Clyde estuary are 12,650 ± 200, 12,610 ± 210 and 12,615 ± 230 B.P. (Bishop and Dickson, 1970; Peacock, 1971a). These dates are also likely to be minimal for deglaciation since at each locality the dated material is underlain by substantial thicknesses of minerogenic sediment. In the case of the Lockerbie date time must also be allowed for the immigration and growth of trees.

Even though these dates provide evidence for the timing of ice-wastage they do not necessarily relate to the major thermal improvement that led to widespread deglaciation. It is probable that there was a time-lag between the two and this must have been of variable length since deglaciation is time-transgressive. It is also possible that at least initially deglaciation resulted not from thermal improvement but from, for example, changes in precipitation values. It is clear therefore that the dates discussed above must be regarded as minimal for deglaciation and unsatisfactory as indicators of times of climatic change.

Several lines of evidence must be considered in defining when the major climatic improvement marking the beginning of the Lateglacial occurred. At the global scale it is of interest to note that Mercer (1972) suggested that the main global warming marking the end of the Pleistocene occurred at about 14,500 - 14,000 B.P. In the vicinity of Britain Ruddiman and McIntyre (1973) found that the change from dominantly polar to non-polar foraminifera and from low to high fine carbonate content of marine sediments occurred at about 13,500 B.P., though this is an extrapolated date based on sedimentation rates. Recent work, including absolute pollen analysis, at Lake Windermere, north-west England (Pennington, this volume) supports earlier conclusions from research at Blelham Bog and Cam Loch (Pennington, 1975b) in indicating that the woody plant expansion took place at about 13,000 B.P., thus suggesting that this continental chronozone boundary may be an important climatic one in north Britain. Coleopteran studies, however, suggest that the transition from an arctic climate to the thermal maximum of the Lateglacial was very rapid (Coope and Brophy, 1972) and occurred before 13,000 B.P. At Roberthill, for example, the beetles indicate that summer temperatures were already as high as at present by about 13,000 B.P. and at Lake Windermere arctic beetle species were absent and temperate ones were present for some time before the woody plant expansion (G. R. Coope, pers. comm.).

It appears, therefore, that the apparently rapid thermal improvement that resulted in final widespread decay of the Late Devensian ice sheet and in major biotic displacements in north Britain was initiated before, possibly well before, the chronostratigraphic boundary of 13,000 B.P. Unfortunately, there is as yet insufficient data for defining the time of the beginning of this thermal improvement but current evidence suggests that it occurred between about 14,000 and 13,000 B.P. (G. R. Coope, pers. comm.).

Mangerud *et al.* (1974) and Pennington (1975b) have subdivided the Late Weichselian/Lateglacial into four chronozones (Fig. 1). However, as explained earlier, there is at present little evidence for a significant climatic oscillation in Scotland comparable to the Older Dryas of continental north-west Europe. Thus only two climatostratigraphic subdivisions of the Scottish Lateglacial appear justified at present, *viz.* the Lateglacial Interstadial and the Loch Lomond Stadial.

Since climate appears to have deteriorated throughout the later part of the Lateglacial Interstadial it is difficult to define a climatostratigraphic Lateglacial Interstadial/ Loch Lomond Stadial boundary. Though the Alleröd/Younger Dryas boundary, dated at 10,800 to 11,000 radiocarbon years B.P. (Iversen, 1953; Godwin and Willis, 1959b; Berglund, 1966; Tauber, 1970; Mangerud *et al.*, 1974) has been generally assumed to define the lower boundary of the Lateglacial revertence phase in British pollen diagrams, this has yet to be tested critically. Pennington (1975b) has accepted the chronozone boundary 11,000 B.P., but the published Lateglacial radiocarbon dates from Cam Loch are non-sequential and are not critically related to the important biostratigraphic horizon indicating climatic deterioration.

Six radiocarbon dates are available from Scotland that are thought to be on or close to important biostratigraphic horizons indicating the beginning of a major vegetational revertence phase implying transition to cold conditions (Table 1 (B)). In addition to the ranges of the standard deviations of these dates and the possibility of errors associated with the radiocarbon-dating method, the thickness of the sediment slices used in the assay may again be important, though in this case this factor will tend to make the date too old. Of the six dates, four are based on slices of 5 or 10 cm and only two, those from Tynaspirit

(2 cm) and Loch Etteridge (1.5 cm), are for thinner horizons. These latter two might be regarded as more critical dates, but there is a marked discrepancy between them with no overlap in the ranges of the standard deviations. Vasari (this volume) has suggested that there were regional differences in Scotland in the time of vegetational response to climatic deterioration. From glaciological and other considerations Sissons (1974c, p.320) concluded that "the Loch Lomond Readvance began before, perhaps well before 10,800. It may also be inferred, in view of the variations in relief and precipitation, that glaciers began to develop at different times in different places." It is probable too that other environmental effects of the continuous climatic deterioration since about 12,500 to 12,000 B.P. (break-up of closed vegetation, soil erosion, gelifluction) were time-transgressive. In spite of the early onset of climatic deterioration there is at present insufficient justification for placing the Lateglacial Interstadial/Loch Lomond Stadial climatostratigraphic boundary any earlier than the chronostratigraphic boundary of 11,000 B.P. Bishop and Coope (this volume) suggest that climatic deterioration became more intense at about this time.

The term Loch Lomond Stadial has been used in place of Younger Dryas or Zone III, since, as a climatostratigraphic unit, it may eventually have a different definition. It is named after the Loch Lomond Readvance, the now well-established glacial advance in Scotland (see above) that resulted from the climatic deterioration that followed the Lateglacial Interstadial.

A number of radiocarbon dates are now available from Scotland (Table 1 (C)) for biostratigraphic boundaries indicating an end to the major vegetational revertence phase, implying climatic improvement. In most cases they coincide with a maximum of *Juniperus* pollen in the pollen diagrams. Clearly there is some disparity in the dates, but in comparing them a number of factors must be considered. First, there is the possibility that the oldest dates are subject to some "hard water error", though there are no obvious reasons to suspect this at any of the sites. Secondly, the youngest dates may have been subject to contamination, though again there is no direct evidence for this; rather the younger dates may well indicate continued minerogenic in-wash after the close of the Loch Lomond Stadial as a result of local site factors. Thirdly, many of these dates are based on sample slices of about 10 cm thickness, and there is a strong possibility that this has resulted in incorporation of organic material that is significantly younger than the climatostratigraphic boundary and the precise biostratigraphic boundary.

A number of the dates are based on very thin sediment slices, *viz.* Amulree (10,770 ± 90), Tynaspirit (10,420 ± 160), Mollands (10,670 ± 85), Lochan Coir' a' Ghobhainn (10,254 ± 220), Loch Etteridge (9,405 ± 260) and Blackness (9,490 ± 160) and a date of 10,520 ± 330 has recently been obtained from a thin basal gyttja layer from a site in Rannoch Moor (Lowe and Walker, in preparation). The date from Loch Etteridge is interpreted as a result of continued solifluction at this site (Sissons and Walker, 1974), though both this date and that from Blackness may relate to contaminated sediments (Walker, 1975a, 1975b). Of the remaining 5 dates all except Lochan Coir' a' Ghobhainn suggest a climatostratigraphic boundary that is earlier than the lower Flandrian boundary of 10,000 B.P., a concept that may also hold for southern Scandinavia (Mangerud *et al.*, 1974). According to the dates listed in Table 1 (C), the possibility cannot be discounted that biostratigraphic changes as a result of climatic improvement may be quite variable in age throughout Scotland.

The chronozone scheme outlined by Mangerud *et al.* (1974) and Pennington (1975b) provides useful boundaries for comparisons, but it is essential that precise regional climatostratigraphic boundaries are defined and dated in order to ascertain whether or not climatic changes are synchronous throughout north-west Europe. Current evidence suggests that in Scotland the recently-proposed chronostratigraphic boundaries, particularly for the beginning and end of the Lateglacial do not correspond exactly with climatostratigraphic boundaries. Unfortunately, however, there is at present insufficient evidence to define precisely when the climatic changes occurred. Until suitable evidence is forthcoming it has been found necessary to adopt temporary working definitions for the Lateglacial Interstadial and Loch Lomond Stadial for use in this volume. The term Lateglacial Interstadial is used for the period between the start of the apparently rapid thermal improvement that occurred between about 14,000 and 13,000 B.P. and the start of the marked thermal decline that took place around 11,000 B.P. The term Loch Lomond Stadial is used for the period between the start of this thermal decline and the internationally accepted chronostratigraphic boundary for the beginning of the Flandrian, *viz.* 10,000 B.P. This scheme, employing both climatostratigraphic and chronostratigraphic boundaries, is felt to be unsatisfactory but unavoidable at the present time.

It seems probable that a problem will eventually arise in dealing with the period between the beginning of the Loch Lomond Stadial and the international boundary of 10,000 B.P. If the early dates for biostratigraphic boundaries marking the end of stadial conditions are confirmed by further research it may be necessary to subdivide the above period into two, as Pennington (this volume) has done. The earlier part would be the period between the thermal decline that apparently took place around 11,000 B.P. and the beginning of the thermal improvement that occurred before 10,000 B.P. The later part would be the period between the start of this thermal improvement and the beginning of the Flandrian chronozone. It is proposed that the term Loch Lomond Stadial should eventually be applied to the earlier period but a problem arises in naming the later one.

Finally, it is argued that the formal acceptance of dates for particular stratigraphic boundaries is perhaps premature at this stage in view of errors in the radiocarbon dating method, the relatively few dates available, and the standard deviations associated with them. The fact that there is only a 68 per cent probability that the "true" age lies within the range quoted is often ignored. Many more dates are essential to reduce the uncertainty currently associated with stratigraphic interpretation of the Lateglacial in Scotland and to establish to what extent climatostratigraphic boundaries are synchronous over Europe. As evidence increases the chronology and intensity of climatic changes will become clearer and dates for the boundaries of the Lateglacial Interstadial and Loch Lomond Stadial will be defined with more confidence. The climatostratigraphic boundaries shown in Fig. 1 should be regarded as extremely tentative.

FUTURE RESEARCH

Finally it is worth outlining where the gaps in our knowledge of the Scottish Lateglacial environment exist or where further information is required or may soon be forthcoming. For example, much more research is needed to determine the pattern and timing of the decay of the Late Devensian ice sheet. In coastal areas this could partly be achieved by careful delimitation of the marine limit and radiocarbon dating of marine shells, although there is the problem of "apparent age" associated with the latter (Mangerud, 1972; Mangerud and Gullikson, 1975). Further dating of basal organic material from Lateglacial lake sites might eventually allow the construction of isochrones of ice-sheet wastage, but both meticulous sampling and careful interpretation of terrestrial organic sediments are necessary (see above). Radiocarbon dating and measurement of the marine limit in Buchan is required to test Clapperton and Sugden's (this volume) suggestion that part of this area was deglaciated by about 16,000 B.P. Sedimentological studies are also required in order to gain insight into the glacial and fluvioglacial processes operative during deglaciation.

If boreholes through the base of the large kettle holes at Almondbank penetrate the fine grained rhythmites below the surface outwash a "Perth Readvance" would be implied, though this need not have been a major forward movement (R. A. Cullingford, this volume; pers. comm.). The mapping of the extent of the Loch Lomond Readvance may well be completed in the forseeable future, though in some areas morphological evidence is poor, and the implementation of palynological, erratic and soil studies is urgently required in problem areas such as the south-west Highlands (Gray and Sutherland, this volume). When this work has been carried out it should be possible to reconstruct in detail the climatology of Scotland during the Loch Lomond Stadial in the way Sissons and Sutherland (1976, pers. comm.) have done on the basis of evidence from the south-east Grampians and elsewhere in Scotland.

As far as sea-level changes are concerned, verification of the Lochgilphead relative sea-level curve (Peacock *et al.*, this volume) must be a research priority in view of its implications for the timing of deglaciation. Dating of basal lake deposits in the Lochgilphead area would add significant evidence to this problem. Little is known about the details and timing of Lateglacial sea-level changes over most of the Scottish coastline. In view of the general lack of datable quantities of Lateglacial marine shells that can be related accurately to specific sea-levels, the most promising way of determining the chronology of Lateglacial sea-level changes is likely to be by radiocarbon dating of the horizons in coastal peat bogs at which the pollen assemblages change from saltmarsh to freshwater reedswamp and fen communities or *vice versa*. Future work on sea-levels might also profitably include submarine investigation of the underwater continuation of the Main Lateglacial Shoreline (*cf.* Sissons, 1976a). It is one of the best developed glacioisostatically deformed shorelines in the northern Hemisphere and as such should yield a great deal of information on isostatic rebound as well as on the processes that formed it.

Despite the importance of Lateglacial periglaciation few detailed studies have been carried out. There is a particular need for further dating of organic matter below solifluction deposits in order to determine which were the most important periglacial periods.

It is now apparent that a number of problems in the interpretation of Scottish Lateglacial pollen profiles cannot be resolved until more absolute pollen data is available. In particular it would seem imperative to employ this kind of analysis at sites in north-east Scotland (Vasari, this volume) to establish whether the complex stratigraphical variation in some of the sites reflects general climatic variation or local site factors only. In addition, interpretation of pollen diagrams remains to a large extent subjective and inferential. The use of modern pollen spectra derived from surface samples from modern plant communities (Birks, 1973; Pennington, this volume) provides a more objective means of assessing fossil pollen assemblages, but no modern analogues have been found for a number of Lateglacial pollen assemblage types. Additional information on the relationship of modern pollen spectra to present-day European and Arctic plant communities may enhance the interpretation of fossil pollen spectra and closer analogues to "exceptional" spectra may be discovered. Further, more insight may be gained into the degree of variation of ground cover and into plant community types in existence during the Lateglacial in Scotland.

Extremely little is known of terrestrial animal populations in the Lateglacial, and the data from south-west Scotland (Bishop and Coope, this volume), though restricted mainly to coleopteran remains, is a significant step forward. This work is valuable not only for the climatic parameters derived from the faunal species assemblages, but also for their important implications for physical characteristics of sites of deposition (*e.g.* depth of water, water-flow, sediment load), for the nature of surrounding soils, and for the local plant life, species of which may not be recorded in microfossil and macrofossil assemblages of plant remains. It is to be hoped that other detailed coleopteran studies are to follow, and that information can be gained of other animals such as ostracoda and cladocera that are known to be abundant in Lateglacial sediments. It would be useful to discover further faunal remains and obtain additional radiocarbon dates from the stratigraphy of west central Scotland (see Sissons, 1976b).

Lists of marine faunal assemblages also need to be expanded and many more radiocarbon dates are required to determine precisely their environmental and chronological implications. In addition to the problem of "apparent age" there is the fundamental difficulty that the presence of cold water assemblages need not relate to regional climatic deterioration, but could result from influxes of cold water into estuaries and near-shore coastal localities following rapid ice-sheet wastage (Ruddiman and McIntyre, 1973). The reconciliation of implications based on marine faunal assemblages with those derived from terrestrial evidence remains problematical.

Many of our concepts regarding soil history and diversity during the Lateglacial are derived indirectly from changing base-status conditions of lake waters, based on changing aquatic flora (Vasari and Vasari, 1968) and on actual chemical measurements (Pennington *et al.*, 1972; Pennington and Lishman, 1974; Pennington, this volume), from physical changes in lithostratigraphy and variation in the preservation of pollen and spores (*e.g.* Lowe and Walker, this volume), or from the inferred character of plant communities represented by pollen spectra (*e.g.* Birks, 1973). It would be especially valuable to discover preserved Lateglacial Interstadial soil horizons, particularly from areas that supported closed vegetation with perhaps tree birch present, for it is difficult to assess the maturity of Lateglacial Interstadial soils and the amount of leaching that may have taken place prior to the Loch Lomond Stadial. Under warm Lateglacial Interstadial conditions leaching may have been rapid in those areas that remained tree-less and thus where there was little nutrient recycling, and rock weathering, as an alternative source of minerals, was at a minimum until the more active climatic regime of the Loch Lomond Stadial.

In the preceding section of this paper the need for more research on the chronology of Lateglacial climatic changes was emphasised. In addition to establishing more confident dates for boundaries, priority should be given to testing the validity of the two-fold climatostratigraphic subdivision of the Lateglacial, for it may be that the climatic recession identified by Pennington (1975b, this volume) and correlated with the Older Dryas was more pronounced than current stratigraphic evidence suggests. Further, a temperature fluctuation within Alleröd time has been suggested on the basis of pollen percentage changes in sediments from several sites in northern Scotland (Pennington *et al.*, 1972), and Vasari (this volume) has suggested the possibility of the recognition of a climatic deterioration within the early Flandrian equivalent to the Piottino Oscillation of continental north-west Europe. These require to be substantiated.

It is clear that although much has been learnt recently about the Scottish Lateglacial environment there are aspects about which we still know very little. This paper has concentrated on providing a *résumé* and synthesis of those areas, both regional and systematic, where recent research has been concentrated. Though more is known about the Lateglacial environment than any other part of the Scottish Pleistocene, significant enhancement of our appreciation of the complexity of Lateglacial environmental variation awaits the availability of more absolute dates and the closer scrutiny of those regions and topics so far neglected.

ACKNOWLEDGEMENTS

We are most grateful to Dr. H. J. B. Birks, Dr. G. R. Coope, Dr. J. T. Greensmith, Mr. A. R. Gunson, Dr. R. J. Price, Dr. J. B. Sissons, Mr. D. G. Sutherland and Dr. M. J. C. Walker for their critical comments on drafts of this paper.

References

ANDERSEN, B. G. 1965. The Quaternary of Norway. *In* RANKAMA, K. (ed.) *The Quaternary*, vol. 1, 91-138.

ANDERSEN, S. Th. 1961. Vegetation and its environment in Denmark in the Early Weichselian Glacial. *Danm. Geol. Unders.*, Ser. II, 75, 1-175.

ANDREWS, J. T. 1970. A geomorphological study of post-glacial uplift with particular reference to Arctic Canada. *Inst. Br. Geogr. Special Publication*, 2.

ARMSTRONG, M., PATERSON, I. B. and BROWNE, M. E. A. 1975. Late-glacial ice limits and raised shorelines in east central Scotland. *In* GEMMELL, A. M. D. (ed.) *Quaternary Studies in North East Scotland*, 39-44, Aberdeen.

ASHWORTH, A. C. 1972. A Late-glacial insect fauna from Red Moss, Lancashire, England. *Entomol. Scand.*, 3, 211-24.

ASHWORTH, A. C. 1973. The climatic significance of a Late Quaternary insect fauna from Rodbaston Hall, Staffordshire, England. *Entomol. Scand.*, 4, 191-205.

ATHERTON, N. M., CRANWELL, P. A., FLOYD, A. J. and HAWORTH, R. D. 1967. Humic acid - 1 ESR spectra of humic acids. *Tetrahedron*, 23, 1653-67.

BAILEY, E. B. *et al.* 1924. Tertiary and post-Tertiary geology of Mull, Loch Aline and Oban. *Mem. Geol. Surv. Gt. Br.*

BALD, R. 1822. Notices regarding the Fossil Elephant in Scotland. *Mem. Wernerian Nat. Hist. Soc. Edinb.*, IV, 63.

BARTLEY, D. D. 1962. The stratigraphy and pollen analysis of lake deposits near Tadcaster, Yorkshire. *New Phytol.*, 61, 277-87.

BARTLEY, D. D. 1966. Pollen analysis of some lake deposits near Bamburgh in Northumberland. *New Phytol.*, 65, 141-56.

BAXTER, M. S., ERGIN, M. and WALTON, A. 1969. Glasgow University Radiocarbon Measurements, 1. *Radiocarbon*, 11, 43-52.

BEHRE, K.-E. 1967. The Late Glacial and early Postglacial history of vegetation and climate in Northwestern Germany. *Rev. Palaeobotan. Palynol.*, 4, 149-61.

BENNIE, J. 1885. Note on the contents of two bits of clay from the elephant bed at Kilmaurs in 1817. *Proc. R. Phys. Soc. Edinb.*, VIII, 451.

BENNINGHOFF, W. S. 1962. Calculations of pollen and spore density in sediments by addition of exotic pollen in known quantities (Abstract). *Pollen Spores*, 4, 232.

BERGLUND, B. E. 1966. Late-Quaternary vegetation in eastern Blekinge, south-eastern Sweden. I. Late-glacial time. *Opera Botanica*, 12, 1-180.

BINNS, P. E., HARLAND, R., and HUGHES, J. J. 1974. Glacial and post-glacial sedimentation in the Sea of the Hebrides. *Nature, Lond.*, 248, 751-4.

BIRKS, H. H. 1970. Studies in the vegetational history of Scotland. I. A pollen diagram from Abernethy Forest, Inverness-shire. *J.Ecol.*, 58, 827-46.

BIRKS, H. H. 1972. Studies in the vegetational history of Scotland. II. Two pollen diagrams from the Galloway Hills, Kirkcudbrightshire. *J. Ecol.*, 60, 183-217.

BIRKS, H. J. B. 1968. The identification of *Betula nana* pollen. *New Phytol.*, 67, 309-14.

BIRKS, H. J. B. 1970. Inwashed pollen spectra at Loch Fada, Isle of Skye. *New Phytol.*, 69, 807-20.

BIRKS, H. J. B. 1973. *Past and Present Vegetation on the Isle of Skye - a palaeoecological study.* Cambridge.

BIRKS, H. J. B., DEACON, J. and PEGLAR, S. 1975. Pollen maps for the British Isles 5000 years ago. *Proc. R. Soc. Lond.*, B, 189, 87-105.

BIRKS, H. J. B. and SAARNISTO, M. 1975. Isopollen maps and principal components analysis of Finnish pollen data for 4000, 6000 and 8000 years ago. *Boreas*, 4, 77-96.

BISHOP, W. W. 1963. Lateglacial deposits near Lockerbie, Dumfriesshire. *Trans. J. Proc. Dumfries. Galloway Nat. Hist. Antiq. Soc.*, 40, 117-32.

BISHOP, W. W. and DICKSON, J. H. 1970. Radiocarbon dates related to the Scottish Late-Glacial sea in the Firth of Clyde. *Nature, Lond.*, 227, 480-2.

BLACK, R. F. 1976. Periglacial features indicative of permafrost: ice and soil wedges. *Quaternary Res.*, 6, 3-26.

BLOOM, A. L. 1964. Peat accumulation and compaction in a Connecticut coastal marsh. *J. sediment. Petrol.*, 34, 599-603.

BONNY, A. P. 1972. A method for determining absolute pollen frequencies in lake sediments. *New Phytol.*, 71, 391-403.

BOULTON, G. S. 1972. Modern Arctic glaciers as depositional models for former ice sheets. *Q. J. geol. Soc. Lond.*, 128, 361-93.

BRADY, G. S., CROSSKEY, H. W. and ROBERTSON, D. 1874. The post-Tertiary Entomostraca of Scotland. *Mon. Pal. Soc. Lond.*, 229 pp.

BREMNER, A. 1915. Problems in the glacial geology of Northeast Scotland and some fresh facts bearing on them. *Trans. Edinb. geol. Soc.*, 10, 334-47.

BREMNER, A. 1921. *The physical geology of the Dee and Don valleys.* Aberdeen.

BREMNER, A. 1932. Further problems in the glacial geology of Northeastern Scotland. *Trans. Edinb. geol. Soc.*, 12, 147-64.

BRETT, D. W. and NORTON, T. A. 1969. Late glacial marine algae from Greenock and Renfrew. *Scot. J. Geol.*, 5, 42-8.

BRETTUM, P. 1972. The phytoplankton of Lake Øvre Heimsdalvatn, central South Norway. *Norw. J. Bot.*, 19, 79-90.

BROECKER, W. S. 1966. Glacial rebound and the deformation of the shorelines of proglacial lakes. *J. Geophys. Res.*, 71, 4777-83.

BRÖGGER, W. C. 1900-01. Om de senglaciale og postglaciale nivaforandringer i Kristianiafeltet (Molluskfaunan). *Norges geol. Unders.*, 31, 528 pp.

BROWN, A. P. 1971. The *Empetrum* pollen record as a climatic indicator in the Late Weichselian and Early Flandrian of the British Isles. *New Phytol.*, 70, 841-9.

BROWN, R. 1870. On the physics of Arctic ice, as explanatory of the glacial remains in Scotland. *Q. J. geol. Soc. Lond.*, 26, 671-701.

BROWN, T. 1867. On the arctic shell-clay of Elie and Errol, viewed in connection with our other glacial and more recent deposits. *Trans. R. Soc. Edinb.*, 24, 617-33.

BRYCE, J. 1865. On the occurrence of beds in the west of Scotland beneath the boulder clay. *Q. J. geol. Soc. Lond.*, 21, 213.

BRYSON, R. A., WENDLAND, W. M., IVES, J. D. and ANDREWS, J. T. 1969. Radiocarbon isochrones on the disintegration of the Laurentide Ice Sheet. *Arct. Alp. Res.*, 1, 1-14.

BÜDEL, J. 1944. Die morphologischen Wirkungen des Eiszeitklimas im gletscherfreien Gebiet. *Geol. Rdsch.*, 34, 482-519.

BURNETT, J. H. 1964. *The Vegetation of Scotland.* Edinburgh.

CHARLESWORTH, J. K. 1955. The late-glacial history of the Highlands and Islands of Scotland. *Trans. R. Soc. Edinb.*, 62, 769-928.

CLAPPERTON, C. M. 1969. Physiography and terrain analysis. *In* WALTON, K. (ed.) *Royal Grampian Country*, 1-4, Aberdeen.

CLAPPERTON, C. M. 1971. The location and origin of glacial meltwater phenomena in the eastern Cheviot hills. *Proc. Yorks. geol. Soc.*, 38, 361-80.

CLAPPERTON, C. M. and SUGDEN, D. E. 1972. The Aberdeen and Dinnet glacial limits reconsidered. *In* CLAPPERTON, C. M. (ed.) *North-East Scotland: geographical essays*, 5-11, Aberdeen.

CLAPPERTON, C. M. and SUGDEN, D. E. 1975. The glaciation of Buchan: a reappraisal. *In* GEMMELL, A. M. D. (ed.) *Quaternary Studies in North East Scotland*, 19-22, Aberdeen.

CLAPPERTON, C. M., GUNSON, A. R., and SUGDEN, D. E. 1975. Loch Lomond Readvance in the eastern Cairngorms. *Nature*, 253, 710-12.

CONOLLY, A. P. 1961. Some climatic and edaphic indications from the Late-glacial flora. *Proc. Linn. Soc. Lond.*, 172, 56-62.

COOPE, G. R. 1962. Coleoptera from a peat interbedded between two boulder clays at Burnhead, near Airdrie. *Trans. geol. Soc. Glasg.*, 24, 279-86.

COOPE, G. R. 1968. Fossil beetles collected by James Bennie from Late Glacial silts at Corstorphine, Edinburgh. *Scot. J. Geol.*, 4, 339-48.

COOPE, G. R. 1970. Climatic interpretations of Late Weichselian Coleoptera from the British Isles. *Rev. Geogr. Phys. Geol. Dynam.*, 12, 149-155.

COOPE, G. R. 1975. Climatic fluctuations in northwest Europe since the last Interglacial, indicated by fossil assemblages of Coleoptera. *In* WRIGHT, A. E. and MOSELEY, F. (eds.) *Ice Ages: Ancient and Modern*, 153-68, Liverpool.

COOPE, G. R. and BROPHY, J. A. 1972. Late Glacial environmental changes indicated by a coleopteran succession from North Wales. *Boreas*, 1, 97-142.

COOPE, G. R., MORGAN, A. and OSBORNE, P. J. 1971. Fossil Coleoptera as indicators of climatic fluctuations during the Last Glaciation in Britain. *Palaeogeogr. Palaeoclimatol. Palaeoecol.*, 10, 87-101.

CRAIG, A. 1973. *Studies on the ecological history of south-east Ireland, using pollen influx analysis and other methods.* Ph.D. thesis, Univ. of Dublin.

CRAIG, R. 1888. On the post-Pliocene beds of the Irvine valley, Kilmaurs and Dreghorn Districts. *Trans. geol. Soc. Glasg.*, 8, 213.

CRANWELL, P. A. 1973. Chain-length distribution of n-alkenes from lake sediments in relation to post-glacial environmental change. *Freshwat. Biol.*, 3, 259-65.

CRANWELL, P. A. 1974. Monocarboxylic acids in lake sediments: indicators, derived from terrestrial and aquatic biota, of palaeoenvironmental trophic levels. *Chem. geol.*, 14, 1-14.

CROSSKEY, H. W. 1867. Glacial deposits of the Clyde district. *Trans. geol. Soc. Glasg.*, 2, 45-51.

CROSSKEY, H. W. 1874. Post Tertiary fossiliferous beds of Scotland. VII. Garvel Park, New Dock, Greenock. *Trans. geol. Soc. Glasg.*, 4, 32-45.

CROSSKEY, H. W. and ROBERTSON, D. 1871. On the post-Tertiary fossiliferous beds of Scotland. *Trans. geol. Soc. Glasg.*, 3, 118-25, 328-31.

CULLINGFORD, R. A. 1972. *Lateglacial and postglacial shoreline displacement in the Earn-Tay area and eastern Fife.* Ph.D. thesis, Univ. of Edinb.

CULLINGFORD, R. A. and SMITH, D. E. 1966. Late-glacial shorelines in eastern Fife. *Trans. Inst. Br. Geogr.*, 39, 31-51.

CURRAY, J. R. 1965. Late Quaternary history, continental shelves of the United States. *In* WRIGHT, H. E. and FREY, D. G. (eds.) *The Quaternary of the United States*, 723-36, Princeton.

CUSHING, E. J. 1964. Redeposited pollen in Late-Wisconsin pollen spectra from east-central Minnesota. *Am. J. Sci.*, 262, 1075-88.

CUSHING, E. J. 1967. Evidence for differential pollen preservation in Late Quaternary sediments in Minnesota. *Rev. Palaeobotan. Palynol.*, 4, 87-101.

DAVIDSON, C. F. 1932. The arctic clay of Errol, Perthshire. *Trans. Perthsh. Soc. Nat. Sci.*, 9, 55-68.

DAVIDSON, C. F. 1935. Geology. *In* MELVILLE, L. *Errol: its legends, lands and people*, 31-3, Perth.

DAVIS, M. B. 1963. On the theory of pollen analysis. *Am. J. Sci.*, 261, 897-912.

DAVIS, M. B. 1967. Pollen accumulation rates at Rogers Lake, Connecticut, during late- and post-glacial time. *Rev. Palaeobotan. Palynol.*, 2, 219-30.

DAVIS, M. B., BRUBAKER, L. B. and WEBB, T. 1973. Calibration of absolute pollen influx. *In* BIRKS, H. J. B. and WEST, R. G. (eds.) *Quaternary Plant Ecology*, 9-26, Oxford.

DAVIS, M. B. and DEEVEY, E. S. 1964. Pollen accumulation rates: estimates from late-glacial sediment of Rogers Lake. *Science*, 145, 1293-95.

DEACON, J. 1974. The location of refugia of *Corylus avellana* L. during the Weichselian glaciation. *New Phytol.*, 73, 1055-63.

DICKSON, J. H. 1973. *Bryophytes of the Pleistocene*. Cambridge.

DICKSON, J. H., JARDINE, W. G. and PRICE, R. J. 1976. Three late-Devensian sites in West-central Scotland. *Nature, Lond.*, 262, 43-4.

DONNER, J. J. 1957. The geology and vegetation of Late-glacial retreat stages in Scotland. *Trans. R. Soc. Edinb.*, 63, 221-64.

DONNER, J. J. 1958. Loch Mahaick, a Lateglacial site in Perthshire. *New Phytol.*, 57, 183-86.

EMILIANI, C. *et al.* 1975. Palaeoclimatological analysis of Late Quaternary cores from the Northeastern Gulf of Mexico. *Science*, 189, 1083-88.

FAEGRI, K. 1963. Problems of immigration and dispersal of the Scandinavian Flora. *In* LÖVE, A. and LÖVE, D. (eds.) *North Atlantic Biota and their History*, Oxford.

FAEGRI, K. and IVERSEN, J. 1964. *Textbook of Pollen Analysis*. Oxford.

FITZPATRICK, E. A. 1965. An interglacial soil at Teindland, Morayshire. *Nature, Lond.*, 207, 621-2.

FITZPATRICK, E. A. 1972. The principal Tertiary and Pleistocene events in Northeast Scotland. *In* CLAPPERTON, C. M. (ed.) *North-east Scotland: geographical essays*, 1-4, Aberdeen.

FITZPATRICK, E. A. 1975. Particle size distribution and stone orientation patterns in some soils of Northeast Scotland. *In* GEMMELL, A. M. D. (ed.) *Quaternary Studies in North-East Scotland*, 49-60, Aberdeen.

FLEET, H. 1938. Erosion surfaces in the Grampian Highlands of Scotland. *Rapport de la Commission pour la Cartographic des Surfaces d'Appl. Tertiares*, 91-94, Paris.

FOWLER, W. W. 1913. *The Coleoptera of the British Isles*, vols. I-IV, London.

FRANCIS, E. H. *et al.* 1970. The geology of the Stirling district. *Mem. Geol. Surv. Gt. Br.*

FREDSKILD, B. 1973. Studies in the Vegetational History of Greenland: Palaeobotanical investigations of some Holocene lake and bog deposits. *Meddr. Grønland*, 198, 1-245.

FREUDE, H., HARDER, K. W. and LOHSE, G. A. 1966. *Die Käfer Mitteleuropas*, vol. 9, Krefeld.

GEIKIE, A. 1863. On the phenomena of the Glacial Drift of Scotland. *Trans. geol. Soc. Glasg.*, 1, 1-190.

GEMMELL, A. M. D. 1973. The deglaciation of the Island of Arran, Scotland. *Trans. Inst. Br. Geogr.*, 59, 25-39.

GLENTWORTH, R., MITCHELL, W. A. and MITCHELL, B. D. 1964. The red glacial drift deposits of Northeast Scotland. *Clay Min. Bull.*, 5, 373-81.

GODARD, A. 1965. *Recherches de géomorphologie en Ecosse du Nord-Ouest*. Paris.

GODWIN, H. 1960. Radiocarbon dating and Quaternary history in Britain. *Proc. R. Soc.* B, 153, 287-320.

GODWIN, H. 1975. *The History of the British Flora*. Cambridge.

GODWIN, H. and SWITSUR, V. R. 1966. Cambridge University Natural Radiocarbon Measurements VIII. *Radiocarbon*, 8, 390-400.

GODWIN, H. and WILLIS, E. H. 1959a. Cambridge University Natural Radiocarbon Measurements I. *Radiocarbon*, 1, 63-75.

GODWIN, H. and WILLIS, E. H. 1959b. Radiocarbon dating of the Late-glacial Period in Britain. *Proc. R. Soc.* B, 150, 199-215.

GODWIN, H. and WILLIS, E. H. 1960. Cambridge University Natural Radiocarbon Measurements II. *Radiocarbon*, 2, 62-72.

GODWIN, H. and WILLIS, E. H. 1961. Cambridge University Natural Radiocarbon Measurements III. *Radiocarbon*, 3, 60-76.

GODWIN, H. and WILLIS, E. H. 1962. Cambridge University Natural Radiocarbon Measurements V. *Radiocarbon*, 4, 57-70.

GODWIN, H. and WILLIS, E. H. 1964. Cambridge University Natural Radiocarbon Measurements VI. *Radiocarbon*, 6, 116-37.

GODWIN, H., WILLIS, E. H. and SWITSUR, V. R. 1965. Cambridge University Natural Radiocarbon Measurments VII. *Radiocarbon*, 7, 205-212.

GOLDTHWAIT, J. W. 1908. A reconstruction of water planes of the extinct glacial lakes in the Lake Michigan Basin. *J. Geol.*, 16, 459-76.

GRAY, J. M. 1972a. *The inter-, late- and post-glacial shorelines, and ice-limits of Lorn and eastern Mull.* Ph.D. thesis, Univ. of Edinb.

GRAY, J. M. 1972b. Trends through clusters. *Area*, 4, 275-9.

GRAY, J. M. 1974a. The Main Rock Platform of the Firth of Lorn, western Scotland. *Trans. Inst. Br. Geogr.*, 61, 81-99.

GRAY, J. M. 1974b. Lateglacial and postglacial shorelines in western Scotland. *Boreas*, 3, 129-38.

GRAY, J. M. 1975a. Measurement and analysis of Scottish raised shoreline altitudes. *Dept. of Geogr., Queen Mary College, Occasional Papers*, 2.

GRAY, J. M. 1975b. Some comments on the late-glacial history of the Firth of Clyde. *Trans. Inst. Br. Geogr.*, 64, 129-32.

GRAY, J. M. 1975c. The Loch Lomond Readvance and contemporaneous sea-levels in Loch Etive and neighbouring areas of western Scotland. *Proc. Geol. Ass.*, 86, 227-38.

GRAY, J. M. and BROOKS, C. L. 1972. The Loch Lomond Readvance moraines of Mull and Menteith. *Scot. J. Geol.*, 8, 95-103.

GREGORY, J. W. and CURRIE, E. D. 1928. The vertebrate fossils from the glacial and associated post-glacial beds of Scotland in the Hunterian Museum, University of Glasgow. *Monographs of the Geol. Dept. of the Hunterian Museum Glasg. Univ.*, ii, 1-21.

GUNSON, A. R. 1975. The vegetation history of Northeast Scotland. *In* GEMMELL, A. M. D. (ed.) *Quaternary Studies in North East Scotland*, 61-72, Aberdeen.

HAMMEN, T. van der, MAARLEVELD, G. C., VOGEL, J. C. and ZAGWIJN, W. H. 1967. Stratigraphy, climatic succession and radiocarbon dating of the last glacial in the Netherlands. *Geol. Mijnb.*, 46, 79-95.

HAMMEN, T. van der, WIJMSTRA, T. A. and ZAGWIJN, W. H. 1971. The floral record of the Late Cenozoic of Europe. *In* TUREKIAN, K. K. (ed.) *The Late Cenozoic Glacial Ages*, 391-424, New Haven and London.

HARRISON, S. S. 1975. Turbidite origin of glaciolacustrine sediments, Woodcock Lake, Pennsylvania. *J. sediment. Petrol.*, 45, 738-44.

HARKNESS, D. D. and WILSON, H. W. 1973. Scottish Universities Research and Reactor Centre Radiocarbon Measurements 1. *Radiocarbon*, 15, 554-65.

HARKNESS, D. D. and WILSON, H. W. 1974. Scottish Universities Research and Reactor Centre Radiocarbon Measurements II. *Radiocarbon*, 16, 238-51.

HAVINGA, A. J. 1964. Investigation into the differential corrosion susceptibility of pollen and spores. *Pollen Spores*, 6, 621-35.

HAWORTH, E. Y. 1974. *The late-glacial (late-Devensian) and post-glacial (Flandrian) diatom assemblages from lochs in northern Scotland.* Ph.D. thesis, Univ. of Leicester.

HAYNES, J. R. 1973. Cardigan Bay Recent Foraminifera. *Bull. Br. Mus. (N.H.) Zoology. Supp.*, 4, 245 pp.

HIBBERT, F. A., SWITSUR, V. R. and WEST, R. G. 1971. Radiocarbon dating of Flandrian pollen zones at Red Moss, Lancashire. *Proc. R. Soc.* B, 177, 161-76.

HILL, J. B. *et al.* 1905. The geology of Mid-Argyll. *Mem. Geol. Surv. Gt. Br.*

HORNE, J., PEACH, B. N. and TEALL, J. J. H. 1896.

Explanation of Sheet 5, Kirkcudbrightshire. *Mem. Geol. Surv. Gt. Br.*

HUSTEDT, F. 1937-39. Systematische und ökologische Untersuchungen über die Diatmeen-Flora von Bali, Java und Sumatra. II. *Arch. Hydrobiol. Suppl.*, 16, 1-155.

HUSTEDT, F. 1950. Die Diatomeenflora norddeutscher Seen mit besonderer Berücksichtigung des holsteinischen Seengebiets. V-VII. Seen in Mecklenburg, Lauenburg und Nordost-deutschland. *Arch. Hydrobiol.*, 43, 329-458.

HYVÄRINEN, H. 1975. Absolute and relative pollen diagrams from northernmost Fennoscandia. *Fennia*, 142, 1-23.

IVERSEN, J. 1953. Radiocarbon dating of the Alleröd Period. *Science*, 118, 9-11.

IVERSEN, J. 1954. The late-glacial flora of Denmark and its relation to climate and soil. *Danm. geol. Unders. Ser. II*, 80, 87-119.

IVERSEN, J. 1973. The development of Denmark's Nature since the last glacial. *Danm. geol. Unders. Ser. II*, 7c, 126.

JACKES, M. 1973. *Some aspects of the glaciation of Northwest Sutherland.* Honours Dissertation, Dept. of Geogr., Univ. of Edinb.

JAMIESON, T. F. 1865. On the history of the last geological changes in Scotland. *Q. J. geol. Soc. Lond.*, 21, 161-203.

JAMIESON, T. F. 1906. The glacial period in Aberdeenshire and the southern border of the Moray Firth. *Q. J. geol. Soc. Lond.*, 38, 160-77.

JARDINE, W. G. 1964. Post-Glacial sea-levels in south-west Scotland. *Scott. geogr. Mag.*, 80, 5-11.

JARDINE, W. G. 1968. The "Perth" Readvance. *Scot. J. Geol.*, 4, 185-6.

JARDINE, W. G. 1975. Chronology of Holocene marine transgression and regression in south-western Scotland. *Boreas*, 4, 173-96.

JÓHANSEN, J. 1975. Pollen diagrams from the Shetland and Faroe Islands. *New Phytol.*, 75, 369-87.

JÖRGENSEN, E. G. 1948. Diatom communities in some Danish lakes and ponds. *K. danske Vidensk. Selsk. Skr.*, Bd. VI, 2, 2-140.

KEITH, M. L., ANDERSON, G. M. and EICHLER, R. 1964. Carbon and oxygen isotopic composition of Mollusk shells from marine and fresh-water environments. *Geochimica et Cosmochimica Acta*, 28, 1757-86.

KIRBY, R. P. 1969. Morphometric analysis of glaciofluvial terraces in the Esk basin, Midlothian. *Trans. Inst. Br. Geogr.*, 48, 1-18.

KIRK, W. and GODWIN, H. 1963. A lateglacial site at Loch Droma, Ross and Cromarty. *Trans. R. Soc. Edinb.*, 65, 225-49.

KIRK, W., RICE, R. J. and SYNGE, F. M. 1966. Deglaciation and vertical displacement of shorelines in Wester and Easter Ross. *Trans. Inst. Br. Geogr.*, 39, 65-78.

KÖNIGSSON, L-K. 1969. Pollen dispersion and the destruction degree. *Bull. geol. Instn. Univ. Uppsala N.S.*, 1, 6, 161-5.

KROG, H. 1954. Pollen Analytical Investigation of a C^{14}-dated Alleröd Section from Ruds Vedby. *Danm. geol. Unders. Ser. II*, 80, 120-39.

KROG, H. and TAUBER, H. 1973. C-14 chronology of Late- and Post-glacial marine deposits in North Jutland. *Danm. geol. Unders. Arb.*, 93-105.

KYNASTON, H. *et al.* 1908. The geology of the country near Oban and Dalmally. *Mem. geol. Surv. Gt. Br.*

LICHTI-FEDEROVICH, S. and RITCHIE, J. C. 1968. Recent pollen assemblages from the western interior of Canada. *Rev. Palaeobotan. Palynol.*, 7, 297-344.

LINDROTH, C. H. 1974. *Handbook for the identification of British insects. Coleoptera, Carabidae.* Ent. Soc. London.

MACKERETH, F. J. H. 1958. A portable core sampler for lake deposits. *Limnol. Oceanogr.*, 3, 181-91.

MACKERETH, F. J. H. 1965. Chemical investigations of lake sediments and their interpretation. *Proc. R. Soc.*, B, 161, 295-309.

MACKERETH, F. J. H. 1966. Some chemical observations on post-glacial lake sediments. *Phil. Trans. R. Soc.*, B, 250, 165-213.

MANGERUD, J. 1970. Late Weichselian vegetation and ice-front oscillations in the Bergen district, Western Norway. *Norsk. geogr. Tidsskr.*, 24, 121-48.

MANGERUD, J. 1972. Radiocarbon dating of marine shells, including a discussion of the apparent ages of Recent shells from Norway. *Boreas*, 1, 143-72.

MANGERUD, J., ANDERSEN, S. Th., BERGLUND, B. E. and DONNER, J. J. 1974. Quaternary stratigraphy of Norden, a proposal for terminology and classification. *Boreas*, 3, 109-26.

MANGERUD, J. and GULLIKSON, S. 1975. Apparent radiocarbon ages of recent marine shells from Norway, Spitzbergen and Arctic Canada. *Quaternary Res.*, 5, 263-73.

MANLEY, G. 1959. The late-glacial climate of North-west England. *Liv. and Manch. Geol. J.*, 2, 188-215.

MANLEY, G. 1964. The evolution of the climatic environment. *In* WATSON, J. W. and SISSONS, J. B. (eds.) *The British Isles: a systematic geography*. Edinburgh.

McCANN, S. B. 1966a. The limits of the late-glacial Highland, or Loch Lomond, readvance along the West Highland seaboard from Oban to Mallaig. *Scot. J. Geol.*, 2, 84-95.

McCANN, S. B. 1966b. The Main Postglacial Raised Shoreline of western Scotland from the Firth of Lorne to Loch Broom. *Trans. Inst. Br. Geogr.*, 39, 87-99.

MACGREGOR, A. R. 1968. *Fife and Angus Geology. An excursion guide*. Edinburgh.

McMANUS, J. 1966. Bottom structures of the Tay and other estuaries. *Scott. geogr. Mag.*, 82, 194-7.

McMANUS, J. 1972. Estuarine development and sediment distribution, with particular reference to the Tay. *Proc. R. Soc. Edinb.*, 71, 97-113.

McVEAN, D. N. and RATCLIFFE, D. A. 1962. *Plant Communities of the Scottish Highlands*. London.

MERCER, J. H. 1972. The Lower Boundary of the Holocene. *Quaternary Res.*, 2, 15-24.

MITCHELL, G. F. 1952. Late-glacial deposits of Garscadden Mains, near Glasgow. *New Phytol.*, 50, 277-86.

MITCHELL, G. F., PENNY, L. F., SHOTTON, F. W. and WEST, R. G. 1973. A correlation of Quaternary deposits in the British Isles. *Geol. Soc. Lond., Special Rep.*, 4, 99pp.

MOAR, N. T. 1963. Pollen analysis of four samples from the River Annan, Dumfriesshire. *Trans. J. Proc. Dumfries. Galloway Nat. Hist. Antiq. Soc.*, 40, 133-5.

MOAR, N. T. 1964. *The history of the Late Weichselian and Flandrian vegetation in Scotland*. Ph.D. thesis, Univ. of Cambridge.

MOAR, N. T. 1969a. Late Weichselian and Flandrian pollen diagrams from South-west Scotland. *New Phytol.*, 68, 433-67.

MOAR, N. T. 1969b. Two pollen diagrams from the Mainland, Orkney Islands. *New Phytol.*, 68, 201-8.

MOORE, P. D. and BELLAMY, D. J. 1974. *Peatlands*. London.

MÖRNER, N-A. 1969. Climatic and eustatic changes during the last 15,000 years. *Geol. Mijnb.*, 48, 389-99.

MURDOCH, W. 1975. The geomorphology and glacial deposits of the area around Aberdeen. *In* GEMMELL, A. M. D. (ed.), *Quaternary Studies in North East Scotland*, 14-18, Aberdeen.

MURRAY, J. and PULLAR, L. 1910. *Bathymetrical survey of the Scottish fresh-water lochs*. Edinburgh.

MURRAY, J. W. 1968. Living foraminifers of lagoons and estuaries. *Micropal.*, 14, 435-55.

NEALE, J. and HOWE, H. V. 1973. New cold-water recent and Pleistocene species of the ostracod genus *Cytheropton*. *Crustaceana*, 25, 237-44.

NEWEY, W. W. 1970. Pollen analysis of Late-Weichselian deposits at Corstorphine, Edinburgh. *New Phytol.*, 69, 1167-77.

NICHOLS, H. 1970. Late Quaternary pollen diagrams from the Canadian Arctic Barren Grounds at Pelly Lake, Northwestern Keewatin, N.W.T. *Arct. Alp. Res.*, 2, 43-61.

NICHOLS, R. L. 1961. Characteristics of beaches formed in polar climates. *Am. J. Sci.*, 259, 694-708.

OCKELMANN, W. K. 1958. The Zoology of East Greenland. Marine Lamellibranchiata. *Meddr. Grønland*, 122, 256pp.

O'DELL, A. C. and WALTON, K. 1962. *The Highlands and Islands of Scotland*. Edinburgh.

OLDFIELD, F. 1960. Studies in the Post-glacial history of

British Vegetation: Lowland Lonsdale. *New Phytol.*, 59, 192-217.

OSBORNE, P. J. 1972. Insect faunas of Late Devensian and Flandrian age from Church Stretton, Shropshire. *Phil. Trans. R. Soc.*, B, 263, 327-67.

OSBORNE, P. J. 1974. An insect assemblage of early Flandrian age from Lea Marston, Warwickshire, and its bearing on the contemporary climate and ecology. *Quaternary Res.*, 4, 471-86.

PATERSON, I. B. 1974. The supposed Perth Readvance in the Perth district. *Scot. J. Geol.*, 10, 53-66.

PATERSON, W. S. B. 1969. *The physics of Glaciers.* Oxford.

PEACH, B. N. *et al.* 1913. The geology of the Fannich Mountains and the country around upper Loch Maree and Strath Broom. *Mem. Geol. Surv. Gt. Br.*

PEACOCK, J. D. 1970. Some aspects of the glacial geology of West Inverness-shire. *Bull. Geol. Surv. Gt. Br.*, 33, 125-45.

PEACOCK, J. D. 1971a. Marine shell radiocarbon dates and the chronology of deglaciation in western Scotland. *Nature Phys. Sci.*, 230, 43-5.

PEACOCK, J. D. 1971b. Terminal features of the Creran Glacier of Loch Lomond Readvance age in western Benderloch, Argyll, and their significance in the late-glacial history of the Loch Linnhe area. *Scot. J. Geol.* 7, 349-56.

PEACOCK, J. D. 1971c. A re-interpretation of the coastal deposits of Banffshire and their place in the Late-Glacial history of N.E. Scotland. *Bull. Geol. Surv. Gt. Br.*, 37, 81-9.

PEACOCK, J. D. 1974. Borehole evidence for late- and post-glacial events in the Cromarty Firth, Scotland. *Bull. Geol. Surv. Gt. Br.*, 48, 55-67.

PEACOCK, J. D. 1975. Scottish late and post-glacial marine deposits. *In* GEMMELL, A. M. D. (ed.) *Quaternary Studies in North East Scotland*, 45-8, Aberdeen.

PEACOCK, J. D., BERRIDGE, N. G., HARRIS, A. L. and MAY, F. 1968. The geology of the Elgin district. *Mem. Geol. Surv. Gt. Br.*

PEAKE, D. S. and OSBORNE, P. J. 1971. The Wandle Gravels in the vicinity of Croydon: their age and insect fauna. *Proc. Croydon Nat. Hist. Sci. Soc.*, 14, 145-75.

PECHLANER, R. 1971. Factors that control the production rate and biomass of phytoplankton in high-mountain lakes. *Mitt. int. Verein. theor. angew. Limnol.*, 19, 125-45.

PENNINGTON, W. 1947. Lake sediments: pollen diagrams from the bottom deposits of the north basin of Windermere. *Phil. Trans. R. Soc.*, B, 233, 137-75.

PENNINGTON, W. 1970. Vegetation history in the north-west of England: a regional synthesis. *In* WALKER, D. and WEST, R. G. (eds.) *Studies in the Vegetation History of the British Isles*, 41-80, Cambridge.

PENNINGTON, W. 1973. Absolute pollen frequencies in the sediments of lakes of different morphometry. *In* BIRKS, H. J. B. and WEST, R. G. (eds.) *Quaternary Plant Ecology*, 79-104, Oxford.

PENNINGTON, W. 1975a. An application of principal components analysis to the zonation of two Late-Devensian profiles: Section II: Interpretation of the numerical analyses in terms of Late-Devensian (Late-Weichselian) environmental history. *New Phytol.*, 75, 441-53.

PENNINGTON, W. 1975b. A chronostratigraphic comparison of Late-Weichselian and Late-Devensian subdivisions, illustrated by two radiocarbon-dated profiles from Western Britain. *Boreas*, 4, 157-71.

PENNINGTON, W. and BONNY, A. P. 1970. Absolute pollen diagram from the British late-glacial. *Nature, Lond.*, 226, 871-3.

PENNINGTON, W., HAWORTH, E. Y., BONNY, A. P. and LISHMAN, J. P. 1972. Lake sediments in Northern Scotland. *Phil. Trans. R. Soc.*, B, 264, 191-294.

PENNINGTON, W. and LISHMAN, J. P. 1971. Iodine in lake sediments in northern England and Scotland. *Biol. Rev.*, 46, 279-313.

PENNINGTON, W. and SACKIN, M. J. 1975. An application of Principal Components Analysis to the zonation of two Late-Devensian profiles: Section I: Numerical analysis. *New Phytol.*, 75, 419-41.

PENNY, L. F., COOPE, G. R. and CATT, J. A. 1969. Age and insect fauna of the Dimlington silts, East Yorkshire. *Nature, Lond.*, 224, 65-7.

PÉWÉ, T. L. 1966. Palaeoclimatic significance of fossil ice wedges. *Biul. peryglac.*, 15, 65-73.

POORE, M. E. D. and McVEAN, D. N. 1957. A new approach to Scottish mountain vegetation. *J. Ecol.*, 45, 401-39.

PORTER, S. C. and CARSON, R. J. 1971. Problems of interpreting radiocarbon dates from dead-ice terrain, with an example from the Puget Lowland of Washington. *Quaternary Res.*, 1, 410-4.

von POST, L. 1956. The ancient sea fjord of the Viskan valley. Chapter 11. The shore marks. *Geol. För. Stockh. Förh.*, 78, 101-21.

PRICE, R. J. 1975. The glaciation of west central Scotland - a review. *Scott. geogr. Mag.*, 91, 134-45.

PROCTOR, J. and WOODELL, S. R. J. 1971. The plant ecology of serpentine. I. Serpentine vegetation of England and Scotland. *J. Ecol.*, 59, 397-410.

RAMPTON, V. 1971. Late Quaternary vegetational and climatic history of the Snag-Klutlan area, south western Yukon Territory, Canada. *Bull. Geol. Soc. Am.*, 82, 959-78.

RAMSEY, W. 1924. On relations between crustal movements and variations of sea-level during the late Quaternary time, especially in Fennoscandia. *Fennia*, 44, 1-39.

READ, H. H. 1923. The geology of the country round Banff, Huntly and Turriff. *Mem. Geol. Surv. Gt. Br.*, 186-210.

READ, H. H., PHEMISTER, J. and ROSS, G. 1926. The geology of Strath Oykell and lower Loch Shin. *Mem. Geol. Surv. Gt. Br.*

RICHEY, J. E., ANDERSON, E. M. and MACGREGOR, A. G. 1930. The geology of North Ayrshire. *Mem. Geol. Surv. Gt. Br.*

RITCHIE, J. C. and LICHTI-FEDEROVICH, S. 1967. Pollen dispersal phenomena in arctic-subarctic Canada. *Rev. Palaeobotan. Palynol.*, 3, 255-66.

ROBERTSON, D. 1881. On the post-Tertiary beds of Garvel Park, Greenock. *Trans. geol. Soc. Glasg.*, 7, 1-36.

ROBINSON, H. H. 1908. Ancient waterplanes and crustal deformation. *J. Geol.*, 16, 347-56.

ROSE, J. 1975. Raised beach gravels and ice wedge casts at Old Kilpatrick, near Glasgow. *Scot. J. Geol.*, 11, 15-21.

RUDDIMAN, W. F. and McINTYRE, A. 1973. Time-transgressive deglacial retreat of polar waters from the North Atlantic. *Quaternary Res.*, 3, 117-30.

RYDER, R. H. and McCANN, S. B. 1971. Periglacial phenomena on the island of Rhum in the Inner Hebrides. *Scot. J. Geol.*, 7, 293-303.

SANGSTER, A. G. and DALE, H. M. 1961. A preliminary study of differential pollen grain preservation. *Can. J. Bot.*, 39, 35-43.

SANGSTER, A. G. and DALE, H. M. 1964. Pollen grain preservation of under-represented species in fossil spectra. *Can. J. Bot.*, 42, 437-49.

SARS, G. O. 1878. *Bidrag til Kundshaben om Norges arktiske fauna. 1. Mollusca regionis arcticae norvegiae.* Christiania.

SCHÄFER, W. 1972. *Ecology and Palaeoecology of Marine Environments.* Edinburgh.

SCHOFIELD, J. C. 1964. Postglacial sea-levels and isostatic uplift. *N. Z. J. geol. Geophys.*, 7, 359-70.

SEDDON, B. 1962. Late-glacial deposits at Llyn Dwythwch and Nant Ffrancon, Caernarvonshire. *Phil. Trans. R. Soc.*, B, 244, 459-81.

SEN GUPTA, B. K. 1971. The benthonic foraminifera of the Tail of the Grand Banks. *Micropal.*, 17, 69-98.

SEN GUPTA, B. K. and McMULLEN, R. M. 1969. Foraminiferal distribution and sedimentary facies on the Grand Banks of Newfoundland. *Can. J. Earth Sci.*, 6, 475-87.

SHAW, J. 1971. Mechanism of till deposition related to thermal conditions in a Pleistocene glacier. *J. Glaciol.*, 10, 363-73.

SHEPARD, F. P. 1963. Thirty-five thousand years of sea level. *In* CLEMENTS, T. (ed.) *Essays in Marine Geology in honour of K. O. Emery*, 1-10, Los Angeles.

SHOTTON, F. W., BLUNDELL, D. J. and WILLIAMS, R. E. G. 1967. Birmingham University Radiocarbon Dates I. *Radiocarbon*, 9, 35-7.

SHOTTON, F. W., BLUNDELL, D. J. and WILLIAMS, R. E. G. 1968. Birmingham University Radiocarbon Dates II. *Radiocarbon*,

10, 200-6.

SHOTTON, F. W., BLUNDELL, D. J. and WILLIAMS, R. E. G. 1970. Birmingham University Radiocarbon Dates IV. *Radiocarbon*, 12, 385-9.

SHOTTON, F. W. and WEST, R. G. 1969. Appendix Bl. Stratigraphic table of the British Quaternary. *In* Recommendations on stratigraphical usage. *Proc. Geol. Soc. Lond.*, 1656, 139-66.

SHOTTON, F. W. and WILLIAMS, R. E. G. 1971. Birmingham University Radiocarbon Dates V. *Radiocarbon*, 13, 141-56.

SHOTTON, F. W. and WILLIAMS, R. E. G. 1973. Birmingham University Radiocarbon Dates VII. *Radiocarbon*, 15, 451-68.

SIMPSON, J. B. 1933. The late-glacial readvance moraines of the Highland border west of the River Tay. *Trans. R. Soc. Edinb.*, 61, 687-98.

SISSONS, J. B. 1960. Some aspects of glacial drainage channels in Britain, part I. *Scott. geogr. Mag.*, 76, 131-46.

SISSONS, J. B. 1961a. Some aspects of glacial drainage channels in Britain, part II. *Scott. geogr. Mag.*, 77, 15-36.

SISSONS, J. B. 1961b. The central and eastern parts of the Lammermuir - Stranraer moraine. *Geol. Mag.*, 98, 380-92.

SISSONS, J. B. 1963a. Scottish raised shoreline heights with particular reference to the Forth valley. *Geogr. Annlr.*, 45, 180-5.

SISSONS, J. B. 1963b. The Perth Readvance in central Scotland. Part I. *Scott. geogr. Mag.*, 79, 151-63.

SISSONS, J. B. 1964. The Perth Readvance in central Scotland. Part II. *Scott. geogr. Mag.*, 80, 28-36.

SISSONS, J. B. 1966. Relative sea-level changes between 10,300 and 8,300 B.P. in part of the Carse of Stirling. *Trans. Inst. Br. Geogr.*, 39, 19-29.

SISSONS, J. B. 1967a. *The Evolution of Scotland's Scenery.* Edinburgh.

SISSONS, J. B. 1967b. Comments on the paper by F. M. Synge and N. Stephens in Transactions no. 39. *Trans. Inst. Br. Geogr.*, 42, 163-8.

SISSONS, J. B. 1967c. Glacial stages and radiocarbon dates in Scotland. *Scot. J. Geol.*, 3, 375-81.

SISSONS, J. B. 1969. Drift stratigraphy and buried morphological features in the Grangemouth - Falkirk - Airth area, central Scotland. *Trans. Inst. Br. Geogr.*, 48, 19-50.

SISSONS, J. B. 1972a. The last glaciers in part of the south-east Grampians. *Scott. geogr. Mag.*, 88, 168-81.

SISSONS, J. B. 1972b. Dislocation and non-uniform uplift of raised shorelines in the western part of the Forth valley. *Trans. Inst. Br. Geogr.*, 55, 145-59.

SISSONS, J. B. 1973a. Hypotheses of deglaciation in the eastern Grampians, Scotland. *Scot. J. Geol.*, 9, 96.

SISSONS, J. B. 1973b. Delimiting the Loch Lomond Readvance in the eastern Grampians. *Scott. geogr. Mag.*, 89, 138-9.

SISSONS, J. B. 1974a. Glacial readvances in Scotland. *In* CASELDINE, C. J. and MITCHELL, W. A. (eds.) *Problems of the deglaciation of Scotland*, 5-15, St. Andrews.

SISSONS, J. B. 1974b. A Late-glacial ice-cap in the central Grampians, Scotland. *Trans. Inst. Br. Geogr.*, 62, 95-114.

SISSONS, J. B. 1974c. The Quaternary in Scotland: a review. *Scot. J. Geol.*, 10, 311-37.

SISSONS, J. B. 1974d. Lateglacial marine erosion in Scotland. *Boreas*, 3, 41-8.

SISSONS, J. B. 1975a. A fossil rock glacier in Wester Ross. *Scot. J. Geol.*, 11, 83-6.

SISSONS, J. B. 1975b. The Loch Lomond Readvance in the southeast Grampians. *In* GEMMELL, A. M. D. (ed.) *Quaternary Studies in North East Scotland*, 23-9, Aberdeen.

SISSONS, J. B. 1976a. Lateglacial marine erosion in south east Scotland. *Scott. geogr. Mag.*, 92, 17-29.

SISSONS, J. B. 1976b. *The geomorphology of the British Isles: Scotland.* London.

SISSONS, J. B. 1977a. Former ice-dammed lakes in Glen Moriston, Inverness-shire, and their significance in upland Britain. *Trans. Inst. Br. Geogr.*, (in press).

SISSONS, J. B. 1977b. A remarkable protalus rampart complex in Wester Ross. *Scott. geogr. Mag.* (in press).

SISSONS, J. B., CULLINGFORD, R. A. and SMITH, D. E. 1965. Some pre-carse valleys in the Forth and Tay basins. *Scott. geogr. Mag.*, 81, 115-24.

SISSONS, J. B. and GRANT, A. J. H. 1972. The last glaciers in the Lochnagar area, Aberdeenshire. *Scot. J. Geol.*, 8, 85-93.

SISSONS, J. B., LOWE, J. J., THOMPSON, K. S. R. and WALKER, M. J. C. 1973. Loch Lomond Readvance in the Grampian Highlands of Scotland. *Nature Phys. Sci.*, 244, 75-7.

SISSONS, J. B. and SMITH, D. E. 1965. Raised beaches associated with the Perth readvance in the Forth valley and their relation to glacial isostasy. *Trans. R. Soc. Edinb.*, 66, 143-68.

SISSONS, J. B. and SUTHERLAND, D. G. 1976. Climatic inferences from former glaciers in the south-east Grampian Highlands, Scotland. *J. Glaciol.*, (in press).

SISSONS, J. B. and WALKER, M. J. C. 1974. Lateglacial site in the central Grampian Highlands. *Nature, Lond.*, 249, 822-4.

SMITH, A. G. and PILCHER, J. R. 1973. Radiocarbon dates and vegetational history of the British Isles. *New Phytol.*, 72, 903-14.

SMITH, D. E., SISSONS, J. B. and CULLINGFORD, R. A. 1969. Isobases for the Main Perth Raised Shoreline in south-east Scotland as determined by trend-surface analysis. *Trans. Inst. Br. Geogr.*, 46, 45-52.

SPÄRCK, R. 1937. The Benthonic Animal Communities of the Coastal Waters. *The Zoology of Iceland*, 1 (6). Copenhagen.

SPENCE, D. H. N. 1957. Studies on the vegetation of Shetland. I. The serpentine debris vegetation in Unst. *J. Ecol.*, 45, 917-45.

SUGDEN, D. E. 1968. The selectivity of glacial erosion in the Cairngorm mountains, Scotland. *Trans. Inst. Br. Geogr.*, 45, 79-92.

SUGDEN, D. E. 1969. The age and form of corries in the Cairngorm mountains, Scotland. *Scott. geogr. Mag.*, 85, 34-46.

SUGDEN, D. E. 1970. Landforms of deglaciation in the Cairngorm mountains, Scotland. *Trans. Inst. Br. Geogr.*, 51, 201-19.

SUGDEN, D. E. 1971. The significance of periglacial activity on some Scottish mountains. *Geogr. J.*, 137, 388-92.

SUGDEN, D. E. 1973a. Hypotheses of deglaciation in the eastern Grampians, Scotland. *Scot. J. Geol.*, 9, 94-5.

SUGDEN, D. E. 1973b. Delimiting Zone III glaciers in the eastern Grampians. *Scott. geogr. Mag.*, 89, 63-4.

SUGDEN, D. E. 1974. Deglaciation of the Cairngorms and its wider implications. *In* CASELDINE, C. J. and MITCHELL, W. A. (eds.) *Problems of the deglaciation of Scotland*, 17-28, St. Andrews.

SUGDEN, D. E. and CLAPPERTON, C. M. 1975. The deglaciation of Upper Deeside and the Cairngorm mountains. *In* GEMMELL, A. M. D. (ed.) *Quaternary Studies in North East Scotland*, 30-8, Aberdeen.

SUGDEN, D. E. and JOHN, B. S. 1976. *Glaciers and Landscape*. London.

SUGGATE, R. P. and WEST, R. G. 1967. The substitution of local stage names for Holocene and Post-glacial. *Quarternaria*, 9, 245-6.

SYNGE, F. M. 1956. The glaciation of north-east Scotland. *Scott. geogr. Mag.*, 72, 129-43.

SYNGE, F. M. 1963. The Quaternary succession around Aberdeen, north-east Scotland. *Rep. VIth. Conf. int. Ass. quatern. Res. Geomorph. Sect.*, 3, 353-61.

SYNGE, F. M. 1966. The relationship of the raised strandlines and main end-moraines on the Isle of Mull and in the district of Lorn, Scotland. *Proc. Geol. Ass.*, 77, 315-28.

SYNGE, F. M. and STEPHENS, N. 1966. Late- and post-glacial shorelines, and ice limits in Argyll and north-east Ulster. *Trans. Inst. Br. Geogr.*, 39, 101-25.

TAUBER, H. 1970. The Scandinavian varve chronology and C-14 dating. *In* OLSSON, I. U. (ed.) *Radiocarbon Variations and Absolute Chronology*, 173-96, Stockholm.

TEBBLE, N. 1966. *British Bivalve Seashells*. London.

TERASMAE, J. 1951. On the pollen morphology of *Betula nana*. *Svensk. Bot. Tidsk.*, 45, 358-61.

THOMPSON, H. R. 1950. Some corries of north-west Sutherland. *Proc. Geol. Ass.*, 61, 145-55.

THOMPSON, K. S. R. 1972. *The last glaciers in western Perthshire*. Ph.D. thesis, Univ. of Edinb.

TYLDESLEY, J. B. 1973a. Long-range transmission of tree pollen to Shetland. I. Sampling and trajectories. *New Phytol.*, 72, 175-81.

TYLDESLEY, J. B. 1973b. Long-range transmission of tree pollen to Shetland. II. Calculation of pollen deposition. *New Phytol.*, 72, 183-90.

TYLDESLEY, J. B. 1973c. Long-range transmission of tree pollen to Shetland. III. Frequencies over the last hundred years. *New Phytol.*, 72, 691-7.

VASARI, Y. 1970. The Late-glacial Period in North-east Scotland. *In* JAGER, K-D. (ed.) *Probleme der Weichsel-Spätglazialen Vegetationsentwicklung in Mittel- und Nordeuropa*, 61-77. Herausgegeben vom Quartärkomitee der Deutschen Demokratischen Republik bei der Deutschen Akademie der Wissenschaften zu Berlin in Zussamenarbeit mit der Komission für das Studium des Holozäns in der Internationalen Union für Quartärforschung (INQUA) Frankfurt/Oder.

VASARI, Y. 1975. The vegetation of Northern Finland - past and present. *Inter-Nord*, 13-14, 99-118.

VASARI, Y. and VASARI, A. 1968. Late- and Post-glacial macrophytic vegetation in the lochs of Northern Scotland. *Acta Bot. Fenn.*, 80, 1-120.

WALKER, B. 1967. A diver-operated pneumatic core-sampler. *Limnol. Oceanogr.* 12, 144-6.

WALKER, D. 1966. The Late Quaternary history of the Cumberland lowland. *Phil. Trans. R. Soc.*, B, 251, 1-210.

WALKER, M. J. C. 1974. *Lateglacial and early Postglacial environments in part of the Grampian Highlands of Scotland*. Ph.D. thesis, Univ. of Edinb.

WALKER, M. J. C. 1975a. Two Lateglacial pollen diagrams from the eastern Grampian Highlands of Scotland. *Pollen Spores*, 17, 67-92.

WALKER, M. J. C. 1975b. Late Glacial and Early Postglacial environmental history of the central Grampian Highlands, Scotland. *J. Biogeogr.*, 2, 265-84.

WALTON, W. R. 1964. Recent foraminiferal ecology and Palaeoecology. *In* IMBRIE, J. and NEWELL, N. D. (eds.) *Approaches to Palaeoecology*, 151-337, New York.

WELIN, E., ENGSTRAND, L. and VACZY, S. 1972. Institute of Geological Sciences Radiocarbon Dates II. *Radiocarbon*, 14, 140-4.

WEST, R. G. 1968. *Pleistocene Geology and Biology*. London.

WEST, R. G. 1970. Pollen zones in the Pleistocene of Great Britain and their correlation. *New Phytol.*, 69, 1179-83.

von WEYMARN, J. and EDWARDS, K. J. 1973. Interstadial site on the Island of Lewis, Scotland. *Nature, Lond.*, 246, 473-4.

WHITE, I. D. and MOTTERSHEAD, D. N. 1972. Past and present vegetation in relation to solifluction on Ben Arkle, Sutherland. *Trans. Proc. bot. Soc. Edinb.*, 41, 475-89.

YOUNG, J. and CRAIG, R. 1869. Notes on the occurrence of seeds of fresh water plants and arctic shells along with the remains of the mammoth and reindeer in beds under the boulder clay of Kilmaurs. *Trans. geol. Soc. Glasg.*, 3, 315.

YOUNG, J. A. T. 1974. Ice wastage in Glenmore, upper Spey valley, Inverness-shire. *Scot. J. Geol.*, 10, 147-57.

Index